Lecture Notes in Computer Science 8667

Commenced Publication in 1973
Founding and Former Series Editors:
Gerhard Goos, Juris Hartmanis, and Jan van Leeuwen

T0213702

Marco Dorigo   Mauro Birattari
Simon Garnier   Heiko Hamann
Marco Montes de Oca   Christine Solnon
Thomas Stützle (Eds.)

Swarm Intelligence

9th International Conference, ANTS 2014
Brussels, Belgium, September 10-12, 2014
Proceedings

 Springer

Volume Editors

Marco Dorigo
Mauro Birattari
Thomas Stützle
IRIDIA, CoDE, Université Libre de Bruxelles, Belgium
E-mail: {mdorigo, mbiro, stuetzle}@ulb.ac.be

Simon Garnier
Rutgers University, New Jersey Institute of Technology, Newark, NJ, USA
E-mail: garnier@njit.edu

Heiko Hamann
University of Paderborn, Department of Computer Science, Paderborn, Germany
E-mail: heiko.hamann@uni-paderborn.de

Marco Montes de Oca
University of Delaware, Department of Mathematical Sciences, Newark, DE, USA
E-mail: mmontes@math.udel.edu

Christine Solnon
LIRIS, INSA de Lyon, Villeurbanne Cedex, France
E-mail: christine.solnon@liris.cnrs.fr

ISSN 0302-9743   e-ISSN 1611-3349
ISBN 978-3-319-09951-4   e-ISBN 978-3-319-09952-1
DOI 10.1007/978-3-319-09952-1
Springer Cham Heidelberg New York Dordrecht London

Library of Congress Control Number: 2014945573

LNCS Sublibrary: SL 1 – Theoretical Computer Science and General Issues

Typesetting: Camera-ready by author, data conversion by Scientific Publishing Services, Chennai, India

Printed on acid-free paper

Springer is part of Springer Science+Business Media (www.springer.com)

Preface

These proceedings contain the papers presented at ANTS 2014, the 9th International Conference on Swarm Intelligence, held at IRIDIA, Université Libre de Bruxelles, Brussels, Belgium, during September 10–12, 2014. The ANTS series started in 1998 with the First International Workshop on Ant Colony Optimization (ANTS 1998), which attracted more than 50 participants. Since then ANTS, which is held bi-annually, has gradually become an international forum for researchers in the wider field of swarm intelligence. In 2004, this development was acknowledged by the inclusion of the term "swarm intelligence" (next to "ant colony optimization") in the conference title. Since 2010, the ANTS conference is officially devoted to the field of swarm intelligence as a whole, without any bias toward specific research directions. This is reflected in the title of the conference: "International Conference on Swarm Intelligence."

The papers contained in this volume were selected out of 55 submissions. Of these, 17 were accepted as full-length papers, while nine were accepted as short papers. This corresponds to an overall acceptance rate of 47%. Also included in this volume are seven extended abstracts.

All the contributions were presented as posters. The full-length papers were also presented orally in a plenary session. Extended versions of the best papers presented at the conference will be published in a special issue of the *Swarm Intelligence* journal.

We take this opportunity to thank the large number of people that were involved in making this conference a success. We express our gratitude to the authors who contributed their work, to the members of the International Program Committee, to the additional reviewers for their qualified and detailed reviews, and to the staff of IRIDIA for helping with organizational matters.

We hope the reader will find this volume useful both as a reference to current research in swarm intelligence and as a starting point for future work.

July 2014

Marco Dorigo
Mauro Birattari
Simon Garnier
Heiko Hamann
Marco Montes de Oca
Christine Solnon
Thomas Stützle

Organization

ANTS 2014 was organized by IRIDIA, Université Libre de Bruxelles, Belgium

General Chair

Marco Dorigo   Université Libre de Bruxelles, Belgium

Technical Program Chairs

Simon Garnier   Rutgers University, USA
Marco A. Montes de Oca   University of Delaware, USA
Christine Solnon   Institut National des Sciences Appliquées
                              de Lyon, France

Publication Chair

Heiko Hamann   University of Paderborn, Germany

Organization Chairs

Mauro Birattari   Université Libre de Bruxelles, Belgium
Thomas Stützle   Université Libre de Bruxelles, Belgium

Local Arrangements

Manuele Brambilla   Université Libre de Bruxelles, Belgium
Lorenzo Garattoni   Université Libre de Bruxelles, Belgium
Andreagiovanni Reina   Université Libre de Bruxelles, Belgium

Program Committee

Andy Adamatzky   University of the West of England, UK
Daniel Angus   University of Queensland, Australia
Jacob Beal   BBN Technologies, USA
Tim Blackwell   Goldsmiths University of London, UK
Maria José Blesa Aguilera   Universitat Politècnica de Catalunya, Spain
Christian Blum   University of the Basque Country, Spain
Leticia Cagnina   Universidad Nacional de San Luis, Argentina
Alexandre Campo   Université Libre de Bruxelles, Belgium
Stephen Y. Chen   York University, Canada

Marco Chiarandini	University of Southern Denmark, Denmark
Carlos Coello Coello	CINVESTAV-IPN, Mexico
Oscar Cordon	European Centre for Soft Computing, Spain
Maurice Clerc	Independent Consultant, France
Ana Luisa Custodio	Universidade Nova de Lisboa, Portugal
Sanjoy Das	Kansas State University, USA
Kusum Deep	Indian Institute of Technology Roorkee, India
Jean-Louis Deneubourg	Université Libre de Bruxelles, Belgium
Gianni Di Caro	IDSIA, USI-SUPSI, Switzerland
Luca Di Gaspero	University of Udine, Italy
Karl Doerner	Johannes Kepler Universität Linz, Austria
Leandro Dos Santos Coelho	Pontifical Catholic University of Parana and Federal University of Parana, Brazil
Haibin Duan	Beihang University, China
Mohammed El-Abd	American University of Kuwait, Kuwait
Andries Engelbrecht	University of Pretoria, South Africa
Hugo J. Escalante	Instituto Nacional de Astrofisica, Optica y Electronica (INAOE), Mexico
Susana Esquivel	Universidad Nacional de San Luis, Argentina
Gianni Fasano	Ca' Foscari University, Italy
Namiz Fatès	Loria - Inria, France
Juan L. Fernandez-Martinez	Universidad de Oviedo, Spain
Eliseo Ferrante	University of Leuven, Belgium
Luca Maria Gambardella	IDSIA, USI-SUPSI, Switzerland
Jose M. Garcia-Nieto	Universidad de Malaga, Spain
Deborah Gordon	Stanford University, USA
Roderich Groß	The University of Sheffield, UK
Frédéric Guinand	Université du Havre, France
Walter Gutjahr	Universität Wien, Austria
Saman Halgamuge	University of Melbourne, Australia
Julia Handl	University of Manchester, UK
Richard Hartl	Universität Wien, Austria
Thomas Jansen	Aberystwyth University, UK
Mark Jelasity	University of Szeged, Hungary
Yaochu Jin	University of Surrey, UK
Joshua Knowles	University of Manchester, UK
Xiaodong Li	RMIT University, Australia
Manuel López-Ibáñez	IRIDIA, Université Libre de Bruxelles, Belgium
Simone Ludwig	North Dakota State University, USA
Stephen Majercik	Bowdoin College, USA
Vittorio Maniezzo	Università di Bologna, Italy
Franco Mascia	IRIDIA, Université Libre de Bruxelles, Belgium
Ronaldo Menezes	Florida Institute of Technology, USA
Bernd Meyer	Monash University, Australia
Martin Middendorf	Universität Leipzig, Germany
Nicolas Monmarché	Université de Tours, France

Roberto Montemanni	IDSIA, USI-SUPSI, Switzerland
Radhika Nagpal	Harvard University, USA
Frank Neumann	The University of Adelaide, Australia
Ann Nowé	Vrije Universiteit Brussel, Belgium
Randal Olson	Michigan State University, USA
Ender Özcan	University of Nottingham, UK
Kostantinos Parsopoulos	University of Ioannina, Greece
Paola Pellegrini	French Institute of Science and Technology for Transport, Development and Networks, France
Jorge Peña	Max Planck Institute for Evolutionary Biology, Germany
Marc Reimann	University of Graz, Austria
Katya Rodríguez-Vázquez	IIMAS-UNAM, Mexico
Andrea Roli	Università di Bologna, Italy
Michael Rubenstein	Harvard University, USA
Erol Sahin	Middle East Technical University, Turkey
Thomas Schmickl	Karl-Franzens-Universität Graz, Austria
Kevin Seppi	Brigham Young University, USA
Jurij Silc	Jozef Stefan Institute, Ljubljana, Slovenia
Dirk Sudholt	The University of Sheffield, UK
Jon Timmis	University of York, UK
Colin Torney	University of Exeter, UK
Vito Trianni	Institute of Cognitive Sciences and Technologies, CNR, Italy
Elio Tuci	Aberystwyth University, UK
Kolbjørn Tunstrøm	Princeton Univerisity, USA
Michael N. Vrahatis	University of Patras, Greece
Ling Wang	Tsinghua University, China
Tom Wenseleers	University of Leuven, Belgium
Alan Winfield	University of the West of England, UK
Yanjun Yan	Western Carolina University, USA

Additional Reviewers

Eduardo Feo Flushing	IDSIA, USI-SUPSI, Switzerland
Gianpiero Francesca	IRIDIA, Université Libre de Bruxelles, Belgium
Jawad Nagi	IDSIA, USI-SUPSI, Switzerland
Andreagiovanni Reina	Université Libre de Bruxelles, Belgium
Touraj Soleymani	IRIDIA, Université Libre de Bruxelles, Belgium

Table of Contents

Short Papers

Extended Abstracts

A Novel Ant Colony Algorithm for Building Neural Network Topologies

Khalid Salama[1] and Ashraf M. Abdelbar[2]

[1] School of Computing, University of Kent,
Canterbury, United Kingdom
kms39@kent.ac.uk
[2] Dept. of Mathematics & Computer Science, Brandon University,
Manitoba, Canada
abdelbara@brandonu.ca

Abstract. A re-occurring challenge in applying feed-forward neural networks to a new dataset is the need to manually tune the neural network topology. If one's attention is restricted to fully-connected three-layer networks, then there is only the need to manually tune the number of neurons in the single hidden layer. In this paper, we present a novel Ant Colony Optimization (ACO) algorithm that optimizes neural network topology for a given dataset. Our algorithm is not restricted to three-layer networks, and can produce topologies that contain multiple hidden layers, and topologies that do not have full connectivity between successive layers. Our algorithm uses Backward Error Propagation (BP) as a subroutine, but it is possible, in general, to use any neural network learning algorithm within our ACO approach instead. We describe all the elements necessary to tackle our learning problem using ACO, and experimentally compare the classification performance of the optimized topologies produced by our ACO algorithm with the standard fully-connected three-layer network topology most-commonly used in the literature.

1 Introduction

Pattern classification is a crucial real-world problem, concerned with predicting the class of a given pattern based on its input attributes, using a well-constructed classification model. The classification process consists of two stages. The training stage utilizes a *training set* of labeled patterns, that is a set of patterns along with their correct class labels, that should be sufficiently representative of the domain of interest. A classification algorithm uses the training set to construct an internal model of the relationships between the attributes of the input patterns and their corresponding class labels. Then, during the subsequent operating stage, the classifier uses its internal model to predict the class of unlabeled patterns that it was not presented with during the training stage. One of the most widely-studied and applied methods for pattern classification are artificial neural networks (ANN), which use an internal model consisting of a

M. Dorigo et al. (Eds.): ANTS 2014, LNCS 8667, pp. 1–12, 2014.

vector of real-valued weights associated with inter-neuronal connections, as will be described in Section 2.1.

The most commonly used neural network topology is a three-layer topology, consisting of an input layer, a single hidden layer, and an output layer, with full connectivity between layers. The size of the input and output layers are determined by characteristics of the dataset, while the number of neurons in the hidden layer is often manually determined by practitioners. In this paper, we propose a novel algorithm, called ANN-Miner, for automatically learning the topology of Feed-Forward Neural Networks using Ant Colony Optimization (ACO). Our algorithm is not restricted to three-layer networks, and can produce topologies that contain multiple hidden layers, that do not have full connectivity between successive layers, and that have connections between neurons that are not in successive layers. Our algorithm uses Backward Error Propagation (BP) as a subroutine, but it is possible, in general, to use any neural network learning algorithm within our ACO approach instead.

We begin in Section 2 with a review of ACO and of neural networks. We then present our ACO approach in Section 3. Experimental methodology and results are presented in Section 4 and final remarks are offered in Section 5.

2   Background

2.1   Feed-Forward Neural Networks

Feed-forward neural networks (FFNN) are generally acknowledged as being one of the most widely-applied methods for pattern classification. The most common FFNN topology is a three-layer topology in which neurons are arranged in an input layer, a hidden layer, and an output layer. Commonly, there are connections between every neuron in a layer to all the neurons in the succeeding layer.

Each neuron i is a simple combinatorial circuit which accepts r inputs $o_1, \ldots, o_r$, and produces a single output o_i:

$$net_i = \sum_{i=1}^{r} w_{ij} o_j + \theta_i \ , \ \ o_i = f(net_i) = \frac{1}{1 + e^{-net_i}} \tag{1}$$

where each input o_j is the output of a neuron in the previous layer, the weight w_{ij} represents a real-valued weight between neuron j and neuron i, θ_i represents a weight associated with neuron i itself called the neuron's self-bias, and f is an *activation function* that is most commonly selected to be the sigmoidally-shaped logistic function shown in the equation.

A FFNN with n input neurons and m output neurons computes a mapping $R^n \mapsto R^m$. Given a set of training examples $\mathcal{T}$ where each training example p consists of an n-dimensional input vector x and an m-dimensional output vector y, each input pattern is, in turn, applied to the input layer of the network, the signal is allowed to propagate through the network, and the output of the network, denoted y', is compared to the desired output y to determine the error of the network for that pattern, denoted E_p. A common error function is the simple sum of squared error: $E_p = \frac{1}{2} \sum_{i=1}^{m} (y - y')^2$, where the total error over all patterns is simply $E = \sum_p E_p$.

If the topology and the weights and self-biases of the network are fixed, then the error can seen to be a mathematical function of the training set: $E = f(\mathcal{T})$. On the other hand, if the topology and the training set are held constant, then the error can be seen as a function of the weights and biases: $E = f(w)$, where w is a real-valued *weight vector* that includes all the weights and biases of the network. Therefore, the problem of training a fixed-topology neural network can be viewed as a mathematical multi-dimensional function minimization problem.

Perhaps the most popular neural network training algorithm is the gradient descent based Backward Error Propagation (BP) algorithm which is based on repeatedly applying the training set to the network (each full pass through the training set is called an *epoch*), computing the error E, and then modifying each element of the weight vector according to:

$$ w_i = w_i + \Delta w_i \,, \quad \Delta w_i = -\eta \frac{\partial E}{\partial w_i} \tag{2} $$

where η is an external parameter called the learning rate.

Before applying a neural network to a particular dataset $\mathcal{T}'$ for a pattern classification problem, it is usually necessary to apply some preprocessing to the dataset $\mathcal{T}'$ to obtain a new dataset $\mathcal{T}$. Specifically, every c-category categorical attribute h in $\mathcal{T}'$ is transformed into c numerical attributes $g_1, \ldots, g_c$ in $\mathcal{T}$, where $g_k = 1$ if $h = k$, and $g_k = 0$ if $h \neq k$. If the value of a categorical attribute is missing, then it is replaced with the most-frequently occurring category label for that attribute. Any missing numeric attributes are replaced with the value 0.

If the class label b in $\mathcal{T}'$ has c possible values, then the output patterns in $\mathcal{T}$ will be c-dimensional. If a given pattern has class label u, then the output pattern vector y will be constructed as follows: $y_s = (1 - \epsilon)$ if $u = s$, and $y_s = (0 + \epsilon)$ if $u \neq s$, where we use $\epsilon = 0.1$. The purpose for the ϵ-offset is to avoid the "saturation regions" of the logistic activation function.

2.2   ACO Related Work

Ant colony optimization is a meta-heuristic for solving combinatorial optimization problems, inspired by observations of the behavior of ant colonies in nature. ACO has been successful in tackling the classification problem of data mining. A number of ACO-based algorithms have been introduced in the literature with different classification learning approaches. Ant-Miner [8], proposed by Parpinelli et al., is the first ant-based classification algorithm, which discovers a list of classification rules in the form of IF-Conditions-Then-Class. The algorithm has been followed by several extensions in [5, 6, 8–10].

ACDT [1, 2], proposed by Boryczka and Kozak, and Ant-Tree-Miner [7], proposed by Otero et al., are two different ACO-based algorithms for inducing decision trees for classification. Salama and Freitas have recently employed ACO to learn various types of Bayesian network classifiers, such as Bayesian network augmented naïve-Bayes [13], class-based Bayesian multi-nets [14], cluster-based Bayesian multi-nets [11], and class Markov blankets [12].

As for learning neural networks, the ant-based meta-heuristic was utilized in two works. Liu et. Al proposed ACO-PB [4], a hybrid of the ant colony and back-propagation algorithms to optimize the network weights. It adopts ACO to search the optimal combination of weights in the solution space, and then uses BP algorithm to obtain the accurate optimal solution quickly. Blum and Socha applied $ACO_\mathbb{R}$, an ant colony optimization algorithm for continuous optimization [3, 17], to train feed-forward neural networks [15, 16].

Note that, to the best of our knowledge, ACO has not been utilized to learn the topology of the neural networks prior to the current work.

3   Our Proposed Ant Colony Algorithm

As discussed in Section 2.1, many neural network applications use a simple three-layer network topology, with full connectivity between layers. We allow our ACO method to deviate from this by allowing connections to be generated between hidden neurons and other hidden neurons — under the restriction that the topology remain acyclic — as well as direct connections between input neurons and output neurons. This allows producing networks with a variable number of layers, as well as arbitrary connections that skip over layers.

3.1   The Construction Graph

The core element of any ACO-based algorithm is the construction graph that contains the solution components in the search space, with which an ant constructs a candidate solution. As for the problem at hand, a candidate solution is a network topology, and the solution components are the selected connections between the neurons. More precisely, there are four types of available connections: 1) connections between input and hidden neurons, 2) connections between hidden and output neurons, 3) connections between input and output neurons, and 4) connections between different hidden neurons. Each potential connection $c = i \rightarrow j$, connecting between neurons i and j, has two solution components in the construction graph: D_c^{true}, representing the decision to include connection $i \rightarrow j$ in the current candidate topology being constructed by the ant, and D_c^{false}, representing the decision not to include the connection. Therefore, the construction graph can be represented as a two-dimensional $2 \times |C|$ array, where 2 refers to the Boolean solution components, and C is the set of the available connections.

The number of input neurons and output neurons depends of course on the dataset and the representation that is used for the attributes of the dataset, while the total number of hidden neurons is an external user-supplied parameter. Suppose the total number of neurons is N, with N_i input neurons, N_o output neurons, and N_h potential hidden neurons. $N_i \times N_h$, $N_h \times N_o$ and $N_i \times N_o$ are the number of available connections between input and hidden neurons, hidden and output neurons, and input and output neurons, respectively, in the total available number of connections $|C|$. This means that, for instance, an ant can select (or unselect) a connection between any input neuron and any hidden neuron. The same applies for the two other connection types.

However, the available connections between the hidden neurons N_h are defined as follows. In order to ensure that the topology is acyclic, we impose the restriction that $i \rightarrow j$ is not available if $i \geq j$. In other words, each hidden neuron has a numeric index, and we only allow connections from a given hidden neuron n_i to a higher-numbered neuron n_j. It is well-known that any directed acyclic graph is isomorphic to a graph where the nodes are lexicographically ordered and for all arcs (u, v) in the graph u precedes v in the lexicographic order. Hence, the number of the available connection between the N_h hidden neurons $= (N_h - 1) + (N_h - 2) + ... + 1 + 0 = N_h(N_h - 1)/2$.

The number of input and output neurons, N_i and N_o, respectively, are determined by characteristics of the dataset as described in Section 2.1. In the work described in this paper, we set $N_h = N_i + N_o$.

3.2   The ANN-Miner Overall Algorithm

The overall process of ANN-Miner is illustrated in Algorithm 1. In the initialization step of ANN-Miner (line 3), the amount of pheromone assigned to each solution component D_c^a – where a can be true or false – in the construction graph is initialized with the value 0.5. Hence, for each connection c, the probability of including $i \rightarrow j$ (i.e. selecting D_c^{true}) in the topology equals the probability of not including $i \rightarrow j$ (i.e. selecting D_c^{false}).

The outline of the algorithm is as follows. In the inner for-loop (lines 6-12), each ant_i in the colony creates a candidate solution NN_i, i.e. a complete neural network (line 7). Then the quality of the constructed solution is evaluated (line 8). The best solution NN_{tbest} produced in the colony is selected to update the pheromone trail according to the quality of its solution Q_{tbest}. After that, the algorithm compares the iteration-best solution NN_{tbest} with the best-so-far solution NN_{bsf} (the if statement in lines 14-16) to keep track of the best solution found so far during the algorithm execution.

This set of steps is considered an iteration of the outer *repeat − until* loop (lines 4-18) and is repeated until the same solution is generated for a number of consecutive trials specified by the `conv_iterations` parameter (indicating convergence) or until `max_iterations` is reached. The values of `conv_iterations`, `max_iterations` and `colony_size` are user-specified thresholds. In our experiments (see Section 4), we used 10, 500 and 10 for each of these parameters, respectively.

The best-so-far neural network undergoes an (optional) post-processing step to produce the final neural network NN_{final} to be returned by the algorithm. Basically, the algorithm learns the final weights of the connections in the neural network NN_{bsf} — which is structured with the best topology of connections found during the search process of the ACO algorithm. In the current implementation of the ANN-Miner algorithm, we use the standard Backward Error

Algorithm 1. Pseudo-code of ANN-Miner.

 1: **Begin**
 2: $NN_{bsf} = \phi$; $t = 1$;
 3: $InitializePheromone()$;
 4: **repeat**
 5:       $NN_{tbest} = \phi$; $Q_{tbest} = 0$;
 6:       **for** $i = 1 \rightarrow$ `colony_size` **do**
 7:             $NN_i = ant_i.CreateSolution()$;
 8:             $Q_i = EvaluateQuality(NN_i)$;
 9:             **if** $Q_i > Q_{tbest}$ **then**
10:                   $NN_{tbest} = NN_i$; $Q_{tbest} = Q_i$;
11:             **end if**
12:       **end for**
13:       $UpdatePheromone()$;
14:       **if** $Q_{tbest} > Q_{bsf}$ **then**
15:             $NN_{bsf} = NN_{tbest}$; $Q_{bsf} = Q_{tbest}$;
16:       **end if**
17:       $t = t + 1$;
18: **until** $t =$ `max_iterations` **or** $Convergence($`conv_iterations`$)$;
19: $NN_{final} = PostProcessing(NN_{bsf})$;
20: **return** NN_{final};
21: **End**

Propagation procedure, described in Section 2.1, to train NN_{bsf} and learn its final weights. The only difference is that, instead of having BP working on the conventional three-layer fully-connected network topology, it works on the arbitrary topologies constructed by the ACO algorithms.

In our BP post-processing step, we set the learning rate η to 0.01, and we set the number of epochs to 1000. Using BP for training the final neural network allows for comparing the quality of the conventional topology with the arbitrary topologies constructed by our ACO algorithms, by using the same weight learning procedure. Solution creation, quality evaluation and pheromone update procedures are discussed in the following subsections.

3.3   Solution Creation Procedure

The process of creating a new candidate solution (neural network) is described in Algorithm 2. The procedure starts with an empty (edge-less) neural network (line 2) to be constructed throughout the procedure. In addition, an empty array SLN, which represents the ant trail in the construction graph and its selected solution components, is initialized. This data structure is necessary for the pheromone update procedure, as described later. For each connection c in the available set of connections C, the ant selects D_c^a to decide whether to include this connection in the candidate network NN or not (line 4) — by either selecting

---

Algorithm 2. Pseudo-code of solution creation procedure.

1: **Begin** CreateSolution()
2: $NN \leftarrow \phi$; $SLN \leftarrow \phi$;
3: **for** $c = 1 \rightarrow |C|$ **do**
4: $\quad$ $D_c^a = SelectDecisionComponent()$;
5: $\quad$ $SLN = SLN \cup D_c$;
6: $\quad$ **if** $D_c^a == D_c^{true}$ **then**
7: $\quad\quad$ $NN = NN \cup (i \rightarrow j)_c$;
8: $\quad$ **end if**
9: **end for**
10: $TrainNeuralNetwork(NN)$;
11: **return** NN;
12: **End**

---

solution component D_c^{true} or D_c^{false}. The selection of the solution component at each step is based on the following probabilistic state transition formula

$$p(D_c^a) = \frac{\tau(D_c^a)}{\tau(D_c^{true}) + \tau\left(D_c^{false}\right)}, \tag{3}$$

where $p(D_c^a)$ is the probability of selecting decision D^a for connection c, and $\tau(D_c^a)$ is the current amount of pheromone associated with D_c^a. Every selected solution component D^a (where $a = true$ or $a = false$) is added to the data structure SLN (line 5). However, only if $D_c^a = D_c^{true}$, that is, the ant selected the decision to include connection c in the topology, the corresponding connection $(i \rightarrow j)_c$ is appended to the candidate network NN (the if statement in lines 6-8). After the ant visits all the available connections in the construction graph and performs the include-or-not decision, the network topology of NN is now complete.

The generated topology of NN may contain some hidden neurons that have no incoming edges or no outgoing edges. Such a hidden neuron may be considered "dead" and may be safely removed from the topology. In addition, it is possible that NN may contain an input neuron that does not have a complete path to any output neuron: it may have no outgoing edges, or it may have outgoing edges that connect only to dead hidden neurons. This is interpreted as the algorithm making the decision that that particular external input is irrelevant to the output of the network. If this decision is incorrect, then the network NN will likely receive a poor quality evaluation, and will soon be discarded. On the other hand, if this decision is correct, then the algorithm will have discovered something important. It is also possible for NN to contain an output neuron that does not have any incoming edges, or incoming edges only from dead hidden neurons. This would result in that output neuron having a constant value without regard for the network inputs, and would result in the network NN receiving a poor quality evaluation and being quickly discarded.

We train the neural network NN (line 10) using the Backward Error Propagation procedure (described in Section 2.1), with some optimized parameter values (discussed in the following section), as a "quick and dirty" way to obtain a complete neural network and evaluate its pattern classification quality. We use BP for training the candidate neural network, not because it is the best weight optimization method, but because it is a fast procedure that is going to be repeated several times during the algorithm execution. In addition, we are only interested in the relative quality difference between different topologies trained by the same (even if not very efficient) BP procedure.

3.4   Quality Evaluation and Pheromone Update

A key objective of a pattern classification algorithm is to learn models with good generalization capabilities, i.e., models that are able to accurately predict the class labels of *new* unknown examples. Overfitting occurs when the induced model reflects good classification performance (fit) on the training (in-sample) data used in the learning process, yet shows bad predictive performance (generalization) involving new/testing data.

More precisely, in our neural network case, the network topology and its connection weights are the classification model. A bad approach is to use the training set as the pattern set for learning the model (topology and weights), and then use the same pattern set to evaluate the classification quality of the model. In this case, the generalization ability would not be tested during the training phase, and the quality of the optimized model might be due to a high (undesirable) fit on the training set, and the same model might exhibit poor predictive power on the unseen test (out-of-sample) dataset.

Therefore, we split the training set at the beginning of the algorithm into two mutually exclusive parts: 1) the learning set, which contains 80% of the training set and is used to learn the neural network topology and weights; and 2) the validation set, which contains 20% of the training set and is used to evaluate the quality of the model. The quality Q_i of a candidate solution NN_i is evaluated using the *predictive accuracy* of the model on the validation set. Accuracy is a simple and yet popular predictive performance measure, computed as:

$$Accuracy = \frac{|Correct|}{|ValidationSet|}, \tag{4}$$

where *Correct* is the set of correctly classified validation set instances, and *ValidationSet* is the current validation set.

After Q_i is computed for each candidate solution NN_i constructed by all the ants in the colony at iteration t, the iteration-best solution ant updates the pheromone amounts on the construction graph. In essence, the pheromone amounts are increased on all the solution components D_c^a in the SLN_{tbest} selected by the iteration-best ant during its trail, where D_c^a represents the decision to include ($a = true$) or not to include (where $a = false$) connection c in the topology. This influences the probability for the subsequent ants to include, or not to include connection c. The amount of pheromone deposited is based on Q_{tbest}, the quality of the iteration-best solution NN_{tbest}, as follows:

$$\tau(D_c^a) = \tau(D_c^a) + [\tau(D_c^a) \times Q_{tbest}] \quad \forall D_c^a \in |SLN_{tbest}| \tag{5}$$

To simulate pheromone evaporation, normalization is then applied on each pair of solution components associated with each connection c in the construction graph. This keeps the total pheromone amount on each pair $\tau(D_c^{true})$ and $\tau(D_c^{false})$ equal to 1, as follows:

$$\tau(D_c^a) = \tau(D_c^a)/[\tau(D_c^{true}) + \tau(D_c^{false})] \quad \forall c \in |C| \tag{6}$$

3.5 Variations of the Algorithm

We introduce two variations of the ANN-Miner algorithm for learning neural network topologies, which concern how the connection weights are optimized throughout the algorithm. As previously mentioned, after each ant_i constructs a network topology NN_i, the network is trained using the BP procedure.

— **Randomly Reinitialized Weights:** In the first variation of the algorithm, ANN-Miner, the weights of the network connections are randomly initialized with each candidate NN_i topology constructed by ant_i. This means that the algorithm does not make use of the optimized weights of the previously constructed neural network in the BP procedure for training the current neural network NN_i. Such an approach performs a fair comparison between different candidate topologies, since they all start weight optimization from the same point: random initialization of the weights. In this variation, we perform BP with 20 epochs and 0.1 learning rate. Moreover, the BP process in the post-processing step starts also with randomly initialized set of weights.

— **Retaining the Weights-Based "Wisdom":** By contrast, the second variation makes use of the optimized weights of the neural networks constructed in previous iterations to train the current neural network. More precisely, in wANN-Miner, the colony keeps the weight optimization "wisdom", and accumulates on it throughout the algorithm's execution. In fact, NN_{bsf} does not only keep the topology of the best-so-far neural network, but also keeps its connection weights. After each ant constructs a candidate topology NN_i, the weights of its connection are initialized with the weights present in NN_{bsf} before the BP weight optimization procedure executes. Then, if NN_i produced better classification quality than NN_{bsf}, NN_i will replace NN_{bsf}, and its connection weights will be used as initial values for performing BP on subsequent candidate networks.

Note that some connections in NN_i may not be present in NN_{bsf}, in which case their weights will be randomly initialized; there may also be some connections in NN_{bsf} that are not present at all in NN_i. Such a diversion in the topologies maintains the exploration aspect of the weight learning process, in addition to the exploitation aspect that is realized by building on the best weights learned in previous iterations. In wANN-Miner, we perform BP, for each candidate NN_i, with only 10 epochs, and a lower learning rate of 0.05, making use of the accumulated weight optimization wisdom. Moreover, the BP weight learning procedure in the post-processing step starts also with the weights of the NN_{bsf}.

— **BP as Post-Processing Step:** Furthermore, we introduce two other simple variations of the previous algorithms, in which the first uses the BP weight learning post-processing step, and the other returns the NN_{bsf} without any further weight optimization. The idea behind that is to test the hypothesis that wANN-Miner may not benefit from the BP post processing step to the same extent that the first variation, ANN-Miner, may benefit. This is shown in the results in Section 4.

4   Experimental Methodology and Results

The performance of ANN-Miner was evaluated using 20 benchmark pattern classification datasets, from the well-known UCI (University of California at Irvine) repository. We compare the predictive accuracy of four variations of our proposed ant-based ANN-Miner algorithm (ANN, ANN-BP, wANN, wANN-BP) against the standard three-layer fully connected topology trained with Backward Error Propagation – referred to as 3L-BP – as the baseline for our evaluation.

The experiments were carried out using the *stratified* 10-times 10-fold cross validation procedure. In essence, a dataset is divided into ten mutually exclusive partitions (folds), with approximately the same number of patterns and roughly the same class distribution in each partition. Then, each classification algorithm is run ten times, where each time a different partition is used as the test set and the other nine partitions are used as the training set. The results (accuracy rate on the test set) are then averaged and reported as the accuracy rate of the classifier. Since we are evaluating stochastic algorithms, we run each ten times – using a different random seed to initialize the search each time – for each of the ten iterations of the cross-validation procedure. Table 1 reports the average of the predictive accuracy (in percentage) values obtained by 10-times 10-fold cross validation for the 20 datasets, where the highest accuracy for each dataset is shown in bold face. The last row shows the average rank of each algorithm in terms of predictive accuracy. The average rank for a given algorithm g is obtained by first computing the rank of g on each dataset individually. The individual ranks are then averaged across all datasets to obtain the overall average rank for algorithm g. Note that the lower the value of the rank, the better the algorithm.

As shown in Table 1, two variations (ANN-Miner-BP and wANN-Miner-BP) of our proposed ACO-based algorithm had a higher average accuracy rank than 3L-BP, while a third variation (wANN-Miner) had the same rank as 3L-BP. In details, wANN-Miner-BP, which utilizes the idea of "wisdom", and performs BP as a post-processing step on the best-so-far constructed neural network topology, obtained the best overall average rank with a value of 1.93, and achieved the highest predictive accuracy in 10 datasets. The ANN-Miner variation, which randomly initializes the connection weights each time, and performs the BP post-processing step, obtained second place in the overall average rank with a value of 2.4, and achieved the highest predictive accuracy in 5 datasets. In third place, wANN-Miner, which utilizes the "wisdom" idea without the BP post-processing step, obtained 3.28 as an overall average rank, and achieved the best predictive accuracy in 3 datasets. The standard Backward Error Propagation algorithm with the 3 layer topology tied for third place with 3.28 overall average

Table 1. The table on the left shows predictive accuracy (%) results for each dataset, for each of the five algorithms; the last row shows the average rank of each classifier, where the lower the rank the better the algorithm. The table on the right shows the results of applying a Friedman test with the Holm post-hoc test.

Dataset	3L-BP	ANN	ANN-BP	wANN	wANN-BP
balance	**96.50**	63.17	90.50	91.33	94.00
breast-t	32.64	45.27	**62.28**	57.55	59.37
car	90.29	89.47	93.94	**98.19**	**98.19**
credit-a	**84.35**	83.48	**84.35**	82.75	83.19
credit-g	**74.00**	71.90	72.50	71.90	72.60
ecoli	79.53	81.25	84.86	84.86	**85.76**
glass	46.30	48.87	**57.88**	48.46	50.84
hayes	60.01	63.02	69.31	75.45	**75.59**
heart-c	57.46	54.47	56.1	55.43	56.78
heart-h	57.43	62.63	62.66	60.22	**63.29**
hepatitis	83.79	79.46	80.75	**81.92**	**81.92**
ionosph	89.67	92.52	92.81	93.38	**93.66**
iris	87.28	90.62	94.67	90.67	**95.29**
pima	75.78	74.86	76.04	75.13	**76.17**
pop	**55.71**	39.11	48.04	42.86	44.11
s-heart	81.11	84.45	**85.56**	81.85	83.70
segment	**94.16**	88.81	92.77	92.74	93.10
ttt	76.63	50.32	90.94	97.89	**98.00**
voting	93.89	93.24	94.65	**94.90**	**94.90**
wine	94.41	94.93	**96.04**	93.27	94.38
Avg. rank	3.28	4.13	2.40	3.28	**1.93**

Hypothesis	p	Holm
wANN-BP vs ANN-BP	0.342	0.05
wANN-BP vs 3L-BP	**0.007**	0.025
wANN-BP vs w-ANN	**0.007**	0.166
wANN-BP vs ANN	**1E-5**	0.125

ranking, achieving the highest predictive accuracy in 5 datasets. The ANN-Miner algorithm came in the last place, with 4.13 overall average ranking value.

To determine the level of statistical significance, we applied a non-parametric Friedman test at the 0.05 threshold with the Holm post-hoc test, comparing wANN-BP as the control algorithm to each of the other four. These results, reported in Table 4, show for each comparison the p value and the corresponding Holm critical value. In each case, the result is statistically significant at the 0.05 significance threshold if the p value is less than the corresponding critical value; such statistically significant p values are shown in boldface. The results indicate that wANN-BP is significantly better than wANN, ANN, and 3L-BP, but not significantly better than ANN-BP.

5   Conclusions and Future Work Directions

Our results indicate that ACO can be an effective technique for constructing feed-forward neural network topology, and complement the work of Socha and Blum [15, 16] who found that ACO can be effective in learning neural network weights. In future work, we would like to integrate the $ACO_{\mathbb{R}}$ algorithm [15] to optimize neural network weights within our framework. We would also like to explore the use of different quality evaluation functions, as alternatives to the simple predictive accuracy quality function described in Eq. (4). Furthermore, we would like to apply our ACO approach to the problem of optimizing fuzzy membership functions in Adaptive Neuro-Fuzzy Inference Systems (ANFIS).

References

1. Boryczka, U., Kozak, J.: Ant Colony Decision Trees. In: International Conference on Computational Collective Intelligence, pp. 4373–4382. Springer, Berlin (2010)
2. Boryczka, U., Kozak, J.: An Adaptive Discretization in the ACDT Algorithm for Continuous Attributes. In: Jędrzejowicz, P., Nguyen, N.T., Hoang, K. (eds.) ICCCI 2011, Part II. LNCS, vol. 6923, pp. 475–484. Springer, Heidelberg (2011)
3. Liao, T., Socha, K., de Oca, M.M., Stuetzle, T., Dorigo, M.: Ant colony optimization for mixed-variable optimization problems. IEEE Transactions on Evolutionary Computation (to appear, 2014)
4. Liu, Y.P., Wu, M.G., Qian, J.X.: Evolving neural networks using the hybrid of ant colony optimization and bp algorithms. In: Wang, J., Yi, Z., Żurada, J.M., Lu, B.-L., Yin, H. (eds.) ISNN 2006. LNCS, vol. 3971, pp. 714–722. Springer, Heidelberg (2006)
5. Otero, F., Freitas, A., Johnson, C.: Handling continuous attributes in ant colony classification algorithms. In: IEEE Symposium on Computational Intelligence in Data Mining (CIDM 2009), pp. 225–231 (2009)
6. Otero, F., Freitas, A., Johnson, C.: A New Sequential Covering Strategy for Inducing Classification Rules with Ant Colony Algorithms. IEEE Transactions on Evolutionary Computation 17(1), 64–74 (2013)
7. Otero, F.E.B., Freitas, A.A., Johnson, C.G.: Inducing Decision Trees with an Ant Colony Optimization Algorithm. Applied Soft Computing 12(11), 3615–3626 (2012)
8. Parpinelli, R.S., Lopes, H.S., Freitas, A.A.: Data mining with an ant colony optimization algorithm. IEEE Transactions on Evolutionary Computation 6(4), 321–332 (2002)
9. Salama, K., Abdelbar, A., Freitas, A.: Multiple Pheromone Types and Other Extensions to the Ant-Miner Classification Rule Discovery Algorithm. Swarm Intelligence 5(3-4), 149–182 (2011)
10. Salama, K., Abdelbar, A., Otero, F., Freitas, A.: Utilizing multiple pheromones in an ant-based algorithm for continuous-attribute classification rule discovery. Applied Soft Computing 13(1), 667–675 (2013)
11. Salama, K., Freitas, A.: Clustering-based Bayesian Multi-net Classifier Construction with Ant Colony Optimization. In: IEEE Congress on Evolutionary Computation (IEEE CEC), pp. 3079–3086 (2013)
12. Salama, K.M., Freitas, A.A.: Extending the ABC-Miner Bayesian Classification Algorithm. In: Terrazas, G., Otero, F.E.B., Masegosa, A.D. (eds.) NICSO 2013. SCI, vol. 512, pp. 1–12. Springer, Heidelberg (2014)
13. Salama, K., Freitas, A.: Learning Bayesian Network Classifiers Using Ant Colony Optimization. Swarm Intelligence 7(2-3), 229–254 (2013)
14. Salama, K., Freitas, A.: Ant Colony Algorithms for Constructing Bayesian Multi-net Classifiers. Intelligent Data Analysis (accepted, 2014)
15. Socha, K., Blum, C.: Training feed-forward neural networks with ant colony optimization: An application to pattern classification. In: 5th International Conference on Hybrid Intelligent Systems (HIS 2005), pp. 233–238 (2005)
16. Socha, K., Blum, C.: An ant colony optimization algorithm for continuous optimization: Application to feed-forward neural network training. Neural Computing & Applications 16, 235–247 (2007)
17. Socha, K., Dorigo, M.: Ant colony optimization for continuous domains. European Journal of Operational Research 185, 1155–1173 (2008)

An ACO Algorithm to Solve an Extended Cutting Stock Problem for Scrap Minimization in a Bar Mill

Diego Díaz, Pablo Valledor, Paula Areces, Jorge Rodil, and Montserrat Suárez

ArcelorMittal Global R&D Asturias
P.O. Box 90 – 33400, Avilés, Asturias, Spain
{diego.diaz,pablo.valledor-pellicer,paula.areces,
jorge.rodil-martinez,montserrat.suarez}@arcelormittal.com

Abstract. We introduce the problem of scrap minimization in a bar mill with the capability of cutting several bars at once. We show this problem to be an instance of the cutting stock problem with additional constraints due to the ordering of the layers and relationships between orders spanning more than one layer.

We develop an ACO algorithm with a heuristic based on efficient patterns for search space reduction and for tackling the difficulty of building feasible solutions when the number of blocks is limited. We evaluate the performance on actual mill programs of different characteristics to show the usefulness of the algorithm.

1 Introduction

A bar mill transforms a bloom —a short, thick, square-section block of steel— into bars of specified length, section shape, and section size. The bloom is rolled in several stages, changing the section and stretching it lengthwise, and finally cut to length; normally, one mother beam —a bloom after rolling— can be cut into several bars.

One way to increase productivity in a bar mill is to cut several mother beams at once. This saves time, but makes the scheduling more complex, since all the mother beams that are cut together —a layer— must be cut to the same dimensions.[1]

The remainder of the mother beams after cutting them into bars is scrapped, and re-used in the steelshop; however, the scrap should be minimized, to avoid unnecessary waste. Figure 1 shows the process schematically.

To compound the problem, due to inaccuracies in the process upstream of the bar mill, the length of the input blooms is not constant, but varies a few centimeters from one to another. This translates into length variations of a few meters in the mother beam.

---

[1] The cutting process is slower than rolling. Blooms are rolled into mother beams sequentially, and the mother beams accumulated until a given number n is reached. Then this n mother beams are cut together, while the next n accumulate.

M. Dorigo et al. (Eds.): ANTS 2014, LNCS 8667, pp. 13–24, 2014.
© Springer International Publishing Switzerland 2014

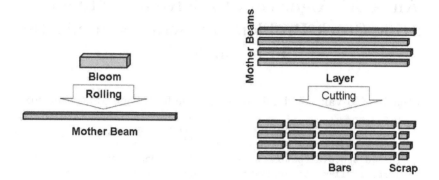

Fig. 1. Schematic of the bar mill process

The processes downstream of the bar mill are not relevant for the problem at hand, except that they impose some additional constraints: the bars are processed in groups called bundles, and there are two different processes fed from the bar mill. Each one must receive the bars one bundle at a time, which translates into a constraint that no more than two bundles can be open at the bar mill at any time. Hence if we start cutting bars for one bundle, then all bars must be for that bundle or at most one other bundle until all bars of the bundle are ready.

The mills up to now tackled this problem by manually scheduling the line, relying on expert knowledge from the line operators. The quality of the manual scheduling is at times very good, but there is an important variance, and the expert knowledge is difficult, if not impossible, to pass on.

We propose to apply a metaheuristic to automate the line scheduling. In this paper we introduce an ACO-based algorithm that solves the core problem of assigning the bars to the blooms, given the bloom processing order and layer structure as inputs, to match the way the mill works. Future extensions may widen the scope to include optimization of one or both of these inputs, starting from just the set of bars to schedule and the available blooms in stock.

We present next the definition of the problem solved, and a review of the relevant literature; in sections 4 and 5 we put forth our solution to it and the experiments we carried out to check the results, respectively. We finish by summarizing the conclusions and examining the possible improvements.

2   Problem Statement

The standard formulation for a cutting stock problem is given in equation (1), where $i = 1, \ldots, n$ are the patterns (combinations of cuts); x_i is the number of times that pattern i is used; c_i is the cost associated to pattern i; Q_j, q_j are the upper and lower limit, respectively, of the amount to be produced of the size for each order $j = 1, \ldots, m$; and a_{ij} is the number of pieces for order j in pattern i.

$$\min \sum_{i=1}^{n} c_i x_i$$

$$\text{s.t. } q_j \leq \sum_{i=1}^{n} a_{ij} x_i \leq Q_j, \quad \forall j = 1, \ldots, m \tag{1}$$

$$x_i \in \mathbb{Z}^+$$

In our problem, the following correspondences hold:

- $j = 1, \ldots, m$ are the bundles, each containing a given number of bars of a specific length.
- c_i is the amount of scrap generated with each pattern for one mother beam.
- Q_j, q_j are the upper and lower limits for the number of bars in the bundle.
- a_{ij} is the number of bars for bundle j in pattern i.

However, we still have additional complications that place our problem apart from the standard cutting stock problem. These arise from the specific constraints of the industrial process, and in one way or another are a matter of ordering, which is not present in the problem definition for cutting stock:

- We cut the mother beams in layers of varying size. This means that the pattern used must be the same for all the bars in the layer, and the actual number of bars of each length obtained depends not only on the pattern itself, but on the layer size. Instead of the number of times a pattern is used we need to identify to which specific layers it is applied.
- Not all mother beams have the same length, so the patterns actually applicable to each layer must result in a minimum scrap generation, corresponding to the length of the shortest mother beam in the layer. The cost function in equation (1) should be adapted, taking into account the individual beam lengths. This also means that not all patterns are applicable to all layers.
- The constraint that at most two open bundles can coexist is difficult, if not impossible, to express in this formulation. Furthermore, it can result in patterns that are not fully used (some of the bars do not belong to a bundle and are therefore considered scrap), and the $q_j \leq \sum_{i=1}^{n} a_{ij} x_i \leq Q_j$ constraint is no longer accurate; it should be updated together with the constraint definition.

This differs considerably from the theoretical problem, and since most of the solution methods proposed rely on these properties, we could not find one that could manage all the extensions and, at the same time, allowing for an efficient solution that could run within the stringent time limit for actual operation in the line.

We overcome this difficulty by re-stating the problem as one of assignment of bars to layers:

$$\min S(\Delta), \quad \Delta = \{\delta_{bl}\}; \; \delta_{bl} = \begin{cases} 1 & \text{if bar } b \text{ is assigned to layer } l \\ 0 & \text{otherwise,} \end{cases} \tag{2}$$

where, $\forall b$ we have $\sum_{l=1}^{n} \delta_{bl} = 1$. $S(\Delta)$ is the scrap generated by assignment Δ.

Since the constraints described above are most easily checked and enforced by proceeding sequentially in the order of processing of the layers, we decided for and ACO approach, taking advantage of the constructive nature of ACO's solution generation mechanism.

When building an assignment, we start at the first layer and select the next bar to assign to it; we repeat this, taking as candidates only feasible bars; when no more bars can be fitted in the layer, we move on to the next layer and repeat the process until all bars are assigned. If we fill up all the layers before all the bars are assigned, the solution is infeasible.

For the calculations of the scrap generated and the remaining capacity of the layer we group all bars of the same length together, and generate the needed number of cuts of each length; if the number of elements is not an exact multiple of the number of mother beams in the layer, the remainder is considered scrap, as is the unassigned length of each beam, as shown in figure 2.

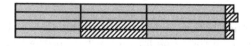

Fig. 2. A layer with 6 bars of length L_1 and 4 bars of length L_2. Scrap is marked with a slanted pattern.

The solution construction procedure may look more like generating a sequence than an assignment, but we opted for the latter as the representation to avoid the many equivalent solutions arising from different ordering of the same bars assigned to a layer. Also, the bundle constraints provide a very good guide in this situation, since only bars from the currently open bundles are candidates for assignment in the layer.

Although we discarded the patterns from the problem formulation due to their number being huge, we retained the idea of patterns to guide the construction phase, as we will explain in section 4.

We must note that even with this approach the need to check all the constraints when defining candidates for each step in the solution construction makes it rather computationally costly, and the number of solutions that can be built is very small due to the time limitation.

Once again, we found ourselves far from the typical methods in the literature, which typically build tens or hundreds of thousands of solutions in every execution. This is utterly impossible for us, with the limitations we face.

3   Literature Review

According to the typology defined by Wäscher to characterize cutting and packing problems [15], the topic analyzed in this paper belongs to the category of One-Dimensional cutting stock problem (1D CSP), and more concretely to the MSSCSP (Multiple Stock Size Cutting Stock Problem), where the objective is to reduce to the minimum the total surface required by the cutting plan considering objects of different sizes and knowing the demand for a specific set of items to be allocated onto the objects.

Multiple techniques have been proposed to solve this type of problem, from exact methods to meta-heuristics going through heuristics based on efficient cutting patterns. Dyckhoff [2] presents two different types of solution approaches. On the one hand, approaches based on cutting patterns (pattern oriented) and, on the other hand, approaches based on profiles and bars (object or item oriented). In the latter case, there is an immediate assignment between bars and profiles; this is an approach used in residual cutting stock problems (RSCP), where the set of profiles is heterogeneous.

Exact methods proposed in the literature are highly dependent on the mathematical formulation of the problem. Within this group, Kantorovich [7] proposed a first mathematical formulation of the 1D CSP problem as an integer linear problem based on binary variables to assign bars with the material available in stock. Also Scheithauer [12] proposed a branch & bound technique to solve exactly the 1D CSP problem. The main disadvantages of the integer linear approaches are the high increase of size when considering more bars in the input and the consideration of equivalent solutions in the search space. These two disadvantages make integer linear approaches impractical for our real world problem.

Other exact methods are based on cutting patterns instead of formulating the problem as an assignment approach. These methods, such as Gilmore and Gomory [4], consider as cutting patterns the feasible combinations of bars for each profile in stock and are focused on decomposing the problem and applying the Dantzig-Wolfe method to calculate the solution (column generation method). Abbas [1], as an example of a real application, proposed an improved linear programming model, applied to the steel industry, with the objective of trim loss reduction (TLR). Although these approaches do not have the problem of symmetric search space, they have the inconvenient of the potentially huge size of the cutting patterns set.

Due to the NP-hardness of the 1D CSP problem and with the objective of reducing the search space size and computing time of the problem, many authors have proposed constructive heuristics to get a trade-off solution between quality and time. Haessler [5] proposed a greedy heuristic, called SHP (Sequential Heuristic Procedure), to schedule cutting patterns, with small material losses and high frequency, until the demand is satisfied. Vasko [13] developed a hierarchical algorithm to generate cutting patterns running a search algorithm repeatedly times based on a branch & bound technique and applied to the steel industry.

Nowadays the presence of meta-heuristic techniques applied to the cutting stock problem has increased importantly. In this group, Wagner [14] developed a genetic algorithm to solve the 1D CSP, Jahromi [6] compared tabu search and simulated annealing meta-heuristics, Nozarian [10] designed an imperialist competitive method focused on trim-loss concentration, Lu [8] developed an integrated method based on a genetic algorithm and a corner arrangement method to solve CSP in the TFT-LCD industry and Pradenas [11] proposed a solution that integrates a genetic algorithm and an integer programming problem solved by CPLEX.

Ant Colony Optimization (ACO) techniques are also applied to the 1D CSP problem. Recently, Eshghi and Javanshir [3] proposed an ACO algorithm where each ant selects probabilistically an item to be cut (a specific length) and the desired cutting pattern to perform that cut by another probabilistic rule. This ACO algorithm has the particularity that the number of ants in each iteration is variable depending on the different cut types remaining to be done.

In other applications of ACO, Qiang [9] developed an ACO applied to the MSSCSP problem, including a mutation operator to avoid the phenomenon of precocity and stagnation emerging, and Yang [16] proposed an ACO algorithm applied to 1D CSP with improvement strategies on part encoding, solution path, state transition probability and pheromone updating rules.

However, we can only generate and evaluate a small number of solutions due to the added complexity and the time limitation. Therefore, most of the techniques proposed in the literature do not fit: all of them rely on evaluating a large number of solutions, either explicitly or by generating several instances under different conditions, competing colonies, or similar approaches. For this reason, we needed to devise our own version of the algorithm.

4   Solution Using ACO with Efficient Patterns

The problem size is on average to schedule 40 bundles, resulting in search space sizes of the order of 10^{48} combinations. Since the model will be used online in a Bar Mill, we have a relatively tight budget of 5 minutes of computing time for each execution, so we made an extra effort to reduce the search space.

An additional problem is that achieving feasible solutions resulted to be the most difficult part of the problem, with negligible differences in cost from one feasible solution to another. This is something we only found out as we managed to achieve solutions, so we could not take it into account during the initial algorithm design phase.

Infeasibility arises when, in the construction phase, an ant finds itself incapable of fitting all bundles into the available space, the given set of layers of beams. We call solutions of this type *partial solutions*.

Therefore, in this section we have focused on two key ideas so that the ACO algorithm is capable of providing good results in terms of quality and performance: efficient patterns and partial solution management.

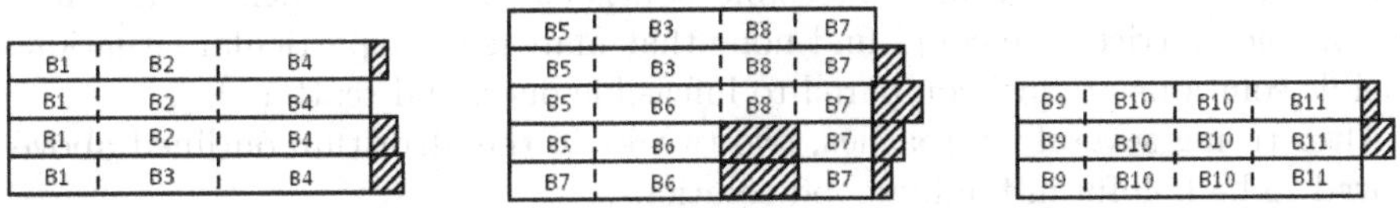

Fig. 3. Example of a candidate solution for 11 bundles across three layers; scrap is marked with a slanted pattern. Bars are identified by bundle.

4.1   Efficient Patterns

Initially, at the beginning of the algorithm, we calculate all patterns (combinations of cut lengths for a mother beam) that can be applied to the longest mother beam. The length of the patterns is bounded below by the beam length minus the shortest bar; in other words, we keep adding cuts until it is not possible to continue. Also, the lengths of the cuts are grouped and ordered, to avoid creating equivalent patterns.

As each layer contains beams of different lengths, the shortest one can be used to discard all patterns that would not fit the layer. Likewise, we focus at this stage on patterns that guarantee a low amount of scrap. Since the scrap generated will be the difference between the pattern length and the lengths of the beams, we filter patterns that are shorter than a given threshold, expressed as a percentage of the longest beam.

Additionally, the assignment of bundles also matters in scrap generation, since a cut might not be filled (*e.g.* see bundle B8 in figure 3).

We use the information provided by the patterns to guide the solution construction phase of the ants. When assigning a bar to a layer, we determine the feasible efficient patterns by filtering out from the list of patterns:

1. patterns longer than the shortest beam of the layer
2. patterns incompatible with the bars already assigned to the layer
3. patterns with lengths not present in the remaining unassigned bars
4. patterns below the threshold length

Since only bars of a length consistent with the efficient patterns are candidates, the search space is reduced. The average scrap generation from the patterns is an estimation of the expected scrap generation if the bar is assigned to the layer, which we use as a heuristic for the ants.

The actual implementation of this scheme involves keeping track of the filtered results for each layer–length combination, updated after each assignment. Also, all bars in a bundle are assigned at once, complicating a little the logic described above, but dramatically decreasing the search space size.

4.2   Partial Solution Management

Because of the difficulty of finding feasible solutions, partial solution management (PSM) becomes mandatory. Although several strategies exist to deal with combinations of feasible and infeasible solutions in ACOs, such as multiple pheromone matrices, we could find none that matched our particular situation: feasible solutions are few compared to infeasible ones, and scattered.

After trying several approaches, two tweaks to the algorithm outlined above proved to be helpful in finding good solutions:

Unification of Cost Function and Pheromone Tables. Partial and complete solutions deposit pheromone in the same pheromone table; this allows to take advantage of the knowledge that is being obtained by partial solutions to promote finding complete solutions. For this to make sense, we have defined a cost function compatible for partial and complete solutions which nudges partial solutions towards feasibility. We add to the scrap generation the length of the bars not assigned, multiplied by a penalty factor. This drives partial solutions towards feasibility, while not affecting the results for feasible solutions.

Assignment of as Many Bundles as Possible. We found out through analysis of runs of the available data sets that the main cause for infeasibility was running out of efficient patterns in the filtering phase. To allow construction of feasible, if less efficient, solutions we changed the last filtering step to relax the scrap threshold if the list of patterns was empty; in this case, the threshold was lowered enough to obtain a minimum number of patterns, and infeasibility only occurs when no pattern matches at all. This provides a trade-off between search space size (too big if all patterns are considered from the beginning) and infeasibility issues. We also included the number of remaining feasible patterns after assignment in the construction heuristic, to promote assignments that have a greater chance of leading to feasible solutions; this term has to be balanced with scrap generation.

An appropriate management of partial solutions yielded an important reduction of the time required to get good results, increasing the likelihood of finding a solution within the time limit.

4.3   Algorithmic Details

We build the pheromone matrix as a 2-dimensional table where each cell b, l corresponds to the assignment of bar b to layer l.

At each step, we evaluate the probability of selecting a bar $b^\star$ to add to the current layer l as $p(b^\star, l)$:

$$p(b^\star, l) = \frac{[\tau_{b^\star, l}]^\alpha [\eta_{b^\star, l}]^\beta}{\sum_{b \in C_l} [\tau_{b, l}]^\alpha [\eta_{b, l}]^\beta}, \qquad \eta_{b, l} = \frac{1}{1 + (N_l^p - N_{l,b}^p)} \tag{3}$$

where $\tau_{b, l}$ is the pheromone associated to the cell b, l and C_l is the set of feasible bundles for layer l, derived as explained above; $\eta_{b, l}$ is the heuristic associated to the assignment of bar b to layer l; N_l^p is the number of feasible patterns for layer l before the assignment, and $N_{l,b}^p$ is the number of feasible patters after assigning bar b. We offset the denominator by one to avoid problems when the difference is zero.

We calculate the cost of a solution Δ as $C(\Delta) = S(\Delta)+kU(\Delta)$, where $S(\Delta)$ is the scrap generated in the solution and $U(\Delta)$ is the unassigned material; both are measured in length.[2] k is a factor to discourage infeasible solutions $(U(\Delta) > 0)$; we calculate it in each instance so that k times the length of the shortest bar is larger than the typical scrap generation.

We otherwise follow typical ACO algorithms with local pheromone update as the inverse of the cost. We use 10 ants per iteration, of which only the best 3 deposit pheromone. We take $\alpha = \beta = 1$ and $\rho = 0.2$.

5   Experimental Results

We ran our tests on a virtual machine emulating a quad-core Intel Xeon processor at 2.9GHz with 16GB of RAM running 64-bit Windows 7 EE, which is representative of the kind of system where the solution will run online.

We restricted the computation time to 5 minutes, the time that will be available during operation. This results in some 200–400 cost function evaluations, depending on the instance.

We used 22 data sets, representative of the operating conditions of the line for different products. The number of bundles in each data set is quite evenly distributed between 5 and 50, with a couple of cases over 90. The results presented here come from running the algorithm ten times on each of the data sets.

The first approach, before adding efficient patterns and partial solution management, was basically unable to reach a feasible solution within the time limit.

Figure 4 shows the improvements we obtained. We focus on the time needed to reach the first feasible solution, since the scrap reduction after that is not significant.

The data sets can have quite different behavior, and the results are not comparable in either time or number of iterations. This happens because different products result in different complexities, arising from the number of bars and bundles, the number of different lengths and so on. We normalized the results to be able to draw conclusions.

For each run of each data set we recorded the number of iterations needed to reach a feasible solution, with and without partial solution management.[3] We made 0% always the first iteration, and 100% the latest recorded one for that data set; this means that both versions of the same data set (with and without partial solution management) share the normalization factor.

Infeasible solutions are arbitrarily represented as 110% to avoid distorting the graph.

---

[2] This is equivalent to weight by means of a constant factor of section area times density.

[3] Since we included efficient patterns first and partial solution management afterwards, we have no results corresponding to using partial solution management without efficient patterns.

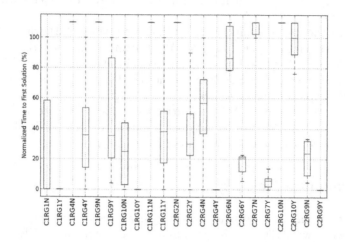

Fig. 4. Comparison of time needed to reach a feasible solution with and without partial solution management

The x axis in figure 4 lists the data sets as the data set identifier, of the form CXRGXXY for the runs with partial solution management, and CXRGXXN for the runs without. Runs without either always fail to find a solution.

The box plot follows the following convention: the line in the middle is the median, the boxes are the second and third quartiles, and the whiskers extend to the minimum and maximum.

We removed several instances from the graph to make it more readable:

- For 40.9% of the data sets (9 out of 22), we always found a feasible solution in the first iteration with PSM, but was always infeasible without.
- For 9.1% of the data sets (2 out of 22), we cannot find a feasible solution within the time limit. Both data sets correspond to a especially problematic product.

For the remaining 11 data sets, shown in the figure, we see clear improvements: out of 5 data sets for which no feasible solution is found without PSM, with it we consistently find a solution for 4, and for the other one (C2RG10) in 7 out of 10 runs. For 4 data sets we go from a variable time to reach a solution to consistently finding it in the first iteration.

The improvement for efficient patterns is that 6 out of 22 (27.3%) data sets become feasible with it, compared to without.

6   Conclusions and Future Work

The extensions to the general cutting stock problem posed by the specific constraints of the Bar Mill had a bigger impact than we expected at first. They distort the feasibility region in a way that renders traditional approaches useless.

This feasibility issue, together with the limitation in running time, were the main challenges of this problem; without them the solution would have been quite straightforward. Instead, we had to prune the search space, removing inefficient patterns without compromising feasibility, even without the ability to look forward.

We now have a system capable of matching or improving best-of-class line operators in the design of the cutting plan. This will allow homogenization of line operation, and we expect it to be easily portable to similar Bar Mills.

The system was recently checked for several hours a day for a week on real-time production data and comparing to operator performance, validating its results overall.

The good results of the combination of efficient patterns and partial solution management means that, even within the time limit allotted to the model, we can now try to improve the quality of feasible solutions. We are starting to work at designing a more complete cost function that goes beyond scrap generation, where there is little potential. Our initial aim is to improve robustness of the solution, considering that the latest beams in the cutting plan have more uncertainty in their lengths.[4]

Finally, we will study the portability to similar lines.

One point we would like to remark is that existing solutions to our knowledge rely on assumptions that do not usually hold in industrial environments. We found it impossible to adapt a method because they relied on properties of the theoretical problem that are not present in reality and/or they typically run for an unaffordable —in this case— number of iterations.

The main points of our case that are not considered in more academic exercises are:

- Variability of the processes, which is unavoidable in real life; we cannot assume a standard beam length or layer size
- Additional constraints; typically an industrial process is limited in certain aspects and can also be influenced by other processes in the supply chain. We can see this for instance in the ordering constraints, which cannot be fitted to the standard problem definition.
- Complex constraint or cost evaluation and running time limitations. Both together tend to result in the possibility of evaluating a number of solutions one or more orders of magnitude smaller than typical in the literature.

We suggest that this type of limitations has been overlooked, and that it could be an interesting field to develop.

---

[4] This is because the blooms are actually measured when they enter the process, and only part of the blooms involved have done so by the time the scheduling has to be done. For the rest, an estimation is available based on bloom production parameters.

References

1. Afshar, A.: An improved linear programming model for one-dimensional cutting stock problem. In: First International Conference on Construction in Developing Countries (ICCIDC-I). Advancing and Integrating Construction Education, Research & Practice, August 4-5 (2008)
2. Dyckhoff, H.: A typology of cutting and packing problems. European Journal of Operations Research 44(2), 145–159 (1990)
3. Eshghi, K., Javanshir, H.: A revised version of ant colony optimization for one-dimensional cutting stock problem. International Journal of Industrial Engineering 15(4), 341–348 (2008)
4. Gilmore, P., Gomoroy, R.: A linear programming approach to the cutting stock problem. Operations Research 11, 863–888 (1963)
5. Haessler, R.W.: Controlling cutting pattern changes in one dimensional trim problems. Operations Resarch 23(3), 483–493 (1975)
6. Jahromi, M.H., Tavakkoli-Moghaddam, R., Makui, A., Shamsi, A.: Solving an one-dimensional cutting stock problem by simulated annealing and tabu search. Journal of Industrial Engineering International 8 (2012)
7. Kantorovic, L.: Mathematical methods of organizing and planning production (1939). Managment Science 6, 366–422 (1960)
8. Lu, H.C., Huang, Y.H., Tseng, K.A.: An integrated algorithm for cutting stock problems in the thin-film transistor liquid crystal display industry. Computers & Industrial Engineering 64, 1084–1092 (2013)
9. Lu, Q., Wang, Z., Chen, M.: An ant colony optimization algorithm for the one-dimensional cutting stock problem with multiple stock lengths. Natural Computation, ICNC 2008 7, 475–479 (2008)
10. Nozarian, S., Jahan, M.V., Jalali, M.: An imperialist competitive algorithm for 1-d cutting stock problem. International Journal of Information Science 3(2), 25–36 (2013)
11. Pradenas, L., Garces, J., Parada, V., Ferland, J.: Genotype-phenotype heuristic approaches for a cutting stock problem with circular patterns. Engineering Applications of Artificial Intelligence 26, 2349–2355 (2013)
12. Scheithauer, G., Terno, J.: A branch & bound algorithm for solving one-dimensional cutting stock problems exactly. Applicationes Mathematicae 23(2), 151–167 (1995)
13. Vasko, F., Newhart, D., Stott, J., Kenneth, L.: A hierarchical approach for one-dimensional cutting stock problems in the steel industry that maximizes yield and minimizes overgrading. European Journal of Operations Research 114(1), 72–82 (1999)
14. Wagner, B.: A genetic algorithm solution for one-dimensional bundled stock cutting. European Journal of Operations Research 117(2), 368–381 (1999)
15. Wäscher, G., Haußner, H., Schumann, H.: An improved typology of cutting and packing problems. European Journal of Operational Research 117(2), 368–381 (2007)
16. Yang, B., Li, C., Huang, L.: Solving one-dimensional cutting-stock problem based on ant colony optimization. Natural Computation, ICNC 2009, 1188–1191 (2009)

An Experiment in Automatic Design of Robot Swarms

AutoMoDe-Vanilla, EvoStick, and Human Experts

Gianpiero Francesca[1], Manuele Brambilla[1], Arne Brutschy[1],
Lorenzo Garattoni[1], Roman Miletitch[1], Gaëtan Podevijn[1],
Andreagiovanni Reina[1], Touraj Soleymani[1], Mattia Salvaro[1,2],
Carlo Pinciroli[1], Vito Trianni[3], and Mauro Birattari[1]

[1] IRIDIA, Université Libre de Bruxelles, Belgium
{gianpiero.francesca,mbiro}@ulb.ac.be
[2] Università di Bologna, Italy
[3] ISTC-CNR, Rome, Italy

Abstract. We present an experiment in automatic design of robot swarms. For the first time in the swarm robotics literature, we perform an objective comparison of multiple design methods: we compare swarms designed by two automatic methods—AutoMoDe-Vanilla and EvoStick—with swarms manually designed by human experts. AutoMoDe-Vanilla and EvoStick have been previously published and tested on two tasks. To evaluate their generality, in this paper we test them without any modification on five new tasks. Besides confirming that AutoMoDe-Vanilla is effective, our results provide new insight into the design of robot swarms. In particular, our results indicate that, at least under the adopted experimental protocol, not only does automatic design suffer from the reality gap, but also manual design. The results also show that both manual and automatic methods benefit from bias injection. In this work, bias injection consists in restricting the design search space to the combinations of pre-existing modules. The results indicate that bias injection helps to overcome the reality gap, yielding better performing robot swarms.

1 Introduction

Automatic design is an appealing way to produce robot control software. So far, evolutionary robotics [1] has been the approach of choice for the automatic design of robot swarms [2,3]. In evolutionary robotics, robots are controlled by a neural network, whose parameters are obtained via artificial evolution in simulation. The main issue of this approach is its difficulty to overcome the *reality gap* [4], that is, the unavoidable difference between simulation and reality.

Recently, Francesca *et al.* [5] proposed a novel approach: AutoMoDe, automatic modular design. AutoMoDe synthesizes control software in the form of a probabilistic finite state machine by selecting, assembling, and fine tuning pre-existing parametric modules. The rationale behind AutoMoDe lies in the machine learning concept of bias–variance tradeoff [6]: Francesca *et al.* [5]

M. Dorigo et al. (Eds.): ANTS 2014, LNCS 8667, pp. 25–37, 2014.

conjectured that the difficulty experienced by evolutionary robotics in overcoming the reality gap bears a resemblance to the generalization problem faced by function approximators in supervised learning. They argued that such difficulty derives from an excessive *representational power* of the control architecture that is typically adopted in evolutionary robotics to be able to fine-tune the dynamics of the robot–environment interaction. Thus, Francesca *et al.* [5] proposed a solution that is reminiscent of the bias injection advocated in the machine learning literature [7] to reduce the representational power of approximators and increase their generalization ability. By synthesizing control software on the basis of pre-existing modules, AutoMoDe reduces the design space. This corresponds to injecting bias, thus decreasing the variance of the design process. As a result, AutoMoDe is expected to overcome the reality gap successfully.

Francesca *et al.* [5] defined, implemented, and tested AutoMoDe-Vanilla (hereafter simply Vanilla) and EvoStick, two specific versions for the e-puck platform [8] of AutoMoDe and evolutionary robotics, respectively. Results obtained on two tasks, aggregation and foraging, indicate that AutoMoDe is a viable and promising approach: Vanilla produced better robot control software than EvoStick, and appeared to better overcome the reality gap [5].

In this paper, we use exactly the same implementations of Vanilla and EvoStick that have been previously published in [5], and we test them on five new tasks. In our analysis, we include also swarms designed manually by human experts and swarms synthesized manually starting from the same modules used by Vanilla. We perform all tests with a swarm of 20 e-puck robots.

In this paper, we give an original contribution to the swarm robotics literature because we perform the first objective assessment of an automatic method for the design of robot swarms. This is indeed the first work in which an automatic design method previously published and tested on some tasks is tested on new tasks strictly without any modification. The new tasks were proposed by researchers that had not been involved in the development of Vanilla and that, at the moment of proposing the tasks, had only a vague idea of its functioning. In particular, they knew that Vanilla assembles pre-existing modules, but they did not have any knowledge of the modules made available to the method. As a consequence, we can claim that the tasks have not been selected to favor or disfavor Vanilla, or any of the other design methods under analysis. This work is also the first one in the domain of swarm robotics in which automatic design and manual design are compared under controlled conditions.

The results presented in this paper confirm that AutoMoDe is a viable approach to the automatic design of robot swarms. They highlight the strengths of Vanilla and also a weakness, for which we suggest a possible solution. More generally, these results provide a new insight into the design of robot swarms. They show that, at least under our experimental protocol, manual design suffers from the reality gap to an extent comparable to that of automatic design. To the best of our knowledge, this has never been discussed in the literature and has never been observed in a controlled empirical study. Finally, contrary to

Table 1. Reference model

Sensor/Actuator	Variables
proximity	$prox_i \in [0, 1]$, with $i \in \{1, 2, \ldots, 8\}$
light	$light_i \in [0, 1]$, with $i \in \{1, 2, \ldots, 8\}$
ground	$gnd_i \in \{black, gray, white\}$, with $i \in \{1, 2, 3\}$
range and bearing	$n \in \mathbb{N}$ and $r_m, \angle b_m$, with $m \in \{1, 2, \ldots, n\}$
wheels	$v_l, v_r \in [-\bar{v}, \bar{v}]$, with $\bar{v} = 0.16\,\mathrm{m/s}$

Period of the control cycle: 100 ms

our original expectations, the results presented in the paper show that human experts produce better swarms when they are constrained to use predefined modules rather then when their creativity is unconstrained.

2   Design Methods Considered

We consider four design methods: Vanilla, EvoStick, U-Human, and C-Human. These methods are intended to design control software for a swarm of e-puck robots [8] with Gumstix Overo Linux board, ground sensor, and range-and-bearing sensor—see [5] for a detailed description of the platform. To be more precise, the design methods are allowed to use a subset of the capabilities of this platform. Such subset of capabilities is formally described by the reference model reported in Table 1. The control software designed by the four methods can access sensors and actuators through suitable variables: $prox_i \in [0, 1]$ are the readings of the eight proximity sensors; $light_i \in [0, 1]$ are the readings of the eight light sensors; $gnd_i \in \{black, gray, white\}$ are the readings of the three ground sensors; n is the number of neighboring robots perceived via the range-and-bearing sensor; r_m and $\angle b_m$ are respectively the range and bearing of the m-th neighbor; finally, $v_l, v_r \in [-\bar{v}, \bar{v}]$, with $\bar{v} = 0.16\,\mathrm{m/s}$ represent the speed of the wheels. All these variables are updated with a period of 100 ms.

Two of the design methods under analysis are automatic methods: Vanilla and EvoStick. The other two are manual methods: U-Human and C-Human. Vanilla and EvoStick have been introduced by Francesca et al. [5] and are an implementation of AutoMoDe and evolutionary robotics, respectively. Concerning U-Human and C-Human, their main difference is that in U-Human the designer is *unconstrained*, that is, he is free to develop the control software in any way he prefers, whereas in C-Human the designer is *constrained* to develop a finite state machine using the same parametric modules available to Vanilla.

In the rest of this section, we introduce the four design methods featured in the experiment. For a thorough description of Vanilla and EvoStick, we refer the reader to the original publication [5].

Vanilla generates control software in the form of a finite state machine starting from a set of twelve pre-existing parametric modules: states are selected from a set of six low-level behaviors and transitions are defined on the basis of six parametric conditions. The low-level behaviors are: exploration, stop, phototaxis, anti-phototaxis, attraction, and repulsion. With the exception of stop, these behaviors include an obstacle avoidance mechanism. The conditions are: black-floor, gray-floor, white-floor, neighbor-count, inverted-neighbor-count, fixed-probability. All these modules are based on the reference model given in Table 1. For a detailed description of the modules and their parameters, see [5]. Vanilla is constrained to create finite state machines composed of at most four states, each with at most four outgoing transitions. As an optimization algorithm, Vanilla adopts F-Race [9,10]. Specifically, it uses the implementation provided by the irace package [11] for R [12]. F-Race can be essentially described as a sample & select algorithm. As already pointed out by Francesca *et al.* [5], F-Race has been adopted in Vanilla due to its simplicity: the authors wished to keep the focus on the control architecture rather than on the optimization process. Within the optimization process, control software candidates are evaluated using the ARGoS multi-robot simulator [13].

EvoStick generates control software in the form of a feed-forward neural network without hidden nodes. Inputs and outputs of the network are defined on the basis of the variables given in the reference model of Table 1. To optimize the parameters of the neural network, EvoStick adopts a standard evolutionary algorithm. Within the optimization process, candidate parameter sets are evaluated using ARGoS.

U-Human is a manual method in which the human designer is left free to design control software as he deems appropriate. Within the control software, sensors and actuators are accessed via an API that implements the reference model given in Table 1. Within the design process, the designer tests his control software using ARGoS.

C-Human is a manual method in which the human designer is constrained to use Vanilla's control architecture and modules. The designer does not directly write the control software: rather, he employs a software tool that allows him to specify a finite state machine, visualize it, and test it using ARGoS. In other words, the human designer takes the role of Vanilla's optimization algorithm. As in Vanilla, the human is constrained to create finite state machines comprised of at most four states, each with at most four outgoing transitions.

3   Experimental Protocol

In the protocol we adopt, five researchers, hereinafter referred to as *experts*, play a central role. The experts are active in swarm robotics, have about two years of experience in the domain, and are familiar with ARGoS. These experts have not been involved in the development of Vanilla and EvoStick and, at the moment of participating in the experiment, they had only a vague idea of Vanilla: they knew that Vanilla operates on pre-existing parametric modules, but they were unaware of what modules are available.

The role of each expert is threefold: i) define a task; ii) solve a task acting as U-Human; and iii) solve a task acting as C-Human. Table 2 summarizes the role

Table 2. Role of the experts, anonymously indicated here by the labels E1 to E5

task		defined by	U-Human	C-Human
SCA	shelter with constrained access	E1	E5	E4
LCN	largest covering network	E2	E1	E5
CFA	coverage with forbidden areas	E3	E2	E1
SPC	surface and perimeter coverage	E4	E3	E2
AAC	aggregation with ambient cues	E5	E4	E3

played by each expert. The criteria and the restrictions that the experts had to follow in the definition of the tasks are presented in Sect. 4. When solving a task either as U-Human or C-Human, each expert worked for four consecutive hours. The involvement of each expert spanned two days: on day one, the expert acted as U-Human on the first task assigned to him; on day two, he acted as C-Human on the second task. In both cases, the expert came to know the definition of the task he had to solve only at the beginning of the four hours. During these four hours, the expert could test the control software in simulation using ARGoS, but was not allowed to test it in reality on the e-pucks.

Regarding the automatic design methods, Vanilla and EvoStick have been allowed a *design budget* of 200,000 simulations for each task. The design process has been conducted on a high performance computing cluster comprised of 400 opteron6272 cores. Vanilla and EvoStick produced the control software for each task in about 2 hours and 20 minutes.

To summarize, all methods under analysis i) produce control software for the same robotic platform formally described by the reference model given in Table 1; ii) complete the design process within 4 hours; and iii) use the same simulator to assess candidate control software during the design process. The protocol of the experiment does not allow any modification of the control software on the basis of its performance in reality on the e-pucks.

We used ARGoS to cross-compile the control software for the e-puck. We then uploaded it to the robots without any modification. We evaluated the control software generated by the four methods for each of the five tasks via 10 runs on the robots. We computed the performance of the robot swarm in an automatic way using a tracking system [14] that acquires images via a ceiling-mounted camera and records the position and orientation of all robots every 100 ms. To assess the impact of the reality gap on the four design methods, we performed also a set of 10 runs per task in simulation.

For each task, we report a notched box-and-whisker plot that summarizes the results: wide boxes represent data gathered with robots, narrow boxes data obtained in simulation. If notches of two boxes do not overlap, the observed difference is significant. We report also the results of a Friedman test [15] that aggregates the data gathered with the robots over the five tasks.

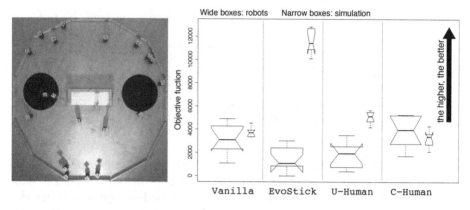

Fig. 1. SCA – arena and 20 e-pucks (left); results of the analysis (right). Robots should aggregate in the white shelter.

4   Tasks and Results

Each of the tasks has been independently defined by one of the experts. Each expert has been provided with the reference model of Table 1 and has been asked to conceive a task that he would be able to solve with a swarm of 20 robots characterized by the given reference model. The expert has also been provided with a list of constraints that the task definition must satisfy: The arena is a dodecagonal area of $4.91\,\mathrm{m}^2$ surrounded by walls. The floor of the arena is gray. The arena can contain up to 3 colored regions in which the floor can be either black or white. These regions can be either circular, with diameter of up to 0.6 m, or rectangular, with sides up to 0.6 m. The setup might include a single light source positioned outside the arena, at 0.75 m from the ground. It might include also up to 5 cuboidal obstacles of size $0.05\,\mathrm{m} \times 0.05\,\mathrm{m} \times L$, where $0.05\,\mathrm{m} \leq L \leq 0.80\,\mathrm{m}$. The swarm comprises 20 e-puck robots in the configuration described in Sect. 2. The duration of each run is $T = 120\,\mathrm{s}$. At the beginning of the run, the robots are randomly distributed in the arena. The task must be formally described by an objective function, which must be either maximized or minimized. The objective function must be defined on the basis of the position of the robots, evaluated with a period of 100 ms.

In the rest of this section, we describe the five tasks and we report the results obtained by the four design methods. For a more detailed description of the tasks and of their objective functions, see the online supplementary material [16].

SCA – Shelter with Constrained Access

In SCA, the goal of the swarm is to maximize the number of robots on an aggregation area. The aggregation area has a rectangular shape, is characterized by a white ground, and is surrounded by walls on three sides. The environment presents also a light source and two black regions that are positioned in front and aside the aggregation area, respectively—see Fig. 1.

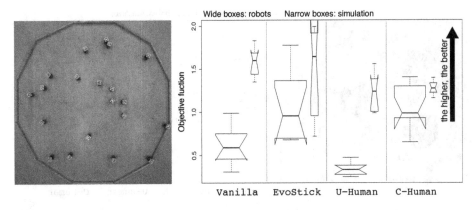

Fig. 2. LCN – arena and 20 e-pucks (left); results of the analysis (right). Robots should cover the largest possible area while maintaining connection with one another.

Formally, the problem is defined as the maximization of $F_{SCA} = \sum_{t=1}^{T} N_a(t)$, where $N_a(t)$ is the number of robots in the aggregation area at time t, and T is the duration of the run.

Results. C-Human and Vanilla perform better than the other methods. In particular, Vanilla is significantly better that EvoStick—this is indicated by the fact that, in Fig. 1, the notches of the respective boxes do not overlap. An interesting result is the inability of EvoStick to overcome the reality gap. The same observation can be made also for U-Human, even though to a far minor extent. In contrast, C-Human and Vanilla overcome the reality gap satisfactorily.

LCN – Largest Covering Network

In LCN, the robots must maintain connection with each other, while trying to cover the largest possible area—see Fig. 2 for a picture of the experimental setting. We assume that i) two robots are connected when their distance is less than 0.25 m, and ii) each robot covers a circular area of radius 0.35 m.

Formally, the problem is defined as the maximization of $F_{LCN} = A_{C(T)}$ where $C(T)$ is the largest group of connected robots at the end T of the run and $A_{C(T)}$ is the surface of the union of the areas covered by the robots in $C(T)$.

Results. C-Human and EvoStick achieve better performance compared to the other methods, with Vanilla performing slightly better than U-Human. The methods performing worse are those that encounter more difficulties in overcoming the reality gap: U-Human and Vanilla.

CFA – Coverage with Forbidden Areas

In CFA, the goal of the swarm is to cover the entire arena except a few forbidden areas characterized by a black ground—see Fig. 3.

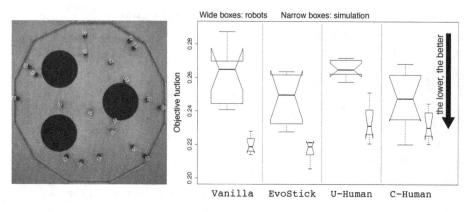

Fig. 3. CFA – arena and 20 e-pucks (left); results of the analysis (right). Robot should cover the arena except the forbidden black areas.

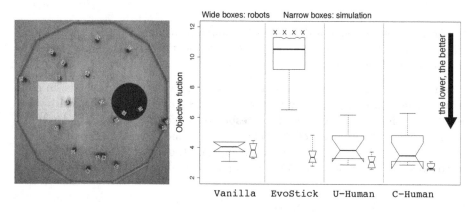

Fig. 4. SPC – arena and 20 e-pucks (left); results of the analysis (right). Robot should cover the surface of the white square and the perimeter of the black circle.

Formally, the problem is defined as the minimization of $F_{CFA} = E[d(T)]$, where $E[d(T)]$ is the expected distance, at time T, between a generic point of the arena and the closest robot that is not in a forbidden area. Distances are measured in meters.

Results. All methods perform more or less similarly. The results are all within a range of few centimeters, that is, less that half of the e-puck's diameter. Concerning the reality gap, for all methods we observe differences between simulation and reality, but these differences are small in absolute terms. Also in this case, they are within a range of few centimeters.

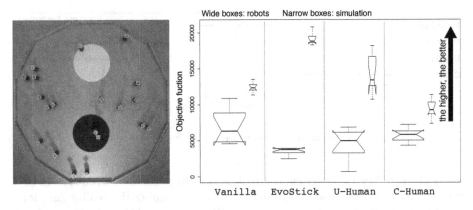

Fig. 5. AAC – arena and 20 e-pucks (left); results of the analysis (right). Robot should aggregate in the black region.

SPC – Surface and Perimeter Coverage

In SPC, the goal of the swarm is to cover the surface of a white square region and the perimeter of a black circular region—see Fig. 4.

Formally, the problem is defined as the minimization of $F_{SPC} = c_a E[d_a(T)] + c_p E[d_p(T)]$, where $E[d_a(T)]$ is the expected distance, at time T, between a generic point of the white region and the closest robot positioned on the white region itself; $E[d_p(T)]$ is the expected distance, at time T, between a generic point of the perimeter of the black region and the closest robot positioned on the perimeter itself; and c_a and c_p are normalization factors [16]. Failing to place at least a robot on the surface of the white region and/or on the perimeter of the black region is a major failure. In this case, $E[d_a(T)]$ and $E[d_p(T)]$ are undefined and we thus assign an arbitrarily large value to F_{SPC}.

Results. The most notable element is that EvoStick is not able to overcome the reality gap and achieves significantly worse results than the other methods. The four Xs marked in the plot indicate four runs that resulted in a major failure. Vanilla, U-Human, and C-Human perform comparably well.

AAC – Aggregation with Ambient Cues

In AAC, the goal of the swarm is to maximize the number of robots on an aggregation area represented by a black region. Besides the black region, the environment comprises a white region and a light source that is placed south of the black region—see Fig. 5.

Formally, the problem is defined as the maximization of $F_{AAC} = \sum_{t=1}^{T} N_b(t)$, where $N_b(t)$ is the number of robots on the black region at time t.

Results. Vanilla performs slightly better than U-Human and C-Human, and significantly better than EvoStick. Concerning the manual methods, C-Human performs slightly better than U-Human. The greatest difference among the methods

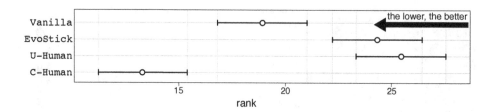

Fig. 6. Friedman test on aggregate data from the five tasks

lies in their ability to overcome the reality gap. In particular, EvoStick is the method that has the most severe difficulty in overcoming the reality gap, followed by U-Human. Vanilla and C-Human still present problems, but to a minor extent compared to the other methods.

5   Analysis and Discussion

To aggregate the results presented in Sect. 4, we perform a Friedman test [15], using the task as a blocking factor and considering 10 replicates per task. The outcome of the test is represented in Fig. 6. The plot represents the expected rank obtained by a design method in the robot experiments, together with a confidence interval. If the confidence intervals of two methods do not overlap, the difference between the expected rank of the two is statistically significant. The test indicates that C-Human perform significantly better than Vanilla, which, in turn, perform significantly better than EvoStick and U-Human.

These results confirm those obtained by Francesca et al. [5]: Vanilla produced swarms with significantly better performance than those produced by EvoStick. However, the results also highlight that Vanilla has a limitation: as already noted in Francesca et al. [5], F-Race, the optimization algorithm adopted in Vanilla, is not particularly powerful and is unable to fully exploit the potential of the available parametric modules—see the results of C-Human, which is based on the same modules. The results clearly suggest that Vanilla can be improved by adopting a more powerful optimization algorithm.

The analysis of the swarms produced by human experts are particularly interesting and informative on their own. First of all, our results show that, when it is not possible to modify the developed control software on the basis of its performance in reality, manual design suffers from the reality gap, as automatic design does. In other words, it is difficult for human experts to foresee whether the developed control software will work in reality as expected or not.

Moreover, we observed that, under the protocol we adopted, human experts produce better swarms when they are constrained to use predefined modules. This result was unexpected and appears counter-intuitive. We expected that the understanding and intuition of human experts would have produced excellent results in case their creativity had been left unconstrained. We expected that the restriction to use predefined modules would have prevented human experts from fully expressing their potential. Our results proved us wrong: although the control software produced by U-Human scored well in simulation, it failed to be effective in reality. Our results clearly indicate that the restriction to use predefined modules enables C-Human to successfully overcome the reality gap.

Concerning the comparison between manual and automatic design, Vanilla produced swarms that are significantly better than those produced by U-Human, but worse that those produced by C-Human. This is a promising result. It proves that the core idea of AutoMoDe is valid: by constraining the design space to the control software that can be obtained assembling predefined modules, one effectively increases the ability to overcome the reality gap. This insight is valid for both automatic and manual design.

As the set of modules used by C-Human are the same used by Vanilla, the performance advantage of C-Human over Vanilla is to be fully ascribed to the limitations of Vanilla's optimization algorithm, as already discussed above. The results obtained by C-Human and by Vanilla show that the set of modules is generally appropriate for tackling swarm robotics tasks with the robotic platform considered: they enabled the synthesis of control software that performed satisfactorily across all the five tasks.

6   Conclusions

In this paper, we presented an experiment in automatic design of robot swarms. This experiment introduces a number of novelties with respect to the literature. In particular, the two automatic methods under analysis—Vanilla and EvoStick—had been previously published [5] and have been used here strictly without any modification. In swarm robotics, this is the first time that i) an automatic design method is tested on as many as five tasks, without adapting it to each of them; ii) the tasks considered are different from the one for which the method has been originally proposed; and iii) the tasks are devised by researchers that had not been involved in the development of the method. Moreover, this is the first time that a comparison between automatic and manual design methods is performed under controlled conditions.

The results of the experiment are encouraging. First of all, they confirm previous results obtained on other tasks [5]: Vanilla performs better that EvoStick. Second, they show that, under the protocol we adopted, human experts produce better swarms when they are constrained to use pre-existing modules: C-Human outperforms U-Human. Together, the superiority of Vanilla over EvoStick and of C-Human over U-Human corroborate the core hypothesis behind AutoMoDe: by introducing a bias in the design process—that is, by restricting the design space—one obtains better robot swarms. Moreover, Vanilla outperformed U-Human, the unconstrained manual design: this is the first clear empirical evidence that the automatic design of robot swarms is a viable reality. On the other hand, we do not consider it a failure that C-Human scored better than Vanilla. As C-Human uses the same modules defined in Vanilla, differences are to be ascribed to the limitations of Vanilla's optimization algorithm. The results indicate that

`Vanilla`'s module set is appropriate to solve swarm robotics tasks with the platform considered in our study.

Our short-term future work will focus on the development of an improved version of `Vanilla` that, taking into account the indications emerged from our results, will adopt a more powerful optimization algorithm. In the medium term, we will develop an instance of AutoMoDe for a more complex reference model.

Acknowledgments. We thank Maria Zampetti and Maxime Bussios for their help with the robots, and Marco Chiarandini for his implementation of the Friedman test. This research has received funding from the European Research Council under the European Union's Seventh Framework Programme—ERC grant agreement n. 246939. It received funding also from the European Science Foundation via the H2SWARM project. Vito Trianni acknowledges support from the Italian CNR. Arne Brutschy and Mauro Birattari acknowledge support from the Belgian F.R.S.-FNRS.

References

1. Nolfi, S., Floreano, D.: Evolutionary Robotics: The Biology, Intelligence, and Technology of Self-organizing Machines. MIT Press, Cambridge (2000)
2. Brambilla, M., Ferrante, E., Birattari, M., Dorigo, M.: Swarm robotics: A review from the swarm engineering perspective. Swarm Intelligence 7(1), 1–41 (2013)
3. Dorigo, M., Birattari, M., Brambilla, M.: Swarm robotics. Scholarpedia 9(1), 1463 (2014)
4. Trianni, V., Nolfi, S.: Engineering the evolution of self-organizing behaviors in swarm robotics: A case study. Artificial Life 17(3), 183–202 (2011)
5. Francesca, G., Brambilla, M., Brutschy, A., Trianni, V., Birattari, M.: AutoMoDe: A novel approach to the automatic design of control software for robot swarms. Swarm Intelligence 8(2), 89–112 (2014)
6. Geman, S., Bienenstock, E., Doursat, R.: Neural networks and the bias/variance dilemma. Neural Computation 4(1), 1–58 (1992)
7. Dietterich, T., Kong, E.B.: Machine learning bias, statistical bias, and statistical variance of decision tree algorithms. Technical report, Department of Computer Science, Oregon State University (1995)
8. Mondada, F., et al.: The e-puck, a robot designed for education in engineering. In: 9th Conf. on Autonomous Robot Systems and Competitions, Portugal, Instituto Politécnico de Castelo Branco, pp. 59–65 (2009)
9. Birattari, M., Stützle, T., Paquete, L., Varrentrapp, K.: A racing algorithm for configuring metaheuristics. In: Proc. of the Genetic and Evolutionary Computation Conference (GECCO 2002), pp. 11–18. Morgan Kaufmann, San Francisco (2002)
10. Birattari, M.: Tuning Metaheuristics. Springer, Berlin (2009)
11. López-Ibáñez, M., Dubois-Lacoste, J., Stützle, T., Birattari, M.: The irace package, iterated race for automatic algorithm configuration. Technical Report TR/IRIDIA/2011-004, IRIDIA, Université Libre de Bruxelles, Belgium (2011)
12. R Development Core Team: R: A language and environment for statistical computing. R Foundation for Statistical Computing (2008)

13. Pinciroli, C., Trianni, V., O'Grady, R., Pini, G., Brutschy, A., Brambilla, M., Mathews, N., Ferrante, E., Di Caro, G., Ducatelle, F., Birattari, M., Gambardella, L.M., Dorigo, M.: ARGoS: A modular, parallel, multi-engine simulator for multi-robot systems. Swarm Intelligence 6(4), 271–295 (2012)
14. Stranieri, A., Turgut, A., Francesca, G., Reina, A., Dorigo, M., Birattari, M.: IRIDIA's arena tracking system. Technical Report TR/IRIDIA/2013-013, IRIDIA, Université Libre de Bruxelles, Belgium (2013)
15. Conover, W.J.: Practical Nonparametric Statistics. Wiley, New York (1999)
16. Francesca, G., et al.: An experiment in automatic design of robot swarms. Supplementary Material (2014), http://iridia.ulb.ac.be/supp/IridiaSupp2014-004

Angle Modulated Particle Swarm Variants

Barend J. Leonard and Andries P. Engelbrecht

Department of Computer Science,
University of Pretoria, South Africa
{bleonard,engel}@cs.up.ac.za

Abstract. This paper proposes variants of the angle modulated particle swarm optimization (AMPSO) algorithm. A number of limitations of the original AMPSO algorithm are identified and the proposed variants aim to remove these limitations. The new variants are then compared to AMPSO on a number of binary problems in various dimensions. It is shown that the performance of the variants is superior to AMPSO in many problem cases. This indicates that the identified limitations may have a significant effect on performance, but that the effects can be overcome by removing those limitations. It is also observed that the ability of the variants to initialize a wider range of potential solutions can be helpful during the search process.

1   Introduction

Angle modulated particle swarm optimization (AMPSO) [12],[13] is an optimization technique that applies particle swarm optimization (PSO) [8] to binary-valued optimization problems [1]. AMPSO makes use of a four-dimensional trigonometric function to generate high-dimensional bit strings. This function is referred to as the *generating function.* The purpose of PSO in AMPSO is to optimize the coefficients of the trigonometric function, such that the resulting bit string is the optimal solution to some binary problem. AMPSO is a generally preferred alternative to Kennedy and Eberhart's discrete binary PSO [9],[13] and has successfully been applied to real-world problems, including power outage prevention [10] and supply chain management [16].

A number of limitations in the original AMPSO model, as presented by Pampara *et al.* [13], potentially inhibit the ability of AMPSO to produce optimal bit strings. The limitations in the generating function include:

- the omittance of a scalable amplitude parameter,
- a pre-defined distance between samples of the bit generating function, and
- a pre-determined starting position when sampling.

In addition to the above limitations, this paper shows how a small initialization domain for PSO also inhibits the search capabilities of AMPSO.

The purpose of this study is to investigate whether some of the limitations that were imposed on AMPSO can effectively be removed or alleviated. This is

---

[1] For the remainder of this paper, referred to as binary problems.

M. Dorigo et al. (Eds.): ANTS 2014, LNCS 8667, pp. 38–49, 2014.
© Springer International Publishing Switzerland 2014

done by introducing three variants of AMPSO that target the limitations listed above. The resulting AMPSO variants are compared to AMPSO to determine whether they perform better on a number of known binary problems.

The remainder of this paper is structured as follows: Section 2 gives an overview of the PSO algorithm. Section 3 explains the AMPSO algorithm, describes the effects of the coefficients of the generating function, and discusses the limitations of AMPSO. Three new AMPSO variants are introduced in section 4. Section 5 discusses the benchmark functions used for this study. Section 6 describes the experimental setup, while the results are presented and discussed in section 7. The paper is concluded in section 8.

2   Particle Swarm Optimization

Particle swarm optimization [8] is a stochastic, population-based optimization technique. The algorithm maintains a population of solutions, known as *particles*. Each particle has a *position* $\mathbf{x}_i$ and a *velocity* $\mathbf{v}_i$, and also keeps track of a *personal best position* $\mathbf{y}_i$ that it has found during the search process. The best personal best position is referred to as the *global best position*, $\hat{\mathbf{y}}$.

At each time step $t + 1$, the velocity of each particle is updated as follows:

$$\mathbf{v}_i(t+1) = \omega\mathbf{v}_i(t) + c_1\mathbf{r}_1(t)[\mathbf{y}_i(t) - \mathbf{x}_i(t)] + c_2\mathbf{r}_2(t)[\hat{\mathbf{y}}(t) - \mathbf{x}_i(t)], \qquad (1)$$

where ω is the inertia weight [14], c_1 and c_2 are acceleration constants, and $r_{1j}(t)$ and $r_{2j}(t)$ are random values, sampled from $U(0,1)$ in each dimension $j = 1, \ldots, n_x$. Then, the position of every particle is updated using

$$\mathbf{x}_i(t+1) = \mathbf{x}_i(t) + \mathbf{v}_i(t+1). \qquad (2)$$

The resulting behaviour is that particles stochastically return to regions of the search space where good solutions have previously been found.

3   Angle Modulated Particle Swarm Optimization

Angle modulated PSO makes use of a particle swarm optimizer to find the optimal coefficients of the following trigonometric bit generating function:

$$g(x) = sin(2\pi(x - a) \times b \times cos(2\pi(x - a) \times c)) + d \qquad (3)$$

The four coefficients, a, b, c and d control the shape of the function. Binary solutions are then generated by sampling the generating function at regular intervals $x = (0, 1, 2, \ldots, n_b)$, where n_b is the length of the required binary solution. The recorded value $g(x)$ at each location is mapped to a binary bit as follows:

$$\begin{aligned} &\text{if } g(x) > 0 \rightarrow 1, \\ &\text{if } g(x) \leq 0 \rightarrow 0. \end{aligned} \qquad (4)$$

An example 5-dimensional solution is constructed in Figure 1.

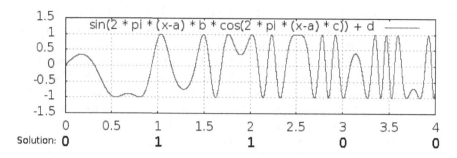

Fig. 1. Bit generating function with $a = 0$, $b = 0.5$, $c = 0.8$, and $d = 0$

3.1   AMPSO Coefficients

During the PSO initialization phase, the particles in the swarm are initialized uniformly in the domain $(-1, 1)^4$, so that the initial values of all four coefficients are between -1 and 1 [12].

The coefficients of the generating function influence the shape of the function in various ways:

- a controls the horizontal shift of the entire function,
- b influences the frequency of the *sin* wave, but also controls the amplitude of the *cos* wave.
- c affects the frequency of the *cos* wave, and
- d controls the vertical shift of the entire function.

3.2   AMPSO Limitations

The first limitation imposed on the generating function is the omittance of a variable amplitude of the *sin* wave. The generating function in AMPSO is a combined *sin* and *cos* wave. While the amplitude of the *cos* wave is controlled by the b coefficient, there is no variable amplitude coefficient that affects the *sin* wave. The result is that, without a sufficiently large vertical shift (at least $d > 1$ or $d < -1$), the generating function will always have regions that produce 1-bits, and other regions that produce 0-bits. This limitation, combined with a small initialization range for PSO, introduces some difficulty in finding binary solutions that consist of a majority of either 1's or 0's. Although this problem is somewhat addressed by the roaming behaviour of particles in PSO (i.e. the tendency of particles to exit the search domain) [4], the introduction of an amplitude parameter may enable AMPSO to find optimal solutions faster.

The second and third limitations, as mentioned in section 1, can be thought of as a single shortcoming on the generating function: sampling from a pre-defined domain at regular intervals. The effect of this is that, if the correct bit string is generated by g in some domain, the algorithm must rely on the horizontal shift parameter a to bring the solution into the sampling domain. In addition, the two frequency parameters b and c must ensure that the frequency of g correctly matches the distance between samples. Therefore, the three coefficients a, b, and c are interdependent. Allowing the algorithm to manipulate the sampling range and sampling interval, independent of the coefficients of g, could potentially enable it to find the solution faster.

4 AMPSO Variations

Three new AMPSO variants are proposed in this section to remove the restrictions that were discussed in section 3.2.

4.1 Amplitude AMPSO

The first variant is called the amplitude angle modulated PSO (A-AMPSO). A-AMPSO augments the generating function with an additional variable e, which controls the amplitude of the sin wave. The generating function is then given by

$$g(x) = e \times sin(2\pi(x - a) \times b \times cos(2\pi(x - a) \times c)) + d, \tag{5}$$

and the position of a particle i becomes a 5-dimensional vector:

$$\mathbf{x}_i = (a, b, c, d, e) \tag{6}$$

The advantage of this approach is that a small amplitude, combined with relatively large vertical shifts (supplied by d), can easily push parts (or all) of the generating function above or below the x-axis. Therefore, it is easier to produce solutions with a majority of 1's or 0's. The A-AMPSO generating function is illustrated in Figure 2.

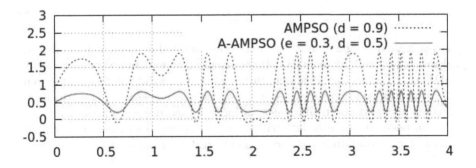

Fig. 2. Bit generating function with $a = 0$, $b = 0.9$, $c = 0.5$

4.2 Increased-Domain AMPSO

The second variant is called the increased-domain AMPSO (ID-AMPSO). ID-AMPSO simply increases the initialization domain of PSO to $(-1.5, 1.5)^4$. Recall from section 3.1 that the original AMPSO had an initialization range of $(-1, 1)^4$.

The main advantage of ID-AMPSO is that a greater variation of solutions are possible during initialization, without any increase in complexity. As an example, consider the bit-string "111111". This string of 1-bits is impossible to create during initialization in the standard AMPSO, because all the coefficients are limited to values between 0 and 1, and there is no amplitude parameter. Since the amplitude of the original generating function is 1, the only way to produce a bit string of 1's is to have a value $d > 1$, which is outside the initialization range. Increasing the initialization range solves this problem.

The ability to consistently initialize bit strings containing only 1-bits is valuable when considering many benchmark problems. The usefulness of this feature in practice must still be investigated.

4.3 Min-Max AMPSO

The final proposed variant is called the min-max AMPSO (MM-AMPSO). This variant augments the position vectors of particles with two additional dimensions. The additional dimensions control the sampling range of the generating function. The position of a particle i is thus a 6-dimensional vector:

$$\mathbf{x}_i = (a, b, c, d, \alpha_1, \alpha_2), \tag{7}$$

where α_1 and α_2 are the bounds of the sampling range.

Let $\alpha_l = min\{\alpha_1, \alpha_2\}$, and $\alpha_u = max\{\alpha_1, \alpha_2\}$. The generating function is then sampled at every δ^{th} position in the range $[\alpha_l, \alpha_u)$, where

$$\delta = \frac{\alpha_u - \alpha_l}{n_b}, \tag{8}$$

ensuring regular sampling intervals within the specified domain. Values are sampled from the standard generating function, given in equation (3).

MM-AMPSO has the advantage that a wider variety of solutions can potentially be generated for given values of a, b, c, and d, because α_1 and α_2 are able to effectively zoom into parts of the generating function. If $\alpha_u - \alpha_l$ is small enough, it is even possible to create bit strings containing only 1's or only 0's, without the need for an amplitude parameter, as illustrated in Figure 3.

5 Benchmark Problems

Five combinatorial benchmark problems were used in this study. Each problem is discussed below.

5.1   N-Queens

For the N-Queens problem, the objective is to place n queens on an $n \times n$ chess board in such a way that no queen is able to capture any other queen, based on the rules of chess. Various solutions to the problem can be found in [3], [11], and [15]. The N-Queens problem is a minimization problem.

For the purpose of this study, the problem was investigated for board sizes of 8, 9, 10, 11, 12, 20, and 25. These board sizes led to solution representations of 64, 81, 100, 121, 144, 400, and 625 bits, respectively.

5.2   Knight's Coverage

The knight's coverage problem [5] is also a chess-board problem and is defined as follows: for any $n \times n$ chess board, use the minimum number of knights to *cover* the maximum number of squares on the chess board. A square on the chess board is covered only if a knight is occupying the square, or a knight may move to the square in a single move from its current location. The allowed movements of knights are defined by the rules of chess. This problem is a minimization problem.

For this study, the problem was optimized for chess board sizes of 8, 9, 10, 11, 12, 15, and 20. The resulting solution representations had 64, 81, 100, 121, 144, 225, and 400 dimensions, respectively.

5.3   Knight's Tour

The knight's tour problem [7] is the third chess board problem that was used in this study. Given an $n \times n$ chess board, the aim is to find a sequence of moves for a single knight, such that every square on the board is visited exactly once. The knight may start on any square, but its movement is restricted by the normal rules of chess. The Knight's Tour is a maximization problem.

From any position on the chess board, a knight has a maximum of eight valid moves. Each move can therefore be encoded as a 3-bit binary value. The complete solution to this problem is then a bit string with $3n$ bits.

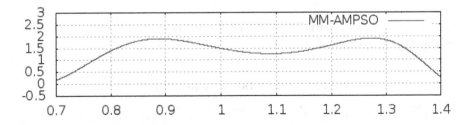

Fig. 3. The exact same generating function as the AMPSO function in Figure 2, except the sampling range is $[0.7, 1.4)$, ensuring that all samples produce 1-bits

This problem was investigated for the following board sizes: 4, 5, 6, 7, 8, 10, and 12. The respective solution representations were 12, 75, 108, 147, 192, 300, and 432 bits in length.

5.4 Deceptive Problems

Finaly, the following two deceptive problems were used in this study: order-3 deceptive, and order-5 deceptive. The concept of deception was introduced by Goldberg [6] and aims to deliberately mislead the evolutionary process, modelled in genetic algorithms. Deceptive problems are designed in such a way that a *deceptive attractor* leads the search away from the global optimum. For example, for an order-3 deceptive problem f, the bits in a candidate solution are grouped into $\frac{n_b}{3}$ three-bit groups. Assuming that f is a maximization problem, with a global maximum at $(1,1,1,\ldots)$ and a global minimum at $(0,0,0,\ldots)$, the relationships in Table 1 must hold for every group of three bits in a candidate solution. Both the order-3 and order-5 deceptive problems are defined as maximization problems.

For this study, the order-3 and order-5 deceptive problems were optimized in the following dimensions: 15, 30, 45, 60, and 75.

6   Experiments

For this study, the three AMPSO variants proposed in section 4 were compared to the original AMPSO on the five benchmark functions described in section 5. The optimal PSO parameter settings for each problem were obtained using iterated F-race [1], [2]. Note that AMPSO was compared to a number of additional algorithms in [12].

Each algorithm executed for 1000 iterations, and average results over 30 runs are reported in section 7. Finally, pair-wise Mann-Whitney U tests were performed to determine significant wins and losses for all algorithms across all problems at a 95% confidence interval.

7   Results and Discussion

Figures 4 to 7 show the fitness profiles of the different algorithms for the lowest and highest dimensions of the various benchmark problems.

Table 1. Order-3 Deceptive Relationships

$f(\ldots 0^{**}\ldots) > f(\ldots 1^{**}\ldots)$	$f(\ldots 00^{*}\ldots) > f(\ldots 11^{*}\ldots), f(\ldots 01^{*}\ldots), f(\ldots 10^{*}\ldots)$
$f(\ldots {}^{*}0^{*}\ldots) > f(\ldots {}^{*}1^{*}\ldots)$	$f(\ldots 0^{*}0\ldots) > f(\ldots 1^{*}1\ldots), f(\ldots 0^{*}1\ldots), f(\ldots 1^{*}0\ldots)$
$f(\ldots {}^{**}0\ldots) > f(\ldots {}^{**}1\ldots)$	$f(\ldots {}^{*}00\ldots) > f(\ldots {}^{*}11\ldots), f(\ldots {}^{*}01\ldots), f(\ldots {}^{*}10\ldots)$

It is observed from Figure 4 that AMPSO, ID-AMPSO and A-AMPSO performed comparably in 64 dimensions, with AMPSO reaching a lower average fitness at times. MM-AMPSO lagged notably behind. This indicates that the two additional dimensions added to the position vector in MM-AMPSO imposed additional complexity that the algorithm was not able to overcome for this particular problem. In 652 dimensions, AMPSO has the steepest initial gradient, but again ID-AMPSO and A-AMPSO eventually reach fitness values close to that of AMPSO.

Figure 5 shows that all algorithms performed comparably in 64 dimensions on the Knight's Coverage problem, with MM-AMPSO obtaining a slightly lower fitness value on average. In 400 dimensions, MM-AMPSO had a lower initial gradient compared to the other algorithms. However, MM-AMPSO overtook all the competitors in the last 100 iterations. The results for the Knight's Coverage problem are in contrast with what was observed for the N-Queens problem. This indicates that the capability to adjust the sampling domain of the generating function during the search process can be beneficial in some problem cases.

For the Knight's Tour problem, Figure 6 shows that AMPSO obtained the highest average fitness in 48 dimensions. All algorithms stagnated at inferior solutions on this problem, with MM-AMPSO having the worst average fitness. In 432 dimensions, MM-AMPSO still had the worst average fitness, but ID-AMPSO obtained the best average fitness after 1000 iterations. Again this result indicates that there are problem cases for which AMPSO suffers some loss in performance due to the limitations of the generating function. In the case of the Knight's Tour problem, the additional complexity of the ID-AMPSO variant is only beneficial in high dimensions, and only towards the end of the 1000-iteration search period.

Figure 7 shows the fitness profiles for the order-5 deceptive problems. Fitness graphs for the order-3 deceptive problems are omitted, but are similar to those in Figure 7. It is immediately obvious that all three proposed variants obtained much better average results than AMPSO in the initial phases of the search process. In fact, the variants consistently found the optimal solutions to these problems during initialization. The reason for this is that the solution to all the deceptive problems is a bit string consisting only of 1-bits. Recall from section 4 that all three the variants are able to create such solutions during initialization. Figure 7 also indicates that dimensionality affects only ID-AMPSO's ability to consistently initialize 1-bit solutions, and the effect only persists through the first few iterations. AMPSO is noticeably affected by an increase in dimensionality in the order-5 deceptive problem. However, due to particles leaving the search space, AMPSO is also able to find the optimal solutions to all the deceptive problems within the first 10 to 70 iterations, regardless of dimensionality.

Tables 2 to 4 report the statistical wins and losses for all algorithms on the N-Queens-, Knight's Coverage-, and Knight's Tour problems after 1000 iterations. The statistical results for the deceptive problems are not reported in tables, because, as discussed above, all the algorithms solved all the deceptive problems, implying no statistical difference in performance after 1000 iterations.

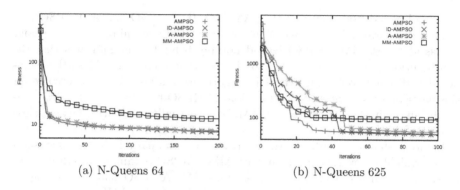

(a) N-Queens 64

(b) N-Queens 625

Fig. 4. Fitness profiles of all algorithms on N-Queens problems

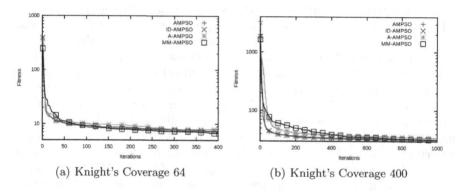

(a) Knight's Coverage 64

(b) Knight's Coverage 400

Fig. 5. Fitness profiles of all algorithms on Knight's coverage problems

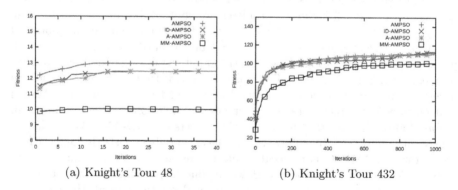

(a) Knight's Tour 48

(b) Knight's Tour 432

Fig. 6. Fitness profiles of all algorithms on Knight's Tour problems

Each table row indicates the wins and losses for all four algorithms on a specific problem. Because a pair-wise comparison was performed, an algorithm can have at most three wins or three losses. In addition, a win for one algorithm necessarily implies a loss for some other algorithm. In the cases where one

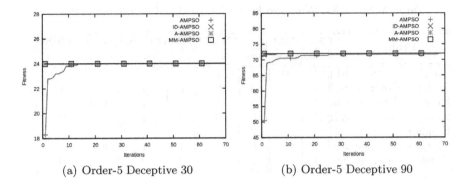

(a) Order-5 Deceptive 30   (b) Order-5 Deceptive 90

Fig. 7. Fitness profiles of all algorithms on Order-5 Deceptive problems

Table 2. Statistical Wins and Losses for N-Queens Problems

Problem	AMPSO		A-AMPSO		ID-AMPSO		MM-AMPSO	
	wins	losses	wins	losses	wins	losses	wins	losses
N-Queens 64	**3**	0	1	1	1	1	0	3
N-Queens 81	1	0	1	1	**2**	0	0	3
N-Queens 100	**2**	0	1	1	1	0	0	3
N-Queens 121	1	0	1	0	1	0	0	3
N-Queens 144	1	1	**2**	0	1	0	0	3
N-Queens 400	**2**	0	1	0	1	1	0	3
N-Queens 625	2	0	1	2	2	0	0	3

algorithm obtained more wins that all the other algorithms, the number of wins for the winning algorithm is printed in bold.

Table 2 shows that AMPSO, A-AMPSO, and ID-AMPSO all obtained statistically superior fitness values to the other three algorithms on some of the N-Queens problems. AMPSO obtained the most wins overall, while MM-AMPSO lost in all cases. The results corroborate the observations made in Figure 4.

Table 3 shows that AMPSO, A-AMPSO, and ID-AMPSO all performed well on the Knight's Coverage problem. However, A-AMPSO was superior in most low-dimensional problems, while MM-AMPSO showed very good performance in high dimensions. This confirms that the gain in performance, due to a little complexity to the AMPSO algorithm, can be significant.

Finally, Table 4 shows that AMPSO is superior when considering low-dimensional Knight's Tour problems, while A-AMPSO and ID-AMPSO perform comparably to AMPSO in 108 to 192 dimensions. However, MM-AMPSO is superior to all other algorithms in high dimensions. This demonstrates how an increase in dimensionality of the Knight's Tour problem affect the performance of the various algorithms. Furthermore, Table 4 illustrates that the enhancements made to the AMPSO algorithm by MM-AMPSO, while having a negative effect on

Table 3. Statistical Wins and Losses for Knight's Coverage Problems

Problem	AMPSO		A-AMPSO		ID-AMPSO		MM-AMPSO	
	wins	losses	wins	losses	wins	losses	wins	losses
Knight's Coverage 64	1	0	1	0	1	0	0	3
Knight's Coverage 81	0	1	3	0	0	1	0	1
Knight's Coverage 100	1	1	3	0	1	1	0	3
Knight's Coverage 121	0	1	3	0	1	1	0	2
Knight's Coverage 144	1	1	2	0	1	0	0	3
Knight's Coverage 225	1	1	0	3	1	1	3	0
Knight's Coverage 400	1	0	0	3	1	0	1	0

Table 4. Statistical Wins and Losses for Knight's Tour Problems

Problem	AMPSO		A-AMPSO		ID-AMPSO		MM-AMPSO	
	wins	losses	wins	losses	wins	losses	wins	losses
Knight's Tour 48	2	0	1	0	1	1	0	3
Knight's Tour 75	3	0	1	1	1	1	0	3
Knight's Tour 108	1	0	1	0	1	0	0	3
Knight's Tour 147	1	0	1	0	1	0	0	3
Knight's Tour 192	1	0	2	0	1	1	0	3
Knight's Tour 300	0	1	0	1	0	1	3	0
Knight's Tour 432	0	1	0	1	0	1	3	0

performance in low dimensions, allow the algorithm to overcome the increased complexity of the problem in higher dimensions.

8 Conclusion

This paper investigated the angle modulated particle swarm optimization (AMPSO) algorithm. A number of limitations due to the algorithm's bit-generating function were discussed. Three variants of the algorithm were proposed to remove these limitations. The three variants that were proposed are: amplitude AMPSO (A-AMPSO), increased-domain AMPSO (ID-AMPSO), and min-max AMPSO (MM-AMPSO). The AMPSO algorithm was then compared to the three variants on a number of binary benchmark problems in various dimensions.

It was observed that, in some problem cases, the limitations imposed on AMPSO affect the performance of the algorithm. Furthermore, it was observed that the additional complexity in the proposed variants was not favourable in low dimensions, but allowed the algorithms to obtain statistically equal or superior performance to AMPSO in higher dimensions.

In addition, it was shown that all three variants allowed for a greater variety of initial solutions to be generated. This ability allowed the variants to solve deceptive problems during initialization, while AMPSO required between 10 to 70 iterations to find optimal solutions.

In future work the problems will be studied in more detail to determine the specific characteristics that affect the performance of the algorithms. The generating function will also be closely studied with respect to an increase in the dimensionality of the binary solution. If shortcomings can be identified, alternative generating functions will be proposed. Furthermore, additional deceptive problems will be derived with optima that are unlikely to be generated during initialization. In this way the algorithms' ability to solve deceptive problems can be tested more thoroughly. Finally, a scalability study, comparing the three variants proposed in this paper to binary PSO [9], will be performed.

References

1. Balaprakash, P., Birattari, M., Stützle, T.: Improvement strategies for the f-race algorithm: Sampling design and iterative refinement. In: Bartz-Beielstein, T., Blesa Aguilera, M.J., Blum, C., Naujoks, B., Roli, A., Rudolph, G., Sampels, M. (eds.) HCI/ICCV 2007. LNCS, vol. 4771, pp. 108–122. Springer, Heidelberg (2007)
2. Birattari, M., Stützle, T., Paquete, L., Varrentrapp, K.: A racing algorithm for configuring metaheuristics. In: Genetic and Evolutionary Computation Conference, pp. 11–18 (2002)
3. Dirakkhunakon, S., Suansook, Y.: Simulated annealing with iterative improvement. In: International Conference on Signal Processing Systems, pp. 302–306 (2009)
4. Engelbrecht, A.: Particle swarm optimization: Velocity initialization. In: IEEE Congress on Evolutionary Computation, pp. 1–8 (2012)
5. Fisher, D.: On the nxn knight cover problem. Ars Combinatoria 69, 255–274 (2003)
6. Goldberg, D.: Simple genetic algorithms and the minimal, deceptive problem. In: Genetic Algorithms and Simulated Annealing, p. 88 (1987)
7. Gordon, V., Slocum, T.: The knight's tour - evolutionary vs. depth-first search. In: IEEE Congress on Evolutionary Computation, vol. 2, pp. 1435–1440 (2004)
8. Kennedy, J., Eberhart, R.: Particle swarm optimization. In: Proceedings of the IEEE International Conference on Neural Networks, vol. 4, pp. 1942–1948 (1995)
9. Kennedy, J., Eberhart, R.: A discrete binary version of the particle swarm algorithm. In: IEEE International Conference on Systems, Man, and Cybernetics. Computational Cybernetics and Simulation, vol. 5, pp. 4104–4108 (1997)
10. Liu, L., Liu, W., Cartes, D., Chung, I.: Slow coherency and angle modulated particle swarm optimization based islanding of large-scale power systems. Advanced Engineering Informatics 23(1), 45–56 (2009)
11. Martinjak, I., Golub, M.: Comparison of heuristic algorithms for the n-queen problem. In: 29th International Conference on Information Technology Interfaces, pp. 759–764 (2007)
12. Pampara, G.: Angle Modulated Population Based Algorithms to solve Binary Problems. Master's thesis, University of Pretoria (2013)
13. Pampara, G., Franken, N., Engelbrecht, A.: Combining particle swarm optimisation with angle modulation to solve binary problems. In: IEEE Congress on Evolutionary Computation, vol. 1, pp. 89–96 (2005)
14. Shi, Y., Eberhart, R.: A modified particle swarm optimizer. In: Proceedings of the IEEE Congress on Evolutionary Computation, pp. 69–73. IEEE (2002)
15. Turky, A., Ahmad, A.: Using genetic algorithm for solving n-queens problem. In: International Symposium in Information Technology, vol. 2, pp. 745–747 (2010)
16. Wang, S., Watada, J., Pedrycz, W.: Value-at-risk-based two-stage fuzzy facility location problems. IEEE Transactions on Industrial Informatics, 465–482 (2009)

Ant Colony Optimization on a Budget of 1000

Leslie Pérez Cáceres, Manuel López-Ibáñez, and Thomas Stützle

IRIDIA, CoDE, Université libre de Bruxelles, Belgium
{leslie.perez.caceres,manuel.lopez-ibanez,stuetzle}@ulb.ac.be

Abstract. Ant Colony Optimization (ACO) was originally developed as an algorithmic technique for tackling NP-hard combinatorial optimization problems. Most of the research on ACO has focused on algorithmic variants that obtain high-quality solutions when computation time allows the evaluation of a very large number of candidate solutions, often in the order of millions. However, in situations where the evaluation of solutions is very costly in computational terms, only a relatively small number of solutions can be evaluated within a reasonable time. This situation may arise, for example, when evaluation requires simulation. In such a situation, the current knowledge on the best ACO strategies and the range of the best settings for various ACO parameters may not be applicable anymore. In this paper, we start an investigation of how different ACO algorithms behave if they have available only a very limited number of solution evaluations, say, 1000. We show that, after tuning the parameter settings for this type of scenario, still the original Ant System performs relatively poor compared to other ACO strategies. However, the best parameter settings for such a small evaluation budget are very different from the standard recommendations available in the literature.

1 Introduction

The first Ant Colony Optimization (ACO) algorithms were introduced more than two decades ago [5]. After the publication of the main journal article describing Ant System (AS) [7], a large number of other ACO algorithms were introduced with the goal of improving over AS's performance and of showing that ACO algorithms could reach highly competitive results for various well-known combinatorial optimization problems. These improved algorithms include Ant Colony System (ACS) [6], Max-Min Ant System (MMAS) [21], rank-based Ant System (RAS) [4] and various others [8]. The main test problems at that time included the Traveling Salesman Problem (TSP) [4,6,7,21] and the Quadratic Assignment Problem (QAP) [7,21], among few others.

When it comes to computational effort, typically a sufficiently large number of solutions constructed or significantly long computation times have been considered. For example, in the 1996 "First International Contest on Evolutionary Optimisation" [3], the competing algorithms could evaluate up to $10\,000 \cdot n$ candidate solutions, where n is the problem dimension, that is, the number of cities in the TSP. Similarly, in most papers large enough computation times have been given to allow a large number of solutions to be generated or expensive local

M. Dorigo et al. (Eds.): ANTS 2014, LNCS 8667, pp. 50–61, 2014.

search methods to be used [9]. In particular, local search usually requires the evaluation of a large number of solutions.

On the other hand, there are situations where an algorithm can generate and evaluate only very few solutions before having to return the best solution found. This is the case when there are very tight real-time constraints even when it is quick to evaluate individual solutions or when the evaluation of solutions itself is very costly and in reasonable computation times only a small number of solutions can be evaluated. Common examples for the latter can be found in the field of simulation-optimization [1,14,23,24]. Moreover, in such a situation the usage of incremental updates to explore neighboring candidate solutions, one of the key factors that make local search algorithms fast [10], is often not applicable. In such situations, an ACO algorithm may only be able to evaluate a few thousand (or even fewer) solutions.

When facing a situation where very few solutions can be evaluated, a first question is how to transfer the knowledge available in the ACO literature, since most experiments on ACO algorithms typically consider many more solution evaluations. Hence, we consider it a valid question to pose which algorithmic ACO variants may be the most promising in such situations and also which values their parameters should take. In this paper, we explore these questions, acknowledging that similar questions have been posed on single ACO algorithms but usually at still much higher computation budgets than we consider here [18]. We do so by comparing the performance of some of the main ACO algorithms, including AS, elitist AS (EAS), RAS, MMAS, and ACS using their default parameter values recommended in the literature [8] and using their parameters tuned by irace [13], an automatic algorithm configuration tool. As benchmark problems, we use the TSP and the QAP, but with the additional limitation that at most 1 000 candidate solutions can be evaluated per run. Our experimental results show that the ACO algorithms that were proposed as improvements over AS still are clearly preferable, even in such a scenario. However, for some of the ACO algorithms this is only the case after re-tuning their parameter settings. In fact, some of the tuned parameter settings differ very strongly from what has been recommended in the literature and, in some cases such as for ACS, from what intuition would dictate as a good setting.

The article is structured as follows. In Sec. 2 we give details on the algorithms we used, the parameter ranges considered and the benchmark problems we used for our tests. Sec. 3 gives the experimental results and we conclude in Sec. 4.

2   Experimental Setting

2.1   Problems

In this article, we consider a scenario where evaluating a solution is costly enough that only a few candidate solutions can be evaluated per run. To allow for a significant number of experiments and to allow for parameter tuning, we evaluate the ACO algorithms on two standard test problems (the TSP and the QAP), but we restrict the number of solution evaluations to 1000. Both the TSP and the

QAP are well-known NP-hard combinatorial optimization problems, often used as benchmarks for heuristic algorithms and in the early literature on ACO [8]. For both problems, we generate a set of benchmark instances to be used in the evaluation of the ACO algorithms.

For the TSP, we generate random uniform Euclidean instances, where points are randomly distributed in a square of dimension 10000×10000. We generate 100 instances for each value of 50, 60, 70, 80, 90, and 100 cities. Half of these instances of each size are used as training instances for the ACO algorithm tuning, while the other half are used as test set for the comparison of the algorithms. For the QAP, we use the instances proposed in [19]. These QAP instances have a structure analogous to the instances that arise in practical applications of the QAP. The instance set comprises 100 instances of each size 60, 80, and 100.

2.2   ACO Algorithms

In our experiments, we use five of the best-known ACO algorithms, namely AS [7], EAS [7], RAS [4], MMAS [21] and ACS [6]. A detailed description of the above algorithms can be found in [8]; here we recall just the main algorithmic rules in the solution construction and the pheromone update so that the parameters we later tune are defined.

ACO algorithms iteratively construct solutions to a problem by using heuristic information and pheromone trails. Most ACO algorithms make use of the random-proportional rule that was introduced with AS: At a decision point i, the next element j is chosen with a probability p_{ij} that is proportional to $\tau_{ij}^{\alpha} \cdot \eta_{ij}^{\beta}$, where τ_{ij} is the pheromone related to a choice of solution component (i, j), η_{ij} is the associated heuristic information and α and β are two parameters that weigh the influence of the pheromone with respect to the heuristic information. ACS used a more deterministic construction, where with a probability q_0, the next element j is chosen deterministically as the one that maximizes $\tau_{ij}^{\alpha} \cdot \eta_{ij}^{\beta}$ (ties being broken randomly). AS-based algorithms may also use this latter rule, that is, make a deterministic choice with probability q_0 and we consider this possibility also in this paper. Once all m ants have constructed a solution, where m is a parameter corresponding to the colony size, the pheromones are updated by evaporating a factor ρ of each pheromone trail, where ρ is a parameter $(0 \leq \rho < 1)$, and depositing an amount of pheromone that is inversely proportional to the solution cost. The various ACO algorithms differ in which ants deposit pheromones and how much they deposit. For example, in AS each ant deposits an amount of pheromone equal to the inverse of the solution cost; in EAS the best solution since the start of the algorithm, the best-so-far solution, deposits additionally a large amount of pheromone; in RAS only some of the best solutions generated in each iteration and the best-so-far solution deposit pheromone; in MMAS only one ant deposits pheromone, which may be either the iteration-best or the best-so-far ant; finally, in ACS typically only the best solution since the start of the algorithm deposits pheromone. As a result, especially the more recent extensions such as RAS, MMAS, and ACS tend to exploit better the best solutions found during the search, but possibly only after longer computation times [8].

For this paper, we have used the implementation of the ACO algorithms given by the ACOTSP software [20]. We do not use candidate sets or local search to avoid biases due to a priori exploitation of specific problem features. For the QAP, we have adapted the ACOTSP software in a straightforward way so that we could use the same implementation. The main difference between the ACO algorithms for the TSP and the QAP is that for the latter we have not derived heuristic information. In the TSP case the avoidance of heuristic information can simply be simulated by setting $\beta = 0$. We also extended ACOTSP such that parameter settings may vary during a single run as described in [15].

2.3 Automatic Configuration

We compare the ACO algorithms using default parameter settings, which were normally derived considering other application scenarios, such as rather large numbers of solution evaluations, and the ACO algorithms after tuning. As tuning tool we use the irace software [13] that implements Iterated F-race and other racing methods for automatic parameter tuning [2]. The details of the various configuration scenarios are described below and the scenario files are available as supplementary material (http://iridia.ulb.ac.be/supp/IridiaSupp2014-006/).

ACOTSP, ACOQAP: These configuration scenarios execute the algorithms using fixed parameter values and they consider a fixed maximum budget for the run of each algorithm of 1 000 evaluations. These scenarios require the configuration of five (QAP) or six (TSP) parameters common to all ACO algorithms, plus one specific parameter in the case of configuring EAS (elitistants: the weight given to the best-so-far solution) or RAS (rasrank: the maximum rank considered corresponding to the maximum number of ants (m) that deposit pheromone). The parameters and their ranges are given in Table 1. The tuning goal is to minimize the solution cost reached after 1 000 solution evaluations.

ACOTSP-V, ACOQAP-V: These scenarios allow the parameters to vary during the algorithm execution by tuning pre-scheduled parameter variations identical to those described in Table 2 of [15]; the other parameters use fixed settings in a range as indicated in Table 1. The pre-scheduled parameter variation is possible for the four parameters; m, β (only in the case of TSP), q_0 and ρ. The rationale for using this alternative tuning scenario is to examine whether parameter variations may improve performance. The configuration goal remains the same as in the ACOTSP and ACOQAP scenarios.

ACOTSP-VA, ACOQAP-VA: These scenarios consider the algorithms with the possibility of pre-scheduled parameter variations as in the previous scenario. However, they consider a different configuration goal: The optimization of the anytime behavior as measured by the normalized hypervolume of the space that is dominated by the pairs of points that describe the development of the best-so-far solution quality over the number of solution evaluations [15]. In this case, the

Table 1. Range of parameters used in the tuning for fixed parameter settings

Common parameters for all scenarios				RAS and EAS		TSP scenario
m	α	ρ	q_0	*rasrank*	*elitistants*	β
$[5, 100]$	$[0, 10]$	$[0.01, 1]$	$[0, 1]$	$[1, 100]$	$[1, 175]$	$[0, 10]$

configuration goal is to minimize the normalized hypervolume. The goal of this tuning setting is to obtain a parameter configuration that is good independent of the specific maximum number of solution evaluation that is given. To still optimize the algorithm behavior for short runs, the maximum execution budget of each run of an ACO algorithms was limited to 5 000 solution evaluations.

In all scenarios, the total configuration budget for each tuning run was 10 000 runs of the algorithm and, as said above, half of the available benchmark instances are used as training set for the tuning.

3   Experimental Results

In this section, we examine the ACO algorithms using different parameter settings. First, we compare the results of the five ACO algorithms when using their default settings in the ACOTSP software. Next, we consider parameter settings that have been tuned following scenarios ACOTSP and ACOQAP. Finally, we consider tuning parameter variation strategies and the anytime behavior of the ACO algorithms, that is, scenarios ACOTSP-V / VA and ACOQAP-V / VA. In the following, each time statistical significance tests are mentioned they refer to Wilcoxon rank-sum tests at the 0.05 significance level with Bonferroni's correction for multiple tests. The experimental results reported here are based on one run on each of the test instances (300 instances for the TSP and 150 instances for the QAP).

3.1   Default Parameter Settings

As a first step, we compare the five ACO algorithms using default parameter settings. For the presentation of the results, we use AS as a baseline, that is, we compute the relative quality deviation obtained by each ACO algorithm with respect to AS on each instance. More concretely, for each test instance i and each ACO algorithm a, we compute the percentage deviation of the result obtained by a on instance i from the result of AS on the same instance i. Figure 1 gives the boxplots of the resulting deviations. A value larger then zero indicates worse performance than AS, while a value lower than zero indicates better performance.

Maybe surprisingly, when using the default settings and limiting the number of evaluations to 1000, some of the ACO algorithms perform much worse than AS. This is particularly striking for MMAS, which generates tours that are about 70% worse than those of AS. Also RAS performs much worse than AS on the

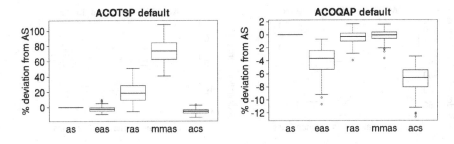

Fig. 1. Boxplots of the percentage deviation of the solution quality obtained by five ACO algorithms from the solution generated by AS, which is taken as reference. The results are given across all test instances.

TSP. The poor performance of these algorithms for short runs is due to the fact the MMAS and RAS parameters were set to allow for a very high final solution quality after a large number of candidate solutions have been generated [8]; for example, both MMAS and RAS use a relatively small evaporation, which does not allow them to bias the search fast enough to focus on the best solutions found. Another reason is the different default setting of the parameter β, which for AS is set to 5 whereas for MMAS and RAS is set to 2. EAS and ACS both show better performance than AS. All differences are statistically significant at the 0.05 significance level. For the QAP, the situation is different from the one of the TSP. None of the other ACO algorithms performs worse than AS. Considering statistical significance, EAS, RAS and ACS perform statistically significantly better than AS, whereas there is no statistically significant difference between MMAS and AS. This difference to the TSP results can be explained by the fact that the ACO algorithms for the QAP do not make use of heuristic information. In fact, if one eliminates heuristic information for the TSP by setting $\beta = 0$, the performance relative to AS follows the same trends as for the QAP (more details are given in the supplementary material).

3.2   Tuned Settings

In a next step, we tuned the parameter settings for both ACOTSP and ACO-QAP scenarios as defined in Section 2.3. After tuning, all algorithms significantly improve the solution quality reached within the limit of 1 000 candidate solutions. Figure 2 shows the relative deviation of each tuned ACO algorithm over its default version. As before, negative values indicate improved quality. The algorithms that most improve their performance are MMAS and RAS, while AS on the QAP is the only ACO algorithm that does not strongly improve its quality after tuning.

Figure 3 compares the performance of the ACO algorithms using AS as a reference in the same way as in the previous section, that is, the relative deviation of the quality obtained by the tuned version of each ACO algorithm with respect

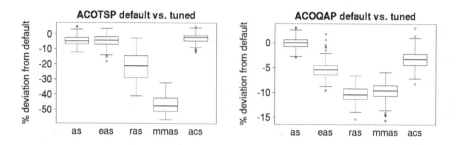

Fig. 2. Boxplots of the observed percentage improvement of each ACO algorithm with tuned parameter settings over the solutions reached with default parameter settings. The reference cost of each tuned ACO algorithm is its respective default parameter setting. For example, the boxplot of MMAS indicates the improvement observed of MMAS with tuned parameter settings over its default parameter settings. The results are given across all test instances.

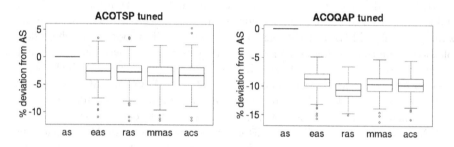

Fig. 3. Boxplots of the percentage deviation of the solution quality obtained by five ACO algorithms from the solution generated by AS, which is taken as reference. The results are given across all test instances.

to the tuned version of AS. This comparison shows that, for both the TSP and QAP, the four ACO algorithms (EAS, RAS, MMAS, and ACS) improve significantly over the performance of AS. (All the differences between AS and the other ACO algorithms are statistically significant.) The overall best performance on the TSP and on the QAP is obtained by MMAS and RAS, respectively.

A main reason for the strong improvements of most ACO algorithms over their default parameter settings is that these settings were designed for scenarios where ample computation time is available, that is, where a large number of candidate solutions may be constructed. To examine the differences between the default and the tuned parameter settings, we performed for each ACO algorithm 20 runs of irace and we analyzed the distribution of the parameter configurations that were obtained. These distributions are given in Figure 4 using boxplots; the red line for each algorithm indicates the default parameter settings.

While the parameter settings selected by irace varied from run to run, we can observe some clear trends. The algorithm for which the tuned parameter settings

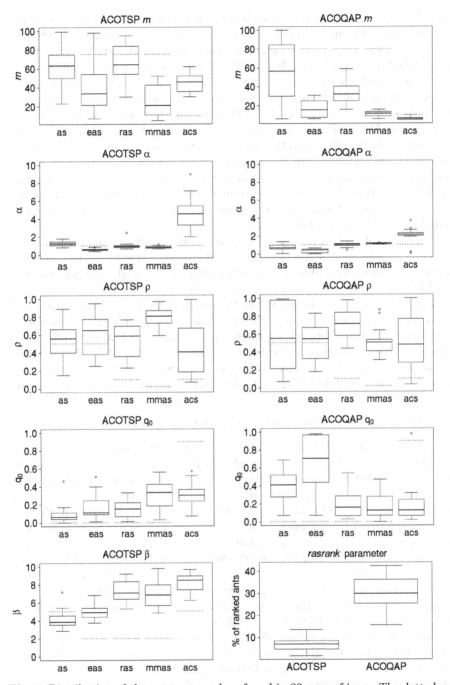

Fig. 4. Distribution of the parameter values found in 20 runs of irace. The dotted red lines indicate the default parameter setting for each algorithm. For the number of ants, the default parameter setting is the instance size; since in our test instances the size varies, we assumed an "average" instance size of 80 to indicate the default setting.

differ the least from the default ones is AS. For the other algorithms, major differences arise. RAS and especially MMAS use much smaller number of ants (m) than in their default version, which allows them to perform more iterations than in the default settings. In the ACOTSP scenario, also the heuristic information gets a much higher emphasis by using a median value for β of around seven instead of the default setting of two. The most noteworthy is probably the much higher evaporation rate ρ in RAS and MMAS than the default evaporation rate for both the ACOTSP and the ACOQAP scenarios. A very high pheromone evaporation has the effect that the search can quickly forget previously obtained worse solutions and focus quickly around the best recent ones. All these differences can be explained by the need of exploiting much more aggressively the search history (and heuristic information if available and helpful) due to the very small number of solutions to be generated biasing in this way the search around the best solutions found so far.

The setting of q_0 larger than zero for all ACO algorithms supports this interpretation, although in the case of ACS the tuned value for q_0 is somewhat surprising: a rather low value of q_0 together with a rather high value for α is proposed by the tuning, thus providing here another means for the exploitation of the search history. If we compare the settings of the ACOTSP and the ACOQAP scenarios, we can observe similar overall trends. However, there are differences in the best parameter settings for specific algorithms. For example, differences in good parameter settings in the two scenarios are evidenced by the fact that the boxplots of the distribution of the tuned parameter settings for the same ACO algorithm in the ACOTSP and ACOQAP scenarios often do not overlap.

3.3 Parameter Variation and Anytime Parameter Tuning

Instead of using fixed parameter settings, another option may be to adapt the parameter settings while running the algorithm. Earlier studies have indicated that pre-scheduled parameter variations [16, 22] may be more promising than self-adaptive schemes for deriving improved ACO parameter settings [17]. In particular, pre-scheduled parameter variation was shown to be particularly successful to improve the anytime behavior of MMAS [15]. In this section, we vary the parameters q_0, β, ρ, and m within a single run using the same variation schemes as proposed in [15]: We consider for all parameters as possible changes either their gradual variation iteration-by-iteration or a single switch of the parameter setting from some value to another one at a particular iteration.

For tuning, we consider the two possibilities that are offered by the scenarios ACOTSP-V / VA and ACOQAP-V / VA, which were described in Section 2.3. In a nutshell, the results do not show a strong difference when tuning for the final quality or for the anytime behavior, when the algorithms are evaluated according to the quality reached at 1 000 evaluations (Fig. 5).

On the other hand, the possibility of tuning the anytime behavior has a positive impact on the behavior of at least some of the algorithms if we look at the solution quality reached for different values of the evaluation budget. The plots of the solution quality development (as measured by the percentage deviation of

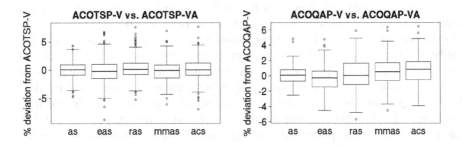

Fig. 5. Performance the ACO algorithm tuned for optimizing their anytime behavior versus the fixed parameter settings tuned for best performance at 1000 evaluations. The results are given for an evaluation of the algorithms on the test set after 1000 evaluated candidate solutions.

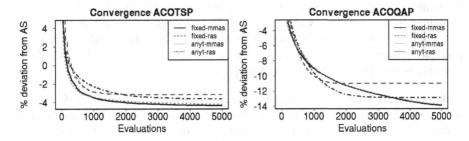

Fig. 6. Plots of the development of the solution quality over the number of solution evaluations for the TSP (left) and the QAP (right). The solution quality is measured as the percentage improvement over the solutions generated by AS across the test instances. The algorithms shown (MMAS and RAS) where tuned either for a fixed evaluation budget (fixed) and for optimizing their anytime behavior (anyt).

the algorithms from the AS solutions) over the number of evaluations (Fig. 6) show that, while MMAS' curve is almost identical independently of how the tuning was done, the curves for RAS show a clear stagnation effect of the RAS tuned for a fixed evaluation budget when compared to the version of RAS tuned for anytime behavior. Since it may be unknown a priori how many evaluations can be done in practice, we would recommend tuning for anytime behavior.

4   Final Remarks and Future Work

In this paper, we have analyzed the performance of five ACO algorithms for very low budgets on the evaluation of candidate solutions. Our computational results showed that EAS, RAS, MMAS, and ACS improve in performance over AS even in such circumstances. However, to make these algorithms reach high performance in short runs, very different parameter settings from the usual default ones have to be used.

There are a number of directions in which this work can be extended. Even though by the analysis of the tuned parameter settings we obtained new insights into the best parameter values for very short runs, the current tuning setting is maybe not the most realistic one. In fact, in an expensive function evaluation setting, the tuning would be rather time-consuming and it may be unrealistic to afford a time-intensive fine-tuning for each different problem being tackled. A next step would be to obtain more general settings. For example, we may tune the ACO algorithms across many different combinatorial problems and in this way derive robust parameter settings. Another question concerns whether to use specific known ACO algorithms or rather consider a framework of ACO algorithms from which known and new ACO algorithms may be instantiated. Such a framework may lead ultimately to a higher performing ACO algorithm than those we know nowadays–and this also holds for the here considered settings of very few solution evaluations. Finally, another possibility is to use surrogate modeling approaches to model fitness landscapes [11, 12]. While such surrogate modeling approaches have widely been applied to black-box continuous function optimization problems, their usage is more rare for combinatorial optimization problems. Hence, the exploration of such models and their integration into ACO algorithms may be a promising next step.

Acknowledgments. This work received support from the COMEX project within the Interuniversity Attraction Poles Programme of the Belgian Science Policy Office and from the European Research Council under ERC grant agreement n. 246939. Manuel López-Ibáñez and Thomas Stützle acknowledge support from the Belgian F.R.S.-FNRS, of which they are a postdoctoral researcher and a senior research associate, respectively. Leslie Pérez Cáceres acknowledges support of CONICYT Becas Chile.

References

1. April, J., Glover, F., Kelly, J., Laguna, M.: Practical introduction to simulation optimization. In: Proceedings of the 2003 Winter Simulation Conference, vol. 1, pp. 71–78 (December 2003)
2. Balaprakash, P., Birattari, M., Stützle, T.: Improvement strategies for the F-race algorithm: Sampling design and iterative refinement. In: Bartz-Beielstein, T., Blesa Aguilera, M.J., Blum, C., Naujoks, B., Roli, A., Rudolph, G., Sampels, M. (eds.) HM 2007. LNCS, vol. 4771, pp. 108–122. Springer, Heidelberg (2007)
3. Bersini, H., Dorigo, M., Langerman, S., Seront, G., Gambardella, L.M.: Results of the first international contest on evolutionary optimisation. In: Bäck, T., Fukuda, T., Michalewicz, Z. (eds.) Proceedings of ICEC 1996, pp. 611–615. IEEE Press, Piscataway (1996)
4. Bullnheimer, B., Hartl, R., Strauss, C.: A new rank-based version of the Ant System: A computational study. Central European Journal for Operations Research and Economics 7(1), 25–38 (1999)
5. Dorigo, M.: Optimization, Learning and Natural Algorithms. Ph.D. thesis, Dipartimento di Elettronica, Politecnico di Milano, Italy (1992) (in Italian)
6. Dorigo, M., Gambardella, L.M.: Ant Colony System: A cooperative learning approach to the traveling salesman problem. IEEE Transactions on Evolutionary Computation 1(1), 53–66 (1997)

7. Dorigo, M., Maniezzo, V., Colorni, A.: Ant System: Optimization by a colony of cooperating agents. IEEE Transactions on Systems, Man, and Cybernetics – Part B 26(1), 29–41 (1996)
8. Dorigo, M., Stützle, T.: Ant Colony Optimization. MIT Press, Cambridge (2004)
9. Gambardella, L.M., Montemanni, R., Weyland, D.: Coupling ant colony systems with strong local searches. European Journal of Operational Research 220(3), 831–843 (2012)
10. Hoos, H.H., Stützle, T.: Stochastic Local Search—Foundations and Applications. Morgan Kaufmann Publishers, San Francisco (2005)
11. Jones, D.R., Schonlau, M., Welch, W.J.: Efficient global optimization of expensive black-box functions. Journal of Global Optimization 13(4), 455–492 (1998)
12. Knowles, J.D., Corne, D., Reynolds, A.P.: Noisy multiobjective optimization on a budget of 250 evaluations. In: Ehrgott, M., Fonseca, C.M., Gandibleux, X., Hao, J.-K., Sevaux, M. (eds.) EMO 2009. LNCS, vol. 5467, pp. 36–50. Springer, Heidelberg (2009)
13. López-Ibáñez, M., Dubois-Lacoste, J., Stützle, T., Birattari, M.: The irace package, iterated race for automatic algorithm configuration. Tech. Rep. TR/IRIDIA/2011-004, IRIDIA, Université Libre de Bruxelles, Belgium (2011), http://iridia.ulb.ac.be/IridiaTrSeries/IridiaTr2011-004.pdf
14. López-Ibáñez, M., Prasad, T.D., Paechter, B.: Ant colony optimisation for the optimal control of pumps in water distribution networks. Journal of Water Resources Planning and Management, ASCE 134(4), 337–346 (2008)
15. López-Ibáñez, M., Stützle, T.: Automatically improving the anytime behaviour of optimisation algorithms. European Journal of Operational Research 235(3), 569–582 (2014)
16. Maur, M., López-Ibáñez, M., Stützle, T.: Pre-scheduled and adaptive parameter variation in $\mathcal{MAX}$–$\mathcal{MIN}$ Ant System. In: Ishibuchi, H., et al. (eds.) Proceedings of CEC 2010, pp. 3823–3830. IEEE Press, Piscataway (2010)
17. Pellegrini, P., Birattari, M., Stützle, T.: A critical analysis of parameter adaptation in ant colony optimization. Swarm Intelligence 6(1), 23–48 (2012)
18. Pellegrini, P., Favaretto, D., Moretti, E.: On $\mathcal{MAX}$–$\mathcal{MIN}$ ant system's parameters. In: Dorigo, M., Gambardella, L.M., Birattari, M., Martinoli, A., Poli, R., Stützle, T. (eds.) ANTS 2006. LNCS, vol. 4150, pp. 203–214. Springer, Heidelberg (2006)
19. Pellegrini, P., Mascia, F., Stützle, T., Birattari, M.: On the sensitivity of reactive tabu search to its meta-parameters. Soft Computing (in press)
20. Stützle, T.: ACOTSP: A software package of various ant colony optimization algorithms applied to the symmetric traveling salesman problem (2002), http://www.aco-metaheuristic.org/aco-code/
21. Stützle, T., Hoos, H.H.: $\mathcal{MAX}$–$\mathcal{MIN}$ Ant System. Future Generation Computer Systems 16(8), 889–914 (2000)
22. Stützle, T., López-Ibáñez, M., Pellegrini, P., Maur, M., Montes de Oca, M.A., Birattari, M., Dorigo, M.: Parameter adaptation in ant colony optimization. In: Hamadi, Y., Monfroy, E., Saubion, F. (eds.) Autonomous Search, pp. 191–215. Springer, Berlin (2012)
23. Teixeira, C., Covas, J., Stützle, T., Gaspar-Cunha, A.: Multi-objective ant colony optimization for solving the twin-screw extrusion configuration problem. Engineering Optimization 44(3), 351–371 (2012)
24. Zeng, Q., Yang, Z.: Integrating simulation and optimization to schedule loading operations in container terminals. Computers & Operations Research 36(6), 1935–1944 (2009)

Application of Supervisory Control Theory to Swarms of e-puck and Kilobot Robots

Yuri K. Lopes[1], André B. Leal[2], Tony J. Dodd[1], and Roderich Groß[1]

[1] Natural Robotics Lab, The University of Sheffield, Sheffield, UK
{y.kaszubowski,t.j.dodd,r.gross}@sheffield.ac.uk
[2] Santa Catarina State University, Joinville-SC, Brazil
andre.leal@udesc.br

Abstract. At present, most of the source code controlling swarm robotic systems is developed in an *ad-hoc* manner. This can make it difficult to maintain these systems and to guarantee that they will accomplish the desired behaviour. Formal approaches can help to solve these issues. However, they do not usually guarantee that the final source code will match the modelled specification. To address this problem, our research explores the application of formal approaches to both synthesise high-level controllers and automatically generate control software for a swarm of robots. The formal approach used in this paper is supervisory control theory. The approach is successfully validated in two experiments using up to 42 Kilobot robots and up to 26 e-puck robots.

1 Introduction

Swarm robotics (SR) studies systems composed of numerous robots that interact and cooperate to achieve certain goals. SR emphasises decentralization of control, limited communication among robots, use of local information, emergence of global behaviour and robustness. Such properties may prove useful in many real-world applications [1].

At present, most of the source code controlling swarm robotic systems is developed in an *ad-hoc* manner, without relying on software engineering methods. This can lead to software that is difficult to maintain. It is also difficult to guarantee that the software will accomplish the desired behaviour.

Formal approaches help to solve or minimise these issues as they require a systematic formalization of the solution. The methods to prove the properties of the system are much more developed to be applied over models expressed by formal approaches than over pure source code. There is a collection of software tools that implement such methods to analyse, validate and even prove properties of systems expressed by formal approaches. Also, the models serve as documentation of the system.

However, even when a project uses formal approaches, it is not guaranteed that the final source code will accomplish its goals. In the context of manufacturing systems, studies have illustrated how control code can be automatically generated using formal approaches [6,12,8,5]. The adaptation of these studies to

M. Dorigo et al. (Eds.): ANTS 2014, LNCS 8667, pp. 62–73, 2014.
© Springer International Publishing Switzerland 2014

the SR field could provide a new approach to engineering SR systems and help their transition to real-world applications.

1.1 Formal Approaches in Swarm Robotics

Works have investigated the use of formal approaches in SR. In [9], a group of probabilistic Finite State Machines is used to describe the structure of a swarm of robots at a microscopic level. This work focuses on modelling and analysis rather than control specification and synthesis. The formal specification, synthesis and verification of swarm robotic systems is an active area of research [3,2].

One of the formal approaches applied to synthesise control logic is Supervisory Control Theory (SCT) [14,13]. SCT is largely applied in manufacturing systems. In this scenario the decomposition of the model and specifications into several "small" subsystems is applied to solve complex problems.

In [4], the dynamics of a robot team is modelled with Discrete Event Systems (DES) using SCT and aiming at the design of a reconfigurable swarm system, which can handle the situation of robots switching off. However, the focus is on recovery control, the implementation of the control is not considered.

In [16], the authors use Deterministic Finite Automata (DFA) based on SCT to address the task allocation problem in a team of mobile robots, which work together as an automated patrolling/inspection system. In [17], Fuzzy-logic-based utility functions are used to quantify the ability of such robots to perform a task. All these works do not use the full Ramadge and Wonham (RW) framework [14,13]; instead, they are only partially based on it. As a consequence, much of the software tools and theory development cannot be applied to them.

While some works address the application of formal approaches in SR, there is a lack of work addressing automatic code generation. Moreover, there is a lack of case studies even in the field of manufacturing coordination, which is a major field of application of SCT.

1.2 Contributions

This paper presents the application of SCT to the domain of swarm robotics. It shows how to design and synthesise controllers for the individual robots in a swarm, and how to automatically generate the control software from the formal specification. Two case studies are presented and the same synthesised controller is applied on two target platforms, the Kilobot [15] and the e-puck [10] miniature mobile robotic systems.

. The main contributions of this work are (1) to adapt the implementation model of the SCT to the SR field; (2) to apply the SCT using the full RW-framework to SR, from the modelling to the software implementation; (3) to develop a software tool that automatically generates the control software for SR; and (4) to present two case studies applying this software tool and the proposed implementation model.

2 Supervisory Control Theory

In this section, SCT is overviewed. A language is defined as a set of words over an alphabet Σ, where the alphabet is a set of symbols. The events of a DES are associated with those symbols, and the words formed by those symbols represent sequences of operations. The control objective is to guarantee that at any time only valid words or prefixes of valid words occur. We are interested in a particular class of languages, namely the regular one.

A generator is a formal representation for a regular language used within the SCT framework. It is a quintuple $G = (Q, \Sigma, \delta, q_0, Q_m)$, where Q is the finite set of states; Σ is the finite set of symbols related to system events; $\delta : Q \times \Sigma \to Q$ is the partial transition function; q_0 is the initial state, where $q_0 \in Q$; and, Q_m is the set of final, marked as accepting states, where $Q_m \subseteq Q$.

The symbols that represent events are of two types: controllable events (Σ_c) and uncontrollable events (Σ_u), where $\Sigma = \Sigma_c \cup \Sigma_u$ and $\Sigma_c \cap \Sigma_u = \emptyset$. The traditional control operates by receiving stimulus signals from the controlled system and then issuing command signals. Thus, uncontrollable events are stimulus signals (e.g. from a sensor) and controllable events are command signals.

The SCT uses generators to represent free behaviour models and control specifications. The *free behaviour models* abstract each subsystem of each robot to be controlled and represent all of the physical possibilities of the system. We use G_i to represent the i-th free behaviour model. The *specifications* represent the desired behaviours of individual robots. We use E_j to represent the j-th specification model. The generators are synthesised to create supervisors that realise the control logic.

The goal of SCT is to obtain a language, realised by a supervisor, that represents valid sequences of events (in particular, they are *minimally restrictive* [14] and *non-blocking* [18]) respecting the control specifications. The supervisor exerts the control over the robot by disabling controllable events.

3 Modelling Swarm Robotic Behaviours with Supervisory Control Theory

In order to explain the SCT modelling processes, we introduce two didactic case studies, one using two robots to illustrate the robot platforms and the second using a proper swarm. In both case studies all robots use the same control logic and synthesised controller.

3.1 Orbit Case Study

The orbit strategy [15] is the first case study and it consists of two robots: the static robot and the orbiting robot. The static robot sends an infra-red (IR) message at a regular interval and the orbiting robot uses this message to estimate the distance and then orbit counterclockwise (CCW) around the static robot. When the orbiting robot is inside the boundary interval (which is the initial

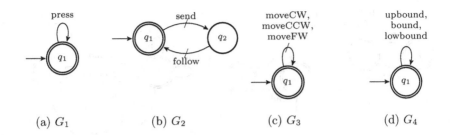

(a) G_1 (b) G_2 (c) G_3 (d) G_4

Fig. 1. Free behaviour models for the orbit strategy. Plain arcs represent uncontrollable events. Arcs with a stroke represent controllable events. Each behaviour model, G_i, has its own set of states $Q_i = \{q_1, ..., q_n\}$.

condition), it moves forward. When the distance of both robots is bigger than a threshold (the upper boundary), the orbiting robot turns CCW around the internal wheel/leg axis and approaches the static robot. When the distance of both robots is smaller than a threshold (the lower boundary), the orbiting robot turns clockwise (CW) around the external wheel/leg axis and moves away from the static robot.

Figure 1 shows the free behaviour models for this experiment. The generator G_1 (Figure 1(a)) represents a device to configure the type of the robot (static or orbiting) where the uncontrollable event *press* occurs when the user activates the configuration device. This device can be implemented as an IR signal received from a remote control. The generator G_2 (Figure 1(b)) represents the configuration of the robot: *orbiting* (in state q_1) or *static* (in state q_2). The controllable events *send* and *follow* start or interrupt the broadcast of the message respectively; the *send* event also stops the robot. This model contains the restriction of both events occuring alternatively.

The generator G_3 (Figure 1(c)) represents the motion capabilities of the robot. The controllable events *moveFW*, *moveCW* and *moveCCW* respectively represent the start of the forward movement, the CW turn and the CCW turn. The robot executes that movement indefinitely. The generator G_4 (Figure 1(d)) represents the boundary sensor; the uncontrollable events *upbound, bound* and *lowbound* are triggered when the orbiting robot is respectively too far, inside the boundary, or too near. As the distance estimation by IR message is not precise, the robot can receive the uncontrollable events in any order and at any time. Those events are continuously triggered after a sampling cycle of $200ms$.

Figure 2 shows the specifications of the orbit strategy. Figure 2(a) shows the specification of the user interaction to configure the robot type. Figure 2(b) shows the specification that enables the movement only for the orbiting robot. Figure 2(c) implements the main rule of the strategy, as previously described. States q_1, q_2 and q_3 (in E_3) specify the motion of the orbiting robot when it is respectively too far, inside the boundary or too near. As the initial state is q_2, it is considered that the robot starts inside the boundary.

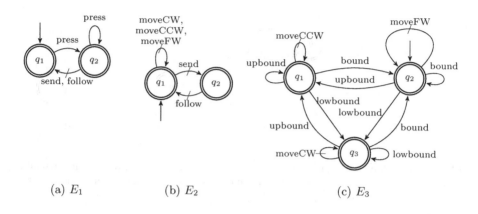

(a) E_1 (b) E_2 (c) E_3

Fig. 2. Specification for the orbit strategy. Each specification, E_j, has its own set of states $Q_j = \{q_1, ..., q_m\}$.

3.2   Segregation Case Study

The second case study addresses a segregation strategy. Each robot is configured as a leader or as a follower at the beginning of the trial. There are three types of leaders, which we refer to as the red, the green and the blue. These leaders are initially spread in the arena. They broadcast by IR a message, at fixed intervals of time, containing their type/colour. The follower robots that are in the signal field of only one type of leader start to belong to that leader. If a robot receives the signal from two or more different types of leaders, this robot starts to move at random until it finds a position where it receives signals from only one type of leader or no signal.

Figure 3 shows the free behaviour models for this experiment. The free behaviour G_1 (Figure 3(a)) represents a device to configure the type of the robot as in the previous experiment. The free behaviour G_2 (Figure 3(b)) represents the message transmission when a robot is a leader and no transmission when it is a follower. Initially, each robot is a follower robot (state q_1 in G_2). The controllable events $sendR$, $sendG$ and $sendB$ start the broadcast of the messages red, green and blue, respectively. The controllable event $sendNothing$ stops the broadcast.

The free behaviour model G_3 (Figure 3(c)) defines the motion behaviour. The controllable events $moveFW$, $moveCW$ and $moveCCW$ start the forward movement, the clockwise turn and counterclockwise turn of the robot for a random period, respectively. An uncontrollable event $moveEnded$ is generated at the end of this period. The controllable event $move Stop$ forces the end of the movement. Initially, the robots do not move.

The free behaviour models G_4, G_5 and G_6 (Figure 3(d)) represent the receiving of a message from the leaders red, green and blue respectively. The uncontrollable event $getX$, $X \in \{R, G, B\}$ occurs if the robot starts to receive a

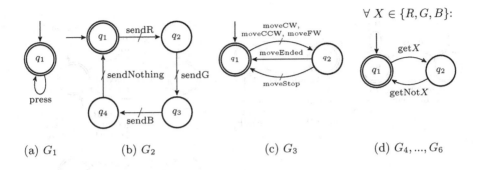

Fig. 3. Free behaviour models for the segregation strategy. Final states Q_m are indicated by double lines.

message from the corresponding type of leader; $getNotX$, $X \in \{R, G, B\}$ occurs when the robot stops receiving messages from the corresponding type of leader. A sampling cycle of $200ms$ is used to check if the robot is receiving the message.

Figure 4 illustrates the specifications to implement the robot type configuration and the segregation strategy. Figure 4(a) shows the specification E_1, which relates the user input with the change of robot type. Each time the user gives an input signal (the *press* event) the event to change the robot type is enabled. The combination with the model G_2 (Figure 3(b)) will guarantee the sequential order of types: follower, red leader, green leader and blue leader. This allows the user to set a robot as one of the three types of leaders or as a follower robot. Figure 4(b) shows the specification E_2 that enables the movement only for the followers and message broadcasts only for the leaders.

Figure 4(c) represents the main rule of the strategy, the specification E_3. When in state q_1 (not receiving any signal) or state q_2 (receiving signals from only one type of leader) the robot is forbidden to move and only the stop event (*moveStop*) is enabled. In state q_3 (receiving signals from two types of leaders) or state q_4 (receiving signals from three types of leaders) the move events are enabled and the stop event (*moveStop*) is disabled. The specification E_3 can be seen as a counter, in state q_1 there is no signal being received, in state q_2 there are signals for one type of leader, in state q_3 there are signals for two types of leaders, and in state q_4 there are signals for all three types of leaders.

The follower robot alternates its movement between the three different modes (forward, CW turn and CCW turn) when receiving signals from more than one type of leader. Each time when the event *moveEnded* occurs, it changes to state q_1 in G_3 where the events *moveFW*, *moveCW* and *moveCCW* are all enabled. As the choice for controllable events is at random, the robot will move randomly.

4   Supervisor Synthesis

The initial approach to synthesise a supervisor is the monolithic approach, where for all free behaviour and specification models only one supervisor S is obtained.

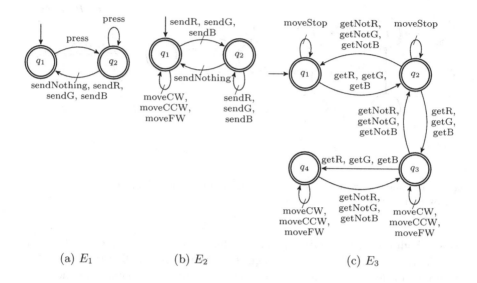

(a) E_1 (b) E_2 (c) E_3

Fig. 4. Specification for the segregation strategy

The parallel composition (represented by $\parallel$) of two generators G_a and G_b is defined as follows:

$$G_a\|G_b = (Q_a \times Q_b, \Sigma_a \cup \Sigma_b, \delta_{a\|b}, (q_{0_a}, q_{0_b}), Q_{m_a} \times Q_{m_b}), \qquad (1)$$

where

$$\delta_{a\|b}((q_a, q_b), e) = \begin{cases} (\delta_a(q_a, e), \delta_b(q_b, e)) & \text{if } \delta_a(q_a, e)! \wedge \delta_b(q_b, e)! \\ (\delta_a(q_a, e), q_b) & \text{if } \delta_a(q_a, e)! \ \text{ only} \\ (q_a, \delta_b(q_b, e)) & \text{if } \delta_b(q_b, e)! \ \text{ only} \\ \text{undefined} & \text{otherwise,} \end{cases} \qquad (2)$$

and $\delta(a)!$ means that function δ is defined for a.

A generator is accessible if all its states are reachable from the initial state q_0. A generator is coaccessible if all its states can reach at least one final state. The $Trim(G)$ operation removes all the non-accessible and non-coaccessible states from G. Let us consider the previous orbit case study with ($n = 4$) free behaviour models and ($m = 3$) specification models. The first step is the parallel composition for all free behaviour models and specifications. That is,

$$G = G_1\|\cdots\|G_4, \qquad (3)$$

$$E = E_1\|E_2\|E_3. \qquad (4)$$

To obtain a supervisor, S, which controls the free behaviour G, it is necessary to calculate the target language K, which is defined as the following parallel composition,

$$K = G\|E. \qquad (5)$$

As a result of the parallel composition, each state q_x^K in K is mapped on a state q_y^G in G. This mapping is not necessarily injective. One says that q_x^K is a composed state $(q_y^G, .)$. An event e is enabled in a state q if $\delta(q, e)!$. If an event e is enabled in q_y^G but not enabled in $q_x^K = (q_y^G, .)$, it means that the event is physically possible to occur, but the control specification denied it to occur. In this case q_x^K is called a bad state and if it is reached, the undesired event e cannot be disabled by the supervisor. Thus, q_x^K and any uncontrollable path to q_x^K must be removed.

To obtain a supervisor S where its supremal controllable sublanguage is $L_m(S/G)$, the iterative removal of bad states and the $Trim$ operation is performed by the $SupC$ operator as shown in Equation 6. This operator removes all bad states and each state that leads to the bad state through an uncontrollable path, that is, each state $q_a : \exists s \in \Sigma_u^+ : \delta(q_a, s) = q_{bad}$. Finally, K is modified by the $Trim$ component and the iterative process restarts until it does not modify K anymore.

$$L_m(S/G) = SupC(G, K). \tag{6}$$

The obtained supervisor, S, represents the control logic according to the designed model and can be used to implement the physical controller. However, the number of states and transitions grows exponentially by the parallel composition, which in some cases may not be feasible. To solve this problem, another approach, called modular supervisors, has been proposed by [18] and it was extended to local modular supervisors by [11].

4.1 Local Modular Supervisors

In this approach one supervisor is created for each control specification and only the free behaviour models that are affected by the control specification are composed in the calculus of each modular supervisor. Thus, each specification has its own local free behaviour model G_j^{loc}. The G_j^{loc} is the parallel composition of each free behaviour G_i which has at least one event in common with E_j. The local free behaviours for the orbit case are:

$$G_1^{loc} = G_1\|G_2, \ G_2^{loc} = G_2\|G_3, \ G_3^{loc} = G_3\|G_4. \tag{7}$$

For the segregation case, the local free behaviours are:

$$G_1^{loc} = G_1\|G_2, \ G_2^{loc} = G_2\|G_3, \ G_3^{loc} = G_3\|...\|G_6. \tag{8}$$

The supervisor for both cases are obtained as,

$$\forall x \in \{1, 2, 3\} : K_x = G_x^{loc}\|E_x, \tag{9}$$

$$S_x : L_m(S_x/G_x^{loc}) = SupC(G_x^{loc}, K_x). \tag{10}$$

In all supervisors for both cases all target languages K_x are already controllable, that is, $K_x = S_x$. This characteristic implies that the case studies are relatively simple. However, the focus of this work is to introduce the application of SCT to swarm robotics. The case studies were chosen to help illustrate the theory.

Applying the local modular approach requires that there is no conflict between supervisors (which could result in deadlocks). To test this, the monolithic system is built and compared with the composition of the local modular supervisors. In the case studies considered here there are no conflicts. If conflicts are present, they can be resolved by replacing the individual supervisors in conflict with a composite supervisor for all the conflicting specifications.

5   Implementation of Supervisory Control in SR

Our implementation is based on the SCT architecture proposed by [12]. However, we use the complete local modular supervisors instead of the reduced ones with the product system applied by [12]. The difference between these are small and are not detailed here. Those supervisors that are assigned to disable the controllable events are stored in the generated data structure proposed by [8].

The core of the controller is the *automata player*, which accesses the generated data to control the flow of the controller logic and it is in charge of evolving the generators. To do so, it stores the current state of each generator. An arbitrary number of supervisors can run in parallel in the same control structure managed by the automata player. Besides the generated data, the automata player evolves according to the occurrence of events defined by the operational procedures.

The *operational procedures* are a low-level interface between the supervisors and the real system that works by generating the control system output and reading the input [12]. As the architecture was designed to be applied mainly in manufacturing coordination the operational procedures are mostly applied to translate the events to the high or low signal in the output pins from Programmable Control Logic devices or to translate its pin signals to events. However, as shown in the following, the operational procedures layer is able to perform more complex tasks.

We associate each event to user defined callback functions to perform the operational procedures as proposed by [8]. We extend the generation tool called Nadzoru [8] to support the e-puck and Kilobot platforms and change the method to insert the operational procedure code. Instead of including the code inside the Nadzoru, now the Nadzoru allows the developer to specify external callback functions. Also, the control of the main loop function is delegated to the user, which must call an update function every cycle. This allows more flexibility to the developer.

The *user code* implements the operational procedures. The developer registers in the initialisation of the code a callback function for each event; this function is called every time that the event happens. Furthermore, the developer must register one callback function for each uncontrollable event to check whether the event occurs. All our implementations, models and the Nadzoru tool can be found in [7].

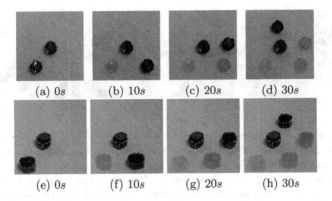

(a) 0s   (b) 10s   (c) 20s   (d) 30s

(e) 0s   (f) 10s   (g) 20s   (h) 30s

Fig. 5. Sequence of (superimposed) snapshots taken from one of the trials where two Kilobots (a-d) and two e-pucks (e-h) perform the orbit task

6   Experiment

This section presents two experiments that validate our application of SCT to swarm robotics. In both experiments the local modular approach is used. Two types of platforms are considered, and for both the same formal approach, free behaviours and specification models are used. This is one advantage of the approach. However, it requires that all devices must have the capability to perform the desired task. Thus, the task is modelled once for all devices and the device differences are abstracted and compensated in the operational procedures layer (user code).

Both experiments take place in a 1.20m × 0.90m two-dimensional arena. The configuration of the orbit experiment consists of two robots. The static robot is placed in the centre of the arena and the orbiting robot is placed inside (or near to) a configured boundary. Ten trials are performed for each type of robot. Each trial is limited to 300s. The experiment evaluates the match of the modelled specifications with the synthesised control logic. We observed that the robots behaved according to the specifications. Figure 5 shows snapshots taken from the experimental trials with the Kilobots and e-pucks.

The configuration of the segregation experiment consists of a group of 39 follower robots in the Kilobot experiment and 20 follower robots in the e-puck experiment. These robots are distributed on a grid. Three leader robots are placed inside the grid in the Kilobot case; in the e-puck case configurations with three pairs of leaders are applied. Ten trials are performed. Each trial runs for 300s or until the robots are segregated, whichever occurs first. The robots are considered to be segregated if they all receive a signal of only one leader or no signal at all (for details, see Figure 6). The experiment evaluates the match of the modelled specifications with the synthesised control logic. We observed that in all trials the robots behaved according to the specifications. Figure 6 shows snapshots taken from two trials.

(a)   (b)   (c)   (d)

Fig. 6. Snapshots from a segregation trial with Kilobots: (a) initial grid formation with three leaders, marked with tags; (b) result after segregation occurred. Trial with e-pucks: (c) initial grid formation with three pairs of leaders marked with tags; (d) result after segregation occurred, tags were added after the experiments for visualisation (based on the robots states as indicated by their Light-Emitting Diodes (LEDs)).

Video recordings from the experiments and additional resources (models, the Nadzoru tool, the source code used) can be found in [7].

7 Conclusions

This paper presented the application of Supervisory Control Theory (SCT) to swarm robotics and validated it through two case studies. First, the basic concepts were presented in a swarm robotics context. Second, an implementation for SCT, based on the architecture proposed by [12], was applied to swarm robotics. Third, the Nadzoru tool was extended to support code generation for two swarm robotics platforms, Kilobot and e-puck. Finally, the synthesised control logic was validated in experiments with these two platforms using the same formally synthesised control logic.

The use of formal approaches brings several advantages for the design of swarm robotic systems. Given a description of a system's capabilities and a set of specifications for the desirable behaviour of the individual robots, the control logic can be obtained. Moreover, the control software can be generated automatically. The control logic models can even be used for code generation in different types of platforms. In this paper, we illustrated this by two case studies with Kilobot and e-puck robot swarms. The only code that had to be manually written are the operational procedures, which is only a small fraction of the amount of code generated when the whole logic was implemented in an *ad-hoc* manner (see additional materials in [7]). Moreover, the use of operational procedures resulted in a more intuitive code that links events with framework functions.

In the future we will investigate how to prove properties of swarm robotic systems modelled by SCT. Also, we will consider more complex case studies.

Acknowledgements. Y.K. Lopes acknowledges support by CAPES - Brazil (grant number: 0462/12-8).

References

1. Brambilla, M., Ferrante, E., Birattari, M., Dorigo, M.: Swarm robotics: a review from the swarm engineering perspective. Swarm Intelligence 7(1), 1–41 (2013)
2. Brambilla, M., Pinciroli, C., Birattari, M., Dorigo, M.: Property-driven design for swarm robotics. In: Proceedings of the 11th International Conference on Autonomous Agents and Multiagent Systems, vol. 1, pp. 139–146 (2012)
3. Dixon, C., Winfield, A., Fisher, M.: Towards temporal verification of emergent behaviours in swarm robotic systems. In: Groß, R., Alboul, L., Melhuish, C., Witkowski, M., Prescott, T.J., Penders, J. (eds.) TAROS 2011. LNCS, vol. 6856, pp. 336–347. Springer, Heidelberg (2011)
4. Gordon-Spears, D., Kiriakidis, K.: Reconfigurable robot teams: modeling and supervisory control. IEEE Transactions on Control Systems Technology 12(5), 763–769 (2004)
5. Leal, A.B., Cruz, D.L.L., Hounsell, M.S.: PLC-based implementation of local modular supervisory control for manufacturing systems. In: Aziz, F.A. (ed.) Manufacturing System, pp. 159–182. InTech (2012)
6. Liu, J., Darabi, H.: Ladder logic implementation of Ramdge-Wonham supervisory controller. In: Proc. of the WODES, pp. 383–392 (2002)
7. Lopes, Y.K., Leal, A.B., Dodd, T.J., Groß, R.: Online supplementary material (2014), http://naturalrobotics.group.shef.ac.uk/supp/2014-001/
8. Lopes, Y.K., Leal, A.B., Rosso, R.S.U., Harbs, E.: Local modular supervisory implementation in micro-controller. In: Proc. of the 9th International Conference of Modeling, Optimization and Simulation (MOSIM 2012), vol. 9 (2012)
9. Martinoli, A., Easton, K., Agassounon, W.: Modeling swarm robotic systems: A case study in collaborative distributed manipulation. Int. Journal of Robotics Research 23(4-5), 415–436 (2004)
10. Mondada, F., Bonani, M., Raemy, X., Pugh, J., Cianci, C., Klaptocz, A., Magnenat, S., Zufferey, J.C., Floreano, D., Martinoli, A.: The e-puck, a robot designed for education in engineering. In: Proceedings of the 9th Conference on Autonomous Robot Systems and Competitions, vol. 1, pp. 59–65 (2009)
11. Queiroz, M.H., Cury, J.E.R.: Modular control of composed system. In: Proceedings of the American Control Conference, Chicago, pp. 4051–4055 (2000)
12. Queiroz, M.H., Cury, J.E.R.: Synthesis and implementation of local modular supervisory control for a manufacturing cell. In: Proceedings of International Workshop on Discrete Event Systems (WODES), pp. 103–110 (2002)
13. Ramadge, P.J., Wonham, W.: The control of discrete event systems. Proceedings of the IEEE 77(1), 81–98 (1989)
14. Ramadge, P.J., Wonham, W.M.: Supervisory control of a class of discrete event process. SIAM J. Control and Optimization 25(1), 206–230 (1987)
15. Rubenstein, M., Ahler, C., Nagpal, R.: Kilobot: A low cost scalable robot system for collective behaviors. In: Proccedings of ICRA 2012, pp. 3293–3298. IEEE (2012)
16. Tsalatsanis, A., Yalcin, A., Valavanis, K.: Optimized task allocation in cooperative robot teams. In: Proc. of the 17th Mediterranean Conference on Control and Automation (MED 2009), pp. 270–275 (2009)
17. Tsalatsanis, A., Yalcin, A., Valavanis, K.P.: Dynamic task allocation in cooperative robot teams. Robotica 30(5), 721–730 (2012)
18. Wonham, W., Ramadge, P.J.: Modular supervisory control of discrete event system. Mathematics of Control, Signals and Systems 1(1), 13–30 (1988)

Can Frogs Find Large Independent Sets in a Decentralized Way? Yes They Can!

Christian Blum[1,2], Maria J. Blesa[3], and Borja Calvo[1]

[1] Department of Computer Science and Artificial Intelligence,
University of the Basque Country UPV/EHU, San Sebastian, Spain
{christian.blum,borja.calvo}@ehu.es
[2] IKERBASQUE, Basque Foundation for Science, Bilbao, Spain
[3] ALBCOM Research Group, Universitat Politécnica de Catalunya, Barcelona, Spain
mjblesa@lsi.upc.edu

Abstract. The problem of identifying a maximal independent (node) set in a given graph is a fundamental problem in distributed computing. It has numerous applications, for example, in wireless networks in the context of facility location and backbone formation. In this paper we study the ability of a bio-inspired, distributed algorithm, initially proposed for graph coloring, to generate large independent sets. The inspiration of the considered algorithm stems from the self-synchronization capability of Japanese tree frogs. The experimental results confirm, indeed, that the algorithm has a strong tendency towards the generation of colorings in which the set of nodes assigned to the most-used color is rather large. Experimental results are compared to the ones of recent algorithms from the literature. Concerning solution quality, the results show that the frog-inspired algorithm has advantages especially for the application to rather sparse graphs. Concerning the computation round count, the algorithm has the advantage of converging within a reasonable number of iterations, regardless of the size and density of the considered graph.

1 Introduction

Given an undirected graph $G = (V, E)$, an independent set is a subset of the nodes of G such that no two nodes of this set are connected by an edge e from E. Furthermore, a *maximal independent set* $V_{\mathrm{MIS}} \subseteq V$ is an independent set such that no other independent set $\hat{V} \subseteq V$ exists with $V_{\mathrm{MIS}} \subseteq \hat{V}$. In other words, it is—per definition—not possible to add an additional node to a maximal independent set without destroying its independent set property. Finally, a *maximum independent set* is a maximal independent set of maximal size. Both the maximum independent set problem and the maximal independent set problem are fundamental in computer science and related fields (see, for example, [4]). From the perspective of centralized algorithms, it is well known that the maximum independent set problem is NP-hard [6], while the maximal independent set problem is in P. In fact, the literature offers various, rather simple, greedy algorithms for the generation of maximal independent sets (see, for example, [7]).

M. Dorigo et al. (Eds.): ANTS 2014, LNCS 8667, pp. 74–85, 2014.
© Springer International Publishing Switzerland 2014

In this work we consider the maximal independent set problem in a distributed setting (henceforth labelled MIS). This problem has applications, for example, in the context of facility location and backbone formation in wireless networks [5]. In particular, we focus on the study of a recently proposed (distributed) algorithm for graph coloring, named FROGSIM (see [8,9,1]). The experimental results show that FROGSIM, which is an algorithm inspired by the self-desynchronization of the calls of male Japanese tree frogs, has the tendency to generate color assignments in which the nodes associated to the most-used color correspond to large independent sets. The results obtained by FROGSIM are compared to one of the most-recently proposed distributed algorithms for the MIS problem, which was initially published in the prestigious journal *Science* [2]. More specifically, the results are compared to an optimized version of this algorithm from [11,10].

The reminder of this article is organized as follows. In Section 2, a short description of the studied algorithm is provided. The experimental evaluation is documented in Section 3. Finally, conclusions and an outlook to future work can be found in Section 4.

2   The FrogSim Algorithm

Even though a description of the FROGSIM algorithm can be found, for example, in [9], in the following we provide a—rather short—description in order for the paper to be self-contained. The following algorithm description is thought for working in a static wireless ad hoc network with n nodes equipped with radio antennas. Depending on the type of antennas and their communication range, a communication graph is implicitly defined in which each edge indicates a pair of nodes for which node-to-node communication is possible.

2.1   Algorithm Preliminaries

A first, preliminary, step requires an a priori organization of the wireless ad hoc network in form of a rooted tree. For the purpose of producing such a tree with a low height, the distributed method described in [3] may be used. The result is an induced tree that includes all the nodes of the network and has minimum diameter. In comparison to the rest of the nodes, the root node (or master node) of the tree will have some additional tasks to fulfill. It runs, for example, a protocol to calculate the height of the tree. Moreover, the master node initiates the start of the FROGSIM algorithm by means of a broadcast message. This message may additionally be used for communicating the height of the tree to the rest of the nodes as well. The induced tree is used during the execution of the FROGSIM algorithm for communicating the node-color assignments to the master node and for calculating the state of convergence which will be used to stop the algorithm.

Algorithm 1. Program of each node $i \in V$

1: $\theta_i :=$ calculateNewThetaValue()
2: $c_i :=$ minimumColorNotUsed()
3: sendColoringMessage()
4: clearMessageQueue()

2.2 Main Algorithm

The main FROGSIM algorithm works as follows. At each communication round (or iteration) each node executes the steps that are shown in Algorithm 1. The precise moment at which a node $i \in V$ starts executing this program depends on the value of variable $\theta_i \in [0, 1)$, which is stored internally. More precisely, assuming that the current communication round starts at time t, node i executes its program at time $t + \theta_i$. Apart from θ_i, each node i also maintains a color, denoted by $c_i \in \mathbb{N}^+$. Note that, for simplicity and without loss of generality, natural numbers greater than zero are used to uniquely identify colors. The execution of the program from Algorithm 1 includes the sending of exactly one message. In order to receive these messages from neighboring nodes, each node i maintains a message queue Q_i. In the following we provide a technical description of the functions of Algorithm 1.

When executing its program, a node i first adapts the value of θ_i in function calculateNewThetaValue(). This is done on the basis of the messages from the message queue Q_i. Only one message from each possible sender node is considered. In the case that Q_i contains two or more messages from the same sender node, the newest one prevails and the others are discarded. A message $m \in Q_i$ has the following format:

$$m = < \text{theta}_m, \text{color}_m > , \tag{1}$$

where $\text{theta}_m \in [0.1)$ contains the θ-value of the emitter and color_m is the color currently used by the emitter. Next, based on the messages in Q_i, function calculateNewThetaValue() calculates a new value for θ_i:

$$\theta_i := \theta_i - \alpha \sum_{m \in Q_i} \frac{\sin(2\pi \cdot (\text{theta}_m - \theta_i))}{2\pi} , \tag{2}$$

where $\alpha \in [0, 1]$ is a parameter used to control the convergence of the system. In general, the lower the value of α the smaller the change applied to θ_i.

Then, node i decides for a (possibly) new color in function minimumColorNotUsed(). Formally, this function computes the following value:

$$c_i := \min\{c \in \mathbb{N}^+ \mid \forall m \in Q_i : \text{color}_m \neq c\} \tag{3}$$

In words, node i chooses the color with the lowest identifier while discarding those colors that appear in messages $m \in Q_i$. Before finalizing its program, node i must communicate its new color to its neighbors. This is done by means of function sendColoringMessage(). This function sends the following message m:

$$m = < \text{theta}_m := \theta_i, \text{color}_m := c_i > \tag{4}$$

To conclude the description of node i's program, the message queue Q_i is cleared by removing all messages (see function clearMessageQueue()).

2.3   Identifying the Best Coloring and Detecting Convergence

The way in which the algorithm identifies a new best coloring and detects convergence is based on the use of the induced tree structure which was generated at the start of the algorithm. In the following we provide a short description of the mechanism. For a complete technical description we refer the interested reader to [8,9].

Henceforth, let h refer to the height—that is, the maximal distance between a leaf and the master node—of the induced tree. Note that h corresponds to the maximum number of communication rounds necessary for the master node to pass information to the rest of the nodes, and vice versa. In the following we assume that the master node knows about the size—in terms of the number of nodes—of the network. At each communication round, each node i is required to communicate the following information to its parent node in the induced tree: (1) a real number corresponding to the sum of the distances between the old theta values and the new ones concerning all nodes included in the subtree of which it acts as root, (2) the index of the largest color used by itself and all nodes included in the subtree of which it acts as root, and (3) an integer indicating the corresponding communication round number. In fact, these values do not need to be sent in extra messages. Instead they may be added to the coloring messages of Algorithm 1. Even though these messages will be received by all neighboring nodes, only the parent nodes in the induced tree will care about this information. Therefore, no additional messages are required by this mechanism.

Note that, in the first communication round, only the leaves of the tree will report the information described above to their parents. This is because the leaves are the only nodes without children. In the second communication round, the parents of the leaves will be able to report the aggregated data to their respective parents. Given the height h of the tree, it takes h communication rounds until all the information regarding a specific communication round has reached the master node. This means that the sensor nodes must store the differences between their old and new theta values, and the information about color use, during h communication rounds. Once the master node has received all the necessary information concerning a specific communication round, it is able to derive the following information. First, it knows the maximum index of any color used at the corresponding communication round. This information can be used to determine if a new best coloring has been found. Second, by dividing the sum of all theta-differences by the size of the network it obtains the average change of the theta-values in the corresponding communication round.

In case this average change is below a certain threshold value (we used 0.001), the master node broadcasts a stopping message to all nodes, which terminates the algorithm.

3   Experimental Evaluation

FROGSIM was implemented in C++ without the use of any external libraries. Experiments were performed by means of discrete event simulation. As mentioned before, in this work we study the ability of the algorithm to generate colorings in which the number of nodes assigned to the most-used color is rather large. In other words, we study if large independent sets may be extracted from the colorings produced by the algorithm.[1] For this purpose we decided to test the algorithm on random geometric graphs, which are commonly used to model wireless ad hoc networks. For comparison we used the optimized version (from [10,11]) of a very recent algorithm published in *Science* [2].

3.1   Generation of the Benchmark Set

Random geometric graphs are arguably the most popular model of wireless ad hoc networks. Therefore, we decided to study the algorithm in the context of this graph type. In order to generate a random geometric graph, one must first choose the number of nodes (n). These nodes are then assigned to random positions in the unit square. Finally, a fixed radius $0 < r < 1$ is used in order to determine the neighbors of each node. In particular, each pair of nodes that are within Euclidean distance of at most r are connected by an edge.

We considered random geometric graphs of sizes $n \in \{100, 500, 1000, 5000\}$. In order to find a reasonable range for the r-values for each n, the following experiments were performed. For each combination of $r \in \{0.01, 0.02, 0.03, \ldots, 0.3\}$ and n we generated 100 random geometric graphs and recorded the probability of these graphs to be connected. The results are graphically presented in Figure 1(a). Based on these results we determined the smallest r-values to be considered for the four graph sizes to be $r = 0.14$ (in case $n = 100$), $r = 0.067$ (in case $n = 500$), $r = 0.049$ (in case $n = 1000$) and $r = 0.024$ (in case $n = 5000$). With these values of r, the generated random geometric graphs have a probability of approx. 5% to be connected. Moreover, the resulting graphs are rather sparse.

In order to find suitable upper ranges for the r-values, we examined the *(relative) average degrees* of the generated graphs. Hereby, the term (relative) average degree refers to the average degree of a node expressed in terms of the fraction of all nodes to which the respective degree corresponds. For example, assume that $n = 100$ and that a node is, on average, connected to 10 neighbors. In this

---

[1] Note that, given a coloring solution, the nodes that are assigned to the same color form an independent set. Henceforth, given a coloring solution, we regard the size of the node set that is assigned to the most-used color as the MIS-value of the corresponding coloring.

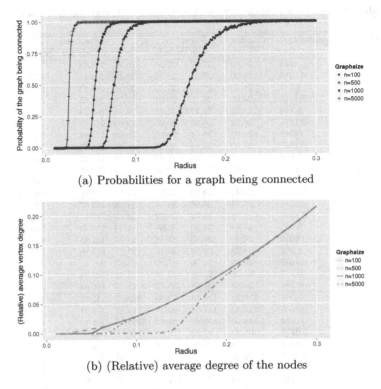

(a) Probabilities for a graph being connected

(b) (Relative) average degree of the nodes

Fig. 1. Test results used for generating the test instances

case the relative average degree is 0.1. In particular, we decided that the densest graphs to be considered in this study should have a (relative) average degree of 0.05. The graphic of Figure 1(b) shows that such graphs can be generated with $r = 0.169$ (in case $n = 100$) and $r = 0.134$ (in case $n \in \{500, 1000, 5000\}$).

For each value of n, numerical tests are performed for 20 values of r equidistantly distributed between the corresponding lower and the upper range.

3.2 Results

The numerical results are shown in Table 1 (graphs with $n \in \{100, 500\}$) and Table 2 (graphs with $n \in \{1000, 5000\}$). Each table row provides—for a certain combination of n and r, as indicated in the first two table columns—average results for 100 random geometric graphs. The column with heading GREEDY provides the results for the most well-known (centralized) greedy algorithm for the MIS problem. This algorithm works as follows: at each iteration, first, it identifies the node with minimal degree and adds it to the maximal independent set under construction. Afterwards, this node—together with all its neighboring nodes—is removed from the input graph. This procedure stops once the input

Table 1. Results for random geometric graphs with 100 and 500 nodes

| #nodes (n) | radius (r) | GREEDY | FRUITFLY | | FROGSIM | | |
			avg.	rounds	avg.	rounds	convergence
	0.14	30.22	27.97	27.86	29.16	244.24	711.35
	0.1415	29.90	27.38	28.92	28.42	203.45	687.60
	0.143	29.47	27.19	27.10	27.96	205.23	691.00
	0.1445	28.99	26.76	29.04	27.58	161.85	678.48
	0.146	28.63	26.17	31.96	26.96	168.29	674.39
	0.1475	28.22	25.86	28.96	26.86	212.75	708.41
	0.149	28.06	25.74	34.30	26.46	179.64	708.32
	0.1505	27.52	25.44	30.48	26.42	221.74	707.78
	0.152	27.27	24.94	31.86	25.89	159.11	700.75
100	0.1535	26.89	24.90	32.34	25.81	202.76	693.56
	0.155	26.69	24.59	32.86	25.51	180.87	708.80
	0.1565	26.41	24.29	33.02	25.19	155.20	699.52
	0.158	25.93	23.68	36.54	24.58	169.57	725.95
	0.1595	25.51	23.72	36.52	24.41	216.11	742.00
	0.161	25.24	23.20	36.68	24.09	159.63	706.00
	0.1625	24.99	22.85	35.58	23.76	167.35	699.47
	0.164	24.79	22.56	36.14	23.67	172.80	713.10
	0.1655	24.63	22.70	37.96	23.57	207.55	736.86
	0.167	24.28	22.21	42.18	23.22	187.38	739.65
	0.169	23.80	21.93	38.72	22.77	166.64	708.97
	0.067	130.20	120.19	52.00	122.52	341.22	726.23
	0.0705	121.42	111.54	62.24	114.19	375.81	740.82
	0.074	113.10	103.57	64.42	106.78	351.93	739.05
	0.0775	105.80	97.99	78.44	100.53	345.00	753.65
	0.081	98.79	91.56	90.98	93.94	312.48	764.53
	0.0845	92.72	86.28	98.70	88.47	305.55	776.57
	0.088	87.38	81.56	126.48	83.22	328.97	778.19
	0.0915	82.40	76.73	151.40	78.46	267.00	771.31
	0.095	77.46	72.31	191.62	73.68	265.35	767.13
500	0.0985	73.06	68.73	219.38	69.96	280.98	788.11
	0.102	69.51	65.37	309.80	66.49	264.84	762.38
	0.1055	65.70	62.38	374.94	62.90	266.77	771.17
	0.109	62.41	59.28	430.32	59.81	236.35	756.69
	0.1125	59.30	56.72	534.84	56.98	259.36	746.62
	0.116	56.47	54.28	626.82	54.39	227.10	770.09
	0.1195	53.69	52.18	777.42	51.91	205.91	757.79
	0.123	51.31	49.65	*1043.28*	49.74	222.74	784.69
	0.1265	49.04	47.43	*1294.66*	47.37	193.61	760.04
	0.13	46.8	46.07	*1663.82*	45.64	206.74	752.46
	0.134	44.71	44.32	*2148.60*	43.49	199.54	724.70

graph is empty. The results of this algorithm are simply given in order to indicate the quality of our algorithm in comparison to a centralized technique.
The next two columns of the result tables provide the results of the most recent
distributed algorithm for the MIS problem [2], which was inspired by the way
in which neural precursors are selected during the development of the nervous
system of the fruit fly *Drosophila*. We implemented an optimized version of this
algorithm—published in [10,11]—which is henceforth referred to as FRUITFLY.
The results of FRUITFLY are given in two columns. The first one, with heading
avg., provides the average result over 100 random geometric graphs. The second
column, with heading **rounds**, indicates the average number of communication

Table 2. Results for random geometric graphs with 1000 and 5000 nodes

#nodes (n)	radius (r)	GREEDY	FRUITFLY		FROGSIM		
			avg.	rounds	avg.	rounds	convergence
	0.049	244.88	225.39	66.66	229.76	416.14	734.08
	0.0534	215.26	199.07	80.78	203.48	442.16	750.92
	0.0578	190.96	176.50	115.30	180.26	414.50	758.72
	0.0622	170.02	158.01	160.02	161.35	366.59	768.56
	0.0666	152.35	142.98	236.02	144.82	325.74	775.16
	0.071	137.45	130.18	315.16	130.68	322.37	767.18
	0.0754	124.53	118.74	480.64	118.82	272.97	770.88
	0.0798	113.26	109.25	721.16	108.32	274.80	767.69
	0.0842	103.82	100.91	1114.90	99.55	248.03	754.76
1000	0.0886	95.35	93.68	1865.20	91.75	268.64	743.73
	0.093	87.82	87.19	2562.90	84.93	279.20	755.75
	0.0974	81.13	81.63	4631.68	78.48	252.60	747.29
	0.1018	75.42	76.72	7014.74	73.16	212.85	750.31
	0.1062	70.55	71.77	11754.56	68.23	216.22	712.49
	0.1106	65.61	67.91	22063.76	64.14	205.49	740.14
	0.115	61.70	63.94	34864.16	60.33	245.18	715.50
	0.1194	57.84	60.60	53523.54	56.55	215.69	684.79
	0.1238	54.67	57.24	96351.26	53.52	178.10	699.82
	0.1282	51.53	54.60	165323.04	50.16	165.77	700.42
	0.134	47.83	50.84	321192.92	46.95	160.96	678.45
	0.024	1040.74	962.55	129.46	976.72	561.54	761.83
	0.0297	736.45	694.10	370.78	694.91	498.26	776.10
	0.0354	543.55	528.05	1211.48	515.79	431.58	771.42
	0.0411	419.17	420.54	4542.54	401.35	380.55	766.95
	0.0468	331.90	345.53	23172.80	323.11	409.65	752.35
	0.0525	269.66	289.20	131172.26	265.49	415.70	740.25
	0.0582	224.68	245.77	749947.06	222.63	353.63	722.16
	0.0639	190.44	75.27	1000000.00	189.05	321.00	718.15
	0.0696	163.88	0.00	1000000.00	163.21	344.87	737.77
5000	0.0753	141.95	0.00	1000000.00	142.00	311.10	731.58
	0.081	124.59	0.00	1000000.00	125.26	331.54	729.67
	0.0867	110.54	0.00	1000000.00	111.30	300.39	703.93
	0.0924	98.90	0.00	1000000.00	99.68	278.69	690.82
	0.0981	89.16	0.00	1000000.00	89.76	262.51	688.26
	0.1038	80.80	0.00	1000000.00	81.28	218.35	676.89
	0.1095	73.94	0.00	1000000.00	74.30	267.05	689.92
	0.1152	67.84	0.00	1000000.00	67.75	191.87	660.59
	0.1209	62.20	0.00	1000000.00	62.50	182.83	640.03
	0.1266	57.66	0.00	1000000.00	57.81	201.70	606.51
	0.134	52.34	0.00	1000000.00	52.33	140.24	590.34

rounds that were performed in order for the algorithm to finish. Finally, the last three table columns contain the results of FROGSIM. As in the case of FRUIT-FLY, the first column provides the average result for the respective 100 random geometric graphs. The second column, with heading **rounds**, provides the average number of communication rounds that were performed in order to achieve the results reported in the column with label **avg.**. The last column, with heading **convergence**, indicates the average number of communication rounds that were performed in order for the algorithm to converge. Finally, note that the best result of each table row is marked with a grey background.

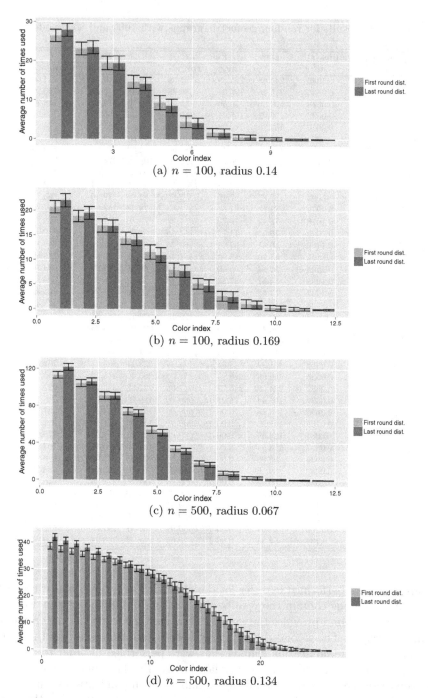

Fig. 2. The graphics show the distribution of the use of the different colors both at the start of the algorithm (see *First round dist.*) and after convergence (see *Last round dist.*) for graphs of sizes 100 and 500, and different values of r

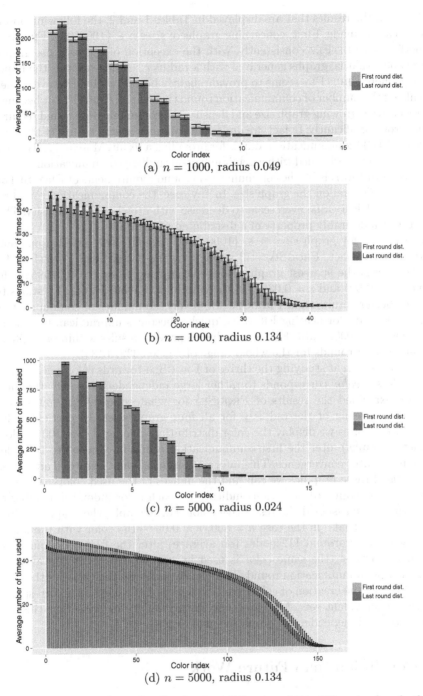

(a) $n = 1000$, radius 0.049

(b) $n = 1000$, radius 0.134

(c) $n = 5000$, radius 0.024

(d) $n = 5000$, radius 0.134

Fig. 3. The graphics show the distribution of the use of the different colors both at the start of the algorithm (see *First round dist.*) and after convergence (see *Last round dist.*) for graphs of sizes 1000 and 5000, and different values of r

Based on the results that are displayed in Tables 1 and 2, the following observations can be made. First, concerning graphs of sizes $n \in \{100, 500\}$, FROGSIM
outperforms FRUITFLY consistently, with the exception of rather dense graphs
of size 500—that is, graphs generated with a radius r tending towards the upper
range—where FRUITFLY seems to provide slightly better results. However, when
consulting the number of communication rounds needed by FRUITFLY it becomes
clear that with growing graph size and density, the communication round requirements grow significantly. In fact, we underlined all cases in which FRUITFLY needs
more than 1000 communication rounds for providing a result. Moreover, all runs of
FRUITFLY were performed with a maximum of one million communication rounds.
In contrast to FRUITFLY, the communication round requirements of FROGSIM do
not seem to depend on the graph size. In any case, the communication round requirements of FROGSIM even seem to decrease with growing graph density. This
is certainly a desirably property of a distributed algorithm.

The results for graph sizes $n \in \{1000, 5000\}$ amplify the observations outlined above. In fact, FROGSIM still seems to work better than FRUITFLY for
what concerns the sparsest graphs. However, starting from $r = 0.0798$ (in the
case of $n = 1000$) and $r = 0.0354$ (in the case of $n = 5000$) FRUITFLY starts to
produce better results than FROGSIM. However, the number of communication
rounds necessary for beating FROGSIM quickly becomes unpractical. For example, when $n = 5000$, FRUITFLY is not able to provide results within one million
communication rounds for the whole range of $r \in [0.0639, 0.134]$.

Finally, we aim at studying the thrive of FROGSIM towards colorings in which
the *most-used color* corresponds to rather large independent sets. For this purpose we examined the results of FROGSIM for what concerns the lowest and
the highest setting of the radius r for all four graph sizes. In particular, for
each of these cases we display the color distribution (averaged over 100 random
geometric graphs) after the first communication round in contrast to the color
distribution after convergence. This information is shown in Figures 2 and 3. On
the x-axis of these graphics we can find the indices of the used colors. The bars
(including the standard deviation) indicate for each color index the number of
nodes that have assigned the respective color. For example, the graphic in Figure 2(c) shows that—in the case $n = 500$, $r = 0.067$—the color with the lowest
index is used by around 112 nodes (on average) after the first communication
round. In contrast, the same color is used by around 125 nodes (on average)
after the last communication round. This clearly indicates the thrive of the algorithm towards the creation of colorings in which the most-used color corresponds
to large independent sets. Moreover, the eight graphics indicate that this is a
general trend, independent of graph size and density.

4   Conclusions and Future Work

In this work we studied a bio-inspired, distributed algorithm—initially introduced for graph coloring—for its ability to generate graph coloring solutions in
which the independent set of nodes assigned to the most-used color is large. The

results, in terms of the size of the independent set that is generated, were compared to the most recent algorithm published in the related literature for the maximal independent set problem. They show that the algorithm performs especially well in the context of sparse graphs. An important additional advantage is to be found in the low number of required communication rounds. The algorithm always converges within a reasonable number of communication rounds, independent of graph size and density.

Future work will focus on the study of the performance of the algorithm on different types of graphs. Moreover, we will study ways for improving the algorithms' performance for dense graphs.

Acknowledgments. This work was supported by projects TIN2012-37930, TIN2010-14931 and TIN2007-66523 of the Spanish Government, and project 2009-SGR1137 of the Generalitat de Catalunya. In addition, support is acknowledged from IKERBASQUE (Basque Foundation for Science) and the Basque Saiotek and Research Groups 2013-2018 (IT-609-13) programs.

References

1. Online scientific news site ScienceDaily: Frog calls inspire a new algorithm for wireless networks (July 2012),
   http://www.sciencedaily.com/releases/2012/07/120717100123.htm
2. Afek, Y., Alon, N., Barad, O., Hornstein, E., Barkai, N., Bar-Joseph, Z.: A biological solution to a fundamental distributed computing problem. Science 331, 183–185 (2011)
3. Bui, M., Butelle, F., Lavault, C.: A distributed algorithm for constructing a minimum diameter spanning tree. Journal of Parallel and Distributed Computing 64(5), 571–577 (2004)
4. Erciyes, K.: Distributed Graph Algorithms for Computer Networks. Springer, London (2013)
5. Erciyes, K., Dagdeviren, O., Cokuslu, D., Yilmaz, O., Gumus, H.: Modeling and Simulation of Mobile Ad hoc Networks, pp. 134–168. CRC Press (2010)
6. Garey, M.R., Johnson, D.S.: Computers and intractability; a guide to the theory of NP-completeness. W. H. Freeman (1979)
7. Halldórsson, M.M., Radhakrishnan, J.: Greedy is Good: Appriximating Independent Sets in Sparse and Bounded-Degree Graphs. Algorithmica 18, 145–163 (1997)
8. Hernández, H., Blum, C.: Distributed graph coloring: An approach based on the calling behavior of japanese tree frogs. Swarm Intelligence 6(2), 117–150 (2012)
9. Hernández, H., Blum, C.: FrogSim: distributed graph coloring in wireless ad hoc networks — an algorithm inspired by the calling behavior of Japanese tree frogs. Telecommunication Systems (2013)
10. Scott, A., Jeavons, P., Xu, L.: Feedback from nature: an optimal distributed algorithm for maximal independent set selection. Tech. rep., ArXiv repository (2012),
    http://arxiv.org/abs/1211.0235
11. Scott, A., Jeavons, P., Xu, L.: Feedback from nature: an optimal distributed algorithm for maximal independent set selection. In: Fatourou, P., Taubenfeld, G. (eds.) Proceedings of PODC 2013 – ACM Symposium on Principles of Distributed Computing, pp. 147–156. ACM Press (2013)

Diversity Rate of Change Measurement for Particle Swarm Optimisers

Phlippie Bosman and Andries P. Engelbrecht

Department of Computer Science, University of Pretoria, Pretoria, South Africa
{pbosman,engel}@cs.up.ac.za

Abstract. The diversity of a particle swarm can reflect the swarm's explorative/exploitative behaviour at a given time step. This paper proposes a diversity rate of change measure to quantify the rate at which particle swarms decrease their diversity over time. The proposed measure is based on a two-piecewise linear approximation of diversity measurements sampled at regular time steps. The proposed measure is the slope of the first of the two lines. It is shown that, when comparing the measure among different algorithms, the measure reflects the differences in the behaviour of algorithms in terms of their exploration-exploitation trade-off. The measure can potentially be used to characterise and classify different algorithms based on algorithm behaviour.

1 Introduction

Particle swarm optimisation (PSO) is a stochastic optimisation algorithm that maintains a swarm of particles, where each particle represents a candidate solution. An important characteristic that describes the search behaviour of a PSO algorithm (and other population-based algorithms) is diversity. The diversity of a swarm is the degree of dispersion of the swarm's particles [15].

Diversity is related to the notions of exploration and exploitation: the more diverse a swarm is, the more its particles are dispersed over the search space, and the more the swarm is exploring. Measuring diversity, then, can give an indication of an algorithm's search behaviour at a certain time step. Considering diversity measures over time can give an indication of the rate at which a swarm converges, or alternatively, the rate at which a swarm moves from an explorative to an exploitative behaviour, which has an impact on the performance of the algorithm.

A single, measurable value that reflects an algorithm's behaviour with regards to the rate at which diversity decreases over time can potentially be used to classify algorithms in different behavioural classes based on the rate at which the algorithms move from exploration to exploitation. Such a measure of diversity rate-of-change can potentially be used to predict performance for the different behavioural algorithm classes. This paper proposes a measure that can be used to differentiate different algorithms in terms of their behaviour with regards to the rate at which diversity decreases. To the knowledge of the authors, this is the first such measure. The proposed diversity rate-of-change (DRoC) measure is based

M. Dorigo et al. (Eds.): ANTS 2014, LNCS 8667, pp. 86–97, 2014.

on a two-piecewise linear approximation of the instantaneous diversity measures, computed at regular time steps: the slope of the first line of the piecewise linear approximation is used as the DRoC measure. A lower negative value for the slope indicates that an algorithm's diversity decreases faster, that the algorithm spends less time exploring, and that its particles converge to a smaller region faster.

The DRoC values for a number of PSO algorithms are computed in this paper for a large set of benchmark functions. These values are then used in the empirical section to see if the DRoC values can be used to characterise the rate at which search behaviour changes from exploration to exploitation, and to see if groups of algorithms can be found that exhibit the same behaviour.

The rest of the paper is organised as follows: Section 2 provides background on PSO, the algorithms used, diversity measures, and linear approximations. Section 3 lists expectations with reference to the rate at which diversity should decrease for the different algorithms. Section 4 presents the proposed measure. Section 5 summarises the experimental procedure. Section 6 provides and discusses the results.

2   Background

This section provides background information on the main concepts used.

2.1   Particle Swarm Optimisers

The basic PSO algorithm, introduced by Kennedy and Eberhart in 1995 [5,10], is a population-based search algorithm inspired from the behaviour of birds in flocks. A PSO algorithm maintains a swarm of particles, where each particle represents a candidate solution to an optimisation problem.

The original (gbest) PSO [10] updates the position of each particle $\mathbf{x}_i$ by adding a velocity, or step size, $\mathbf{v}_i$ to the particle's previous position as follows:

$$\mathbf{x}_i(t+1) = \mathbf{x}_i(t) + \mathbf{v}_i(t+1). \tag{1}$$

The velocity update for each particle consists of three components: the momentum component, which is a fraction of the particle's velocity at the previous time step; the cognitive component, which pulls the particle to a so-called personal best; and a social component, which pulls the particle towards a global best. The velocity update is as follows:

$$\mathbf{v}_i(t+1) = \omega\mathbf{v}_i(t) + c_1\mathbf{r}_1(t)[\bar{\mathbf{y}}_i(t) - \mathbf{x}_i(t)] + c_2\mathbf{r}_2(t)[\hat{\mathbf{y}}(t) - \mathbf{x}_i(t)] , \tag{2}$$

where ω is the inertia weight, c_1 and c_2 are constants, and $\mathbf{r}_1$ and $\mathbf{r}_2$ are vectors of random numbers sampled from the uniform distribution $U(0,1)$.

The cognitive component is a weighted difference between $\bar{y}_i(t)$, the personal best position visited by particle i up to time step t, and its current position. The effect of the component in the update equation is that the particle is drawn towards its personal best position. Similarly, the social component draws the particle in the direction of $\hat{\mathbf{y}}(t)$, the global best position found by the swarm.

2.2   Particle Swarm Optimiser Variations

Many variations of the basic PSO exist. Variations that are used in this study are described in this section.

Some variation arises from introducing a notion of neighbourhoods. The original PSO velocity update (Equation 2) can be changed to

$$\mathbf{v}_i(t+1) = \omega\mathbf{v}_i(t) + c_1\mathbf{r}_1(t)[\bar{\mathbf{y}}_i(t) - \mathbf{x}_i(t)] + c_2\mathbf{r}_2(t)[\hat{\mathbf{y}}_i(t) - \mathbf{x}_i(t)] , \qquad (3)$$

where $\hat{\mathbf{y}}_i(t)$ is particle i's local best position, which is the best position found in that particle's neighbourhood. A neighbourhood is a topology which connects each particle to some other particles; different topologies result in different variations of the PSO. Neighbourhoods have an inhibiting effect on information flow, since particles can, at each iteration of the algorithm, only gain information about local best positions that is already available to their direct neighbours.

The basic gbest PSO uses a star topology where each particle is connected to every other particle. Information flow is not inhibited, and information flow is instant.

The local best (lbest) PSO is a common variation of the gbest PSO that uses a ring topology instead of a star topology [5,19], such that each particle is only connected to two other particles: its index-wise predecessor and successor. The longest path between two particles is half the size of the swarm, so at most $n_s/2$ iterations might be required for information to pass from one particle to another. Information flow is therefore quite slow.

The Von Neumann topology [11] is an intermediate topology where particles are usually logically arranged on a 2-D grid. Information flow in Von Neumann PSO is slower than in gbest PSO but faster than in lbest PSO.

The basic PSO has a potential problem: if $\mathbf{x}_i = \bar{\mathbf{y}}_i = \hat{\mathbf{y}}_i$ for a particle, that particle's update depends only on its previous velocity. This can cause the algorithm to stagnate on the swarm's global best position, even if that position is not a local optimum [6]. The guaranteed convergence PSO (GCPSO) [1] overcomes this problem by using an altered position and velocity update equation for the global best particle, which forces that particle to search for a better position in a confined region around the global best position.

The GCPSO can be used with neighbourhood topologies such as star, ring and Von Neumann. Neighbourhoods have a similar effect in the GCPSO [16] as they do in the standard PSO.

Particles converge to a weighted average between their personal and local best positions [2], referred to in this paper as the theoretical attractor point. Kennedy [9] has proposed that the entire velocity update equation be replaced by a random number sampled from a Gaussian distribution around the theoretical attractor point, with a deviation the magnitude of the distance between the personal and global best. The resultant algorithm is called the barebones PSO (BBPSO). Kennedy also proposed [9] an alternative barebones PSO (aBBPSO), where the particle sampled from the above Gaussian distribution is recombined with the particle's personal best position.

The social PSO (SPSO) is a variation of the gbest PSO where the velocity update does not contain a cognitive component. The particles are only guided by the global best position and their own previous velocity. The particles converge towards the global best position, rather than a weighted average between that and their personal best positions, leading to very fast convergence.

2.3 Swarm Diversity

The diversity of a swarm is the degree of dispersion of its particles. Many existing diversity measures were investigated by Olorunda and Engelbrecht [15]. Note that these measures are instantaneous and thus only measure a swarm's diversity at a single time step. The two measures found in [15] to be the most accurate are the average distance around the swarm centre, and the average distance around all particles in the swarm. The average distance around the swarm centre was used in this study, given by

$$D = \frac{1}{n_s} \sum_{i=1}^{n_s} \sqrt{\sum_{k=1}^{n_x} (x_{ik} - \bar{x}_k)^2} \, , \tag{4}$$

where n_s is the swarm size, n_x is the number of dimensions of the problem, x_{ik} is the k-th dimension of the i-th particle position, and $\bar{x}_k$ is the average of the k-th dimension over all particles.

2.4 Two-Piecewise Linear Approximation

A two-piecewise linear approximation of a function,

$$y(x) \approx f(x) \text{ for } i_0 \leq x \leq i_2 \, , \tag{5}$$

is a mapping of two line segments, taking the form

$$y(x) = \begin{cases} a_1 + b_1 x & \text{for } i_0 \leq x \leq i_1 \\ a_2 + b_2 x & \text{for } i_1 < x \leq i_2 \end{cases} \tag{6}$$

where a_j and b_j are the y intersection and the gradient of the j-th line segment, respectively. The mapping is a minimisation problem aimed at finding optimal values for a_1, a_2, b_1, b_2, and i_1. The goal of the mapping is to minimise the least squares error (LSE) between the function and the linear approximation, given by

$$LSE = \sum_{x=i_0}^{i_2} (f(x) - y(x))^2 \, . \tag{7}$$

3   Algorithm Behaviour

Engelbrecht [7] found that different algorithms exhibit different diversity profiles, with reference to the rate at which diversity is decreased. Preliminary expectations about the rate at which different algorithms reduce diversity, and how this behaviour differs among different algorithms, can be made based on observations published in PSO literature and based on the definitions of position and velocity updates. This section discusses these expectations.

Because neighbourhoods directly influence information flow, it is expected that more connected neighbourhoods will converge more quickly [11,12], and so that their diversity will decrease more quickly. Therefore, when comparing any PSO algorithm that uses a star topology with the same algorithm using a ring or Von Neumann topology, it is expected that the rate of decrease in diversity for the PSO that uses the star topology will be faster than for the other algorithms. It is also expected that algorithms that use a Von Neumann topology will reduce diversity faster than algorithms that use a ring topology.

When comparing the SPSO with any of the other algorithms discussed in Section 2.2, it is expected that the SPSO will reduce diversity at a much faster rate due to the lack of the cognitive component, which is the component that facilitates exploration [7].

Due to the local search around the neighbourhood best that the GCPSO does, it is expected that the GCPSO reduces diversity at a somewhat faster rate than a basic PSO with a corresponding neighbourhood topology, but not at a significantly different rate.

When comparing the BBPSO with the aBBPSO, it is expected that the aBBPSO reduces diversity at a slower rate due to the random combination of the personal best position in the position update. This is expected to strengthen the cognitive-guided behaviour of particles, and weaken their social behaviour, delaying the swarm's convergence towards the theoretical attractor point.

4   Diversity Rate-of-Change Measure

Analysis of diversity measurements taken at regular time steps revealed a common pattern in the diversity profiles of the PSO algorithms studied in this paper, as illustrated in Figure 1: The initial diversity value is very high, due to particles being randomly initialised over the search space. The diversity shows a trend of rapid decrease for a number of iterations, referred to as phase one in this paper. This decrease is due to particles converging on a promising region of the search space. After the first phase, the diversity still generally decreases, though at a slower rate than in phase one, representing exploitation of the promising region in order to locate a good solution. This is referred to as the second phase. The pattern reflects a common behaviour in many PSO algorithms: that exploration is initially high, but then gives way to exploitation. (Of the PSO algorithms included in this study, all find a single solution in a static environment, and none have processes implemented through which the diversity of their swarms are managed.)

A two-piecewise linear approximation of diversity measurements produces one line with a slope that is relatively larger than the second line. An example of such a two-piecewise linear approximation is shown in Figure 2. The slope of the first line, representing phase one, quantifies the rate at which diversity decreases,

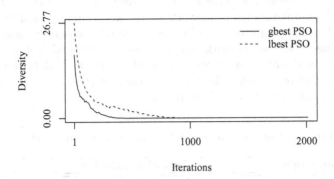

Fig. 1. Average diversity measurements over time for the gbest (solid line) and lbest (dashed line) PSO on Levy's function

i.e. the rate at which the swarm moves from an explorative to exploitative behaviour. The DRoC measure proposed in this paper is therefore simply the slope of this first line.

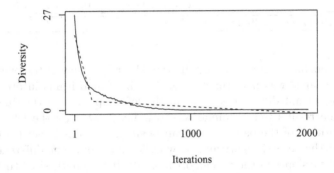

Fig. 2. Average diversity measurements over time for the gbest PSO over Levy's function (solid line) with two-piecewise linear approximation (dashed line)

Because the DRoC measure relies on diversity measurements over the entire run of a simulation, the proposed DRoC measure must be calculated after a simulation has completed.

5   Experimental Procedure

This section describes the experimental procedure followed to evaluate the DRoC measure.

Different PSO algorithms were run on a selection of benchmark functions using CILib,[1] an open-source framework for testing computational intelligence algorithms. The algorithms used, and the corresponding control parameters, are listed in Table 1. The benchmark functions are summarised in table 2. Each algorithm was used with 25 particles. For each function, each algorithm was run 30 times from different random initial conditions, for 2000 iterations. Diversity measures were sampled at every 10 iterations.

Table 1. Algorithms and control parameters used in this study

Algorithm name	Control parameters
Lbest PSO*	Ring neighbourhood topology
Gbest PSO*	Star neighbourhood topology
Von Neumann PSO (V.N. PSO)*	Von Neumann neighbourhood topology
Lbest GCPSO*	Ring neighbourhood topology
Gbest GCPSO*	Star neighbourhood topology
Von Neumann GCPSO (V.N. GCPSO)*	Von Neumann neighbourhood topology
BBPSO*	Star neighbourhood topology
aBBPSO*	Star neighbourhood topology Probability of combination = 0.5
SPSO	Star neighbourhood topology $\omega = 0.729844$, $c_1 = 0$, $c_2 = 1.49618$
* $\omega = 0.729844$, $c_1 = 1.49618$, $c_2 = 1.49618$	

For each benchmark function, a pair-wise Mann-Whitney U test with a 95% level of significance was performed on each pair of algorithms in order to determine if significant differences occur in the left slope among the different algorithms. The results were summarised to indicate whether the left slope for the first algorithm of the pair is significantly smaller (-1) or larger (1) than the left slope for the second algorithm, or whether no significant difference exists between the left slopes of the two algorithms (0). It is hypothesised that, where algorithms are intuitively expected to behave differently in terms of the rate of decrease in diversity, this difference will be reflected by the results of the U tests.

6   Results

The results from all pair-wise U tests are summarised in Table 6. The summarised Mann-Whitney U test result is shown for each algorithm pair (columns) and each function (rows). The final 3 rows respectively provide the number of each result for each algorithm pair.

As expected, the gbest PSO's measures indicate that it usually converged at a faster rate than the lbest PSO. Similarly, the GCPSO with a star topology

---

[1] Availabe at http://www.cilib.net

Table 2. Benchmark functions used in this study

Function name	Domain	Dimensions
Ackley [21]	$x_i\epsilon[-32, 32]$	25
Alpine [18]	$x_i\epsilon[-10, 10]$	25
Eggholder[13]	$x_i\epsilon[-512, 512]$	25
Elliptic [20]	$x_i\epsilon[-100, 100]$	25
Goldstein-Price [21]	$x_i\epsilon[-2, 2]$	2
Griewank [21]	$x_i\epsilon[-600, 600]$	25
Levy [13]	$x_i\epsilon[-10, 10]$	25
Michalewicz [3]	$x_i\epsilon[0, \pi]$	25
Quadric [21]	$x_i\epsilon[-100, 100]$	25
Quartic [21]	$x_i\epsilon[-1.28, 1.28]$	25
Rastrigin [21]	$x_i\epsilon[-5.12, 5.12]$	25
Rosenbrock [21]	$x_i\epsilon[-2.048, 2.048]$	25
Salomon [17]	$x_i\epsilon[-100, 100]$	25
Schwefel 1.2	$x_i\epsilon[-100, 100]$	25
Schwefel 2.22 [21]	$x_i\epsilon[-10, 10]$	25
Schwefel 2.26 [21]	$x_i\epsilon[-500, 500]$	25
Six-hump camel-back [21]	$x_i\epsilon[-5, 5]$	2
Spherical [4]	$x_i\epsilon[-100, 100]$	25
Step [21]	$x_i\epsilon[-20, 20]$	25
Zakharov [8]	$x_i\epsilon[-5, 10]$	25

usually converged faster than the GCPSO with a ring topology. In both the standard PSO and the GCPSO, the Von Neumann variations often produced no significant difference from the gbest variations, but the Von Neumann variations usually converged faster than the lbest variations.

When comparing the standard PSO algorithms with the GCPSO algorithms using the same neighbourhood topologies, no difference was found for most of the functions, as expected. For the few differences found, the GCPSO algorithms decreased diversity at a slower rate than the corresponding standard PSO algorithms, contrary to expectations. This may be an indication that the fitness landscape plays a role in the DRoC; this possibility is also observed by Engelbrecht in [7].

The SPSO usually decreased its diversity at a faster rate than any other algorithm, as expected.

Comparison of the BBPSO with the aBBPSO supports the expectation that the variation to BBPSO should converge more slowly, though there was often no significant difference between the two.

Both the gbest PSO and the gbest GCPSO were usually not significantly different when compared to the BBPSO. For most of the significant differences found in both comparisons, the BBPSO decreased its diversity faster. However, comparison of the gbest PSO and the gbest GCPSO to the aBBPSO produced varying results: there was usually a significant difference, but the aBBPSO was

Table 3. Summarised results of pair-wise Mann-Whitney U tests for each pair of algorithms, for each benchmark function

	Gbest PSO, Lbest PSO	Gbest PSO, V.N. PSO	Gbest PSO, SPSO	Gbest PSO, Gbest GCPSO	Gbest PSO, Lbest GCPSO	Gbest PSO, V.N. GCPSO	Gbest PSO, BBPSO	Gbest PSO, aBBPSO	Lbest PSO, V.N. PSO	Lbest PSO, SPSO	Lbest PSO, Gbest GCPSO	Lbest PSO, Lbest GCPSO	Lbest PSO, V.N. GCPSO	Lbest PSO, BBPSO	Lbest PSO, aBBPSO	V.N. PSO, SPSO	V.N. PSO, Gbest GCPSO	V.N. PSO, Lbest GCPSO
Ackley	-1	0	1	0	-1	0	0	1	1	1	1	0	1	1	1	1	0	-1
Alpine	-1	-1	1	0	-1	-1	0	-1	0	1	1	0	1	1	0	1	1	0
Elliptic	1	1	1	0	0	1	1	1	1	1	-1	-1	1	1	1	1	-1	-1
Eggholder	-1	-1	1	0	-1	-1	0	-1	1	1	1	0	1	1	0	1	1	-1
Goldstein-Price	1	1	1	0	1	1	1	0	0	0	-1	0	0	0	-1	0	-1	-1
Griewank	-1	0	1	0	-1	0	0	1	1	1	1	0	1	1	1	1	0	-1
Levy	1	0	1	0	1	0	1	1	0	1	-1	0	0	1	1	1	-1	0
Michalewicz	-1	-1	1	-1	-1	-1	0	-1	1	1	1	0	1	1	1	1	1	-1
Quadric	-1	-1	1	0	-1	-1	0	-1	1	1	1	0	1	1	0	1	1	-1
Quartic	-1	0	1	0	-1	0	0	1	1	1	0	0	0	1	1	1	-1	-1
Rastrigin	-1	-1	1	0	-1	-1	0	-1	0	1	1	0	0	1	-1	1	1	0
Rosenbrock	-1	0	1	-1	-1	-1	0	0	1	1	1	0	1	1	1	1	-1	-1
Salomon	-1	0	1	0	-1	0	0	0	1	1	0	0	1	0	0	1	0	-1
Schwefel 1.2	-1	-1	1	0	-1	-1	0	-1	1	1	1	0	1	1	0	1	1	-1
Schwefel 2.22	-1	-1	0	-1	-1	-1	1	1	1	1	1	0	1	1	1	1	0	-1
Schwefel 2.26	-1	-1	1	0	-1	-1	0	-1	1	1	1	0	1	1	1	1	1	-1
Sixhump	0	0	1	0	0	-1	0	1	0	1	0	-1	-1	0	1	1	0	0
Spherical	-1	0	1	0	-1	0	0	1	1	1	1	0	1	1	1	1	0	-1
Step	-1	0	1	1	-1	0	1	1	1	1	1	0	1	1	1	1	1	-1
Zakharov	-1	0	1	0	-1	0	-1	-1	1	1	1	0	1	0	-1	1	0	-1
Total -1's	16	8	0	3	16	10	1	8	0	0	3	2	1	0	3	0	5	16
Total 0's	1	10	1	16	2	8	14	3	5	1	3	18	4	4	5	1	7	4
Total 1's	3	2	19	1	2	2	5	9	15	19	14	0	15	16	12	19	8	0

found to decrease its diversity faster than the gbest PSO and the gbest GCPSO as often as the aBBPSO did so at a slower rate than the gbest PSO and gbest GCPSO. This indicates that the rate at which the aBBPSO decreases its diversity can vary widely, possibly depending on the fitness landscape.

	V.N. PSO, V.N. GCPSO	V.N. PSO, BBPSO	V.N. PSO, aBBPSO	SPSO, Gbest GCPSO	SPSO, Lbest GCPSO	SPSO, V.N. GCPSO	SPSO, BBPSO	SPSO, aBBPSO	Gbest GCPSO, Lbest GCPSO	Gbest GCPSO, V.N. GCPSO	Gbest GCPSO, BBPSO	Gbest GCPSO, aBBPSO	Lbest GCPSO, V.N. GCPSO	Lbest GCPSO, BBPSO	Lbest GCPSO, aBBPSO	V.N. GCPSO, BBPSO	V.N. GCPSO, aBBPSO	BBPSO, aBBPSO
Ackley	0	1	1	-1	-1	-1	-1	-1	-1	0	0	1	1	1	1	1	1	0
Alpine	0	1	-1	-1	-1	-1	-1	-1	-1	-1	0	-1	1	1	0	1	-1	-1
Elliptic	0	1	1	-1	-1	-1	-1	-1	1	1	1	1	1	1	1	1	1	0
Eggholder	0	1	-1	-1	-1	-1	-1	-1	-1	-1	0	-1	1	1	0	1	-1	-1
Goldstein-Price	-1	0	-1	-1	0	-1	0	-1	1	1	1	0	0	0	-1	0	-1	-1
Griewank	0	1	1	0	-1	-1	0	0	-1	0	0	0	1	1	1	0	1	0
Levy	0	1	1	-1	-1	-1	0	-1	1	1	1	1	0	1	1	1	1	0
Michalewicz	0	1	0	-1	-1	-1	-1	-1	-1	-1	1	-1	1	1	1	1	0	-1
Quadric	0	1	-1	-1	-1	-1	-1	-1	-1	-1	0	-1	1	1	0	1	-1	-1
Quartic	-1	0	1	-1	-1	-1	-1	-1	-1	0	1	1	1	1	1	1	1	1
Rastrigin	0	1	-1	-1	-1	-1	-1	-1	-1	-1	0	-1	0	1	-1	1	-1	-1
Rosenbrock	0	0	0	-1	-1	-1	-1	-1	-1	0	1	0	1	1	1	1	0	-1
Salomon	0	0	-1	-1	-1	-1	-1	-1	0	0	0	0	1	0	0	0	-1	0
Schwefel 1.2	0	1	-1	-1	-1	-1	-1	-1	-1	-1	0	-1	1	1	0	1	-1	-1
Schwefel 2.22	0	1	1	-1	-1	-1	0	1	-1	0	1	1	1	1	1	1	1	0
Schwefel 2.26	0	1	0	-1	-1	-1	-1	-1	-1	-1	-1	-1	1	1	1	1	0	-1
Sixhump	0	0	1	-1	-1	-1	0	0	0	0	0	1	0	1	1	1	1	0
Spherical	0	0	1	-1	-1	-1	0	0	-1	0	0	1	1	1	1	0	1	0
Step	0	1	1	0	-1	-1	0	0	-1	-1	0	0	1	1	1	1	1	0
Zakharov	0	-1	-1	-1	-1	-1	-1	-1	-1	0	-1	-1	1	0	-1	-1	-1	-1
Total -1's	2	1	8	18	19	20	13	15	15	8	2	8	0	0	3	1	8	10
Total 0's	18	6	3	2	1	0	7	4	2	9	11	5	4	3	5	4	3	9
Total 1's	0	13	9	0	0	0	0	1	3	3	7	7	16	17	12	15	9	1

7 Conclusions

This paper proposed a measure to quantify the rate at which swarms, for different particle swarm optimisation (PSO) algorithms, decrease their diversity. The diversity rate-of-change (DRoC) measure is obtained by fitting two-piecewise linear approximations to diversity measurements taken at regular time steps. The proposed DRoC measure is the slope of the left of those two lines.

The DRoC measure was computed for different PSO algorithms for which there are intuitive expectations about the differences in behaviour in terms of decrease in diversity between the algorithms. The DRoC measure was shown to reflect those expected differences. Firstly, where one algorithm was expected to decrease its diversity faster than a second algorithm, the DRoC measure for the first algorithm was usually a statistically significantly lower negative value than for the second algorithm. Secondly, where no significant difference was expected between the rate at which two algorithms decreased their diversity, no statistically significant difference was usually found between the DRoC measures for the algorithms.

For each comparison, the results for some benchmark functions contradicted expectations. Furthermore, when comparing the alternative barebones PSO to the gbest PSO and the gbest GCPSO, the results varied widely for different benchmark functions. This could indicate that the fitness landscape has an influence on how algorithms decrease their diversity.

Future work will investigate the possible influence that the fitness landscape may have on the behaviour of swarms in terms of decreasing diversity.

Alternative measures can possibly be obtained form the two-piecewise linear approximations that were used to obtain the proposed DRoC measure. For example, using the angle between the slopes of the first and the second lines of the approximation might provide valuable information. Future work will investigate such alternative measures.

The DRoC measure for a simulation must be calculated after the simulation has completed. Methods will be investigated that allow the measure to be calculated in real time. Such methods can be used in algorithms where the diversity of a population is managed in real time, such as attractive-repuslive PSO [14].

References

1. Van den Bergh, F., Engelbrecht, A.P.: A new locally convergent particle swarm optimizer. In: Proceedings of the IEEE International Conference on Systems, Man, and Cybernetics, pp. 96–101 (2002)
2. Van den Bergh, F., Engelbrecht, A.P.: A study of particle swarm optimization particle trajectories. Information Sciences 176(8), 937–971 (2006)
3. Chen, M.R., Li, X., Zhang, X., Lu, Y.Z.: A novel particle swarm optimizer hybridized with extremal optimization. Applied Soft Computing 10(2), 367–373 (2010)
4. De Jong, K.A.: Analysis of the behavior of a class of genetic adaptive systems. Ph.D. thesis, University of Michigan, Ann Arbor, MI, USA (1975)
5. Eberhart, R.C., Kennedy, J.: A new optimizer using particle swarm theory. In: Proceedings of the Sixth International Symposium on Micro Machine and Human Science, New York, NY, vol. 1, pp. 39–43 (1995)
6. Engelbrecht, A.P.: Computational intelligence: an introduction. John Wiley & Sons (2007)
7. Engelbrecht, A.P.: Scalability of a heterogeneous particle swarm optimizer. In: Proceedings of the 2011 IEEE Symposium on Swarm Intelligence, pp. 1–8. IEEE (2011)

8. Fan, S.K.S., Chang, J.M.: Dynamic multi-swarm particle swarm optimizer using parallel PC cluster systems for global optimization of large-scale multimodal functions. Engineering Optimization 42(5), 431–451 (2010)
9. Kennedy, J.: Bare bones particle swarms. In: Proceedings of the 2003 IEEE Swarm Intelligence Symposium, pp. 80–87. IEEE (2003)
10. Kennedy, J., Eberhart, R., et al.: Particle swarm optimization. In: Proceedings of IEEE International Conference on Neural Networks, Perth, Australia, vol. 4, pp. 1942–1948 (1995)
11. Kennedy, J., Mendes, R.: Population structure and particle swarm performance. In: Proceedings of the 2002 IEEE World Congress on Computational Intelligence, vol. 2, pp. 1671–1676. IEEE Computer Society (2002)
12. Kennedy, J.F., Kennedy, J., Eberhart, R.C.: Swarm intelligence. Morgan Kaufmann (2001)
13. Mishra, S.: Some new test functions for global optimization and performance of repulsive particle swarm method. Tech. rep., University Library of Munich, Germany (2006)
14. Monson, C.K., Seppi, K.D.: Adaptive diversity in PSO. In: Proceedings of the 8th Annual Conference on Genetic and Evolutionary Computation, pp. 59–66. ACM (2006)
15. Olorunda, O., Engelbrecht, A.P.: Measuring exploration/exploitation in particle swarms using swarm diversity. In: Proceedings of the 2008 IEEE Congress on Evolutionary Computation (IEEE World Congress on Computational Intelligence), pp. 1128–1134. IEEE (2008)
16. Peer, E.S., Van den Bergh, F., Engelbrecht, A.P.: Using neighbourhoods with the guaranteed convergence PSO. In: Proceedings of the 2003 IEEE Swarm Intelligence Symposium, pp. 235–242. IEEE (2003)
17. Price, K., Storn, R.M., Lampinen, J.A.: Appendix A.1: Unconstrained uni-modal test functions. In: Differential Evolution: a Practical Approach to Global Optimization. Natural Computing Series, pp. 514–533. Springer, Berlin (2006)
18. Rahnamayan, S., Tizhoosh, H.R., Salama, M.: A novel population initialization method for accelerating evolutionary algorithms. Computers & Mathematics with Applications 53(10), 1605–1614 (2007)
19. Suganthan, P.N.: Particle swarm optimiser with neighbourhood operator. In: Proceedings of the 1999 Congress on Evolutionary Computation, vol. 3. IEEE (1999)
20. Tang, K., Yao, X., Suganthan, P.N., MacNish, C., Chen, Y.P., Chen, C.M., Yang, Z.: Benchmark functions for the CEC 2008 special session and competition on large scale global optimization. Tech. rep. (2007)
21. Yao, X., Liu, Y., Lin, G.: Evolutionary programming made faster. IEEE Transactions on Evolutionary Computation 3(2), 82–102 (1999)

Evolutionary Swarm Robotics: Genetic Diversity, Task-Allocation and Task-Switching

Elio Tuci

Computer Science Department, Aberystwyth University, Aberystwyth, UK
elt7@aber.ac.uk

Abstract. The goal of this study is to investigate the role of genetic diversity for engineering more resilient evolutionary swarm robotic systems. The resilience of the swarm is evaluated with respect to the capability of the system to re-distribute agents to tasks in response to changes in operating conditions. We compare the performances of two evolutionary approaches: the *clonal approach* in which the teams are genetically homogeneous, and the *aclonal approach* in which the teams are genetically heterogeneous. We show that the *aclonal approach* outperforms the *clonal approach* for the design of robot teams engaged in two task-allocation scenarios, and that heterogeneous teams tend to rely on less plastic strategies. The significance of this study for evolutionary swarm robotics is discussed and directions for future work are indicated.

1 Introduction

This study draws inspiration from evidence in the study of social insects to generate alternative design principles for swarm of robots [4]. In particular, we target the processes of task-allocation and task-switching.

Recent studies have shown that division of labour in social insects can be guided by emergent circumstances concerning the life of the colony that are independent of the worker age and morphology [8,13]. For example, in [7] the author shows that in a species of harvester ants, foragers can be recruited from workers originally performing other tasks (e.g., nest-patrolling or nest-maintenance) when the quantity of food close to the nest is experimentally manipulated. What mechanisms do insects use to generate this plasticity in division of labour? There are evidence that division of labour in social insects is based on a combination of positive and negative feedback mechanisms and on the variability among the workers to respond to task stimuli [3,5]. Workers with a low response threshold for a task tend to prefer or specialise to that task. Workers with a high response threshold for a task are likely not to perform that task in normal conditions. However, they can be progressively attracted to that task in response to changes in colony conditions that determine an increase level of the stimulus associated to the task. The increase in the number of agents performing a task decreases the stimulus associated to the task. Consequently, those workers with a higher response threshold are likely to abandon the task, reducing the

M. Dorigo et al. (Eds.): ANTS 2014, LNCS 8667, pp. 98–109, 2014.

number of agents performing it. Entomologists have found out that the colony variability in response thresholds to task stimuli is genetically determined, and that colonies with higher genetic variability tend to be more efficient in division of labour [12].

In this paper, we look at dynamic task-allocation (i.e., the autonomous allocation or roles/tasks to robots) from the perspective of evolutionary swarm robotics, where the robot behavioural mechanisms are automatically generated by using evolutionary computation techniques to synthesise artificial neural network controllers [17]. Our long term objective is to identify the elements that facilitate the evolution of neural mechanisms that underpin behavioural responses similar to those observed in insects societies. In other words, we aim to design teams of robots in which at least some of the agents are potentially capable of carrying out multiples tasks, and in which each of these generalist agents engage on a task based on current circumstances. The great majority of the research work on task-allocation in evolutionary swarm robotics refers to scenarios in which each robot is required to specialise on a single role/task for the entire duration of the evaluation [16,18]. In this type of scenario, the designers have generally opted for solutions in which the differentiation of behavioural competencies is obtained using genetically identical (i.e., homogeneous) robots, which manage to autonomously specialise on different functions by exploiting some form of neural plasticity [1]. The genetic homogeneity of the team facilitates the design process by offering alternative solutions to problems and costs that generally emerge with the use of heterogeneous teams.

The originality of this work is based, on the one hand, on a scenario in which, contrary to previous work, some of the team members need to switch task during evaluation. This behaviour is referred to as task-switching. Thus, task-specialisation (i.e., the tendency to carry out a single task for the entire duration of the evaluation phase) is not a valuable option for all the team members. On the other hand, we question the effectiveness of the homogeneity condition in view of the significance of the genetic variability for the plasticity in division of labour in social insects. In particular, we compare two different approaches for the design of individual mechanisms required by a team of robots to re-distribute agents to tasks in response to changes in operating conditions. In the *clonal approach*, teams are formed using a single genotype from the evolving population of genotypes. Thus, the teams are homogeneous because all the team members have a controller derived from the same genotype. In the *aclonal approach*, teams are formed by using multiple genotypes from the evolving population of genotypes. Thus, teams are heterogeneous because each team member has a controller derived from a different genotype. In both approaches, the genes code for the parameters of dynamic neural networks.

We are aware that it is not possible to propose "general recipes" from a single case study. Nevertheless, the results of our study indicate that the *aclonal approach* outperforms the *clonal approach* for the design of teams engaged in scenarios requiring dynamic task-allocation and task-switching behaviour. This suggests that in similar swarm robotic scenarios it may be worth to pay the

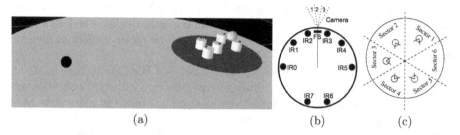

Fig. 1. (a) Experimental scenario (snapshot taken in *Env. A*, during *Phase 1*) showing the nest (dark grey circle) with the light on top, the light indicating the foraging site (black circle), and the five robots (white cylinders in the nest). (b) E-puck body-plan. The black circles refer to the position of the infra-red (IR), the black rectangle refers to the position of the floor sensor (FS). The dotted lines indicate the robot's view with the the three camera's sectors. (c) Robots starting positions within the nest.

costs associated to the evolution of heterogeneous teams to get better results. However, our study also indicates that the *aclonal approach* tends to converge on solutions in which the majority of the members of successful teams are genetically specialised for a task, and a minority (one or two members) display a limited behavioural plasticity. This can be a limit to the robustness of the solutions. These results will be discussed in details in Section 6.

2   The Simulation Environment

In the foraging scenario studied in this paper, the environment is a boundless arena with a nest and a foraging site. The nest is a circular area indicated by a green light, in which the colour of the floor is in shades of grey. The foraging site is also a circular area indicated by a red light. With the exclusion of the nest, the colour of the arena floor is white. The radius of both the nest and the foraging site is randomly defined at the beginning of each trial in the interval [20cm, 30cm]. Both lights, the green one located in the nest and the red one located in the foraging site, are positioned 6cm above the floor and, when turned on, they are visible from everywhere within the arena. In each trial, the green light is placed at the centre of the nest. The red light is placed at a distance from the centre of the nest that varies from 100cm to 110cm (see Fig. 1a).

The robots kinematics are simulated using a modified version of the "minimal simulation" technique described in [9]. Our simulation models a e-puck robot, a 3.5cm radius cylindrical robot. It is provided with eight infra-red sensors (IR^i with $i = \{0, .., 7\}$), which give the robot a noisy and non-linear indication of the proximity of an obstacle (in this task, an obstacle can only be another robot); a linear camera to see the lights; and a floor sensor (FS) positioned facing downward on the underside of the robot (see Fig. 1b). The IR sensor values are extrapolated from look-up tables provided with the Evorobot* simulator [11]. The FS sensor can be conceived of as an IR sensor capable of detecting the intensity of grey of the floor. It returns 0 if the robot is on white floor, 0.5 if it is on light grey floor, and 1 if it is on dark grey floor. The robots camera has a receptive field of 30°, divided in three equal sectors, each of which has three binary sensors (C_i^B for blue, C_i^G for green, and C_i^R for red, with $i = \{1, 2, 3\}$ indicating the sector). Each sensor returns a value which is 0 if no light is detected, 1 when a light is detected. The camera can detect coloured objects up to a distance of 150cm. The robot has left and right motors which can be independently driven forward or reverse, allowing it to turn fully in any direction. The robot maximum speed is 8cm/s.

3   The Task and the Fitness Function

Teams comprising five simulated e-puck robots are evaluated in the context of a dynamic role allocation and role switching behaviour. By taking inspiration from the behaviour of social insects, the roles are nest patrolling and foraging (hereafter, we refer to them as *role P*, and *role F*, respectively). Roughly speaking, *role P* requires a robot to remain within the nest. *Role F* requires a robot to leave the nest for the foraging site, to spend a certain amount of time at the foraging site, and then to come back to the nest. A team is required to execute both roles simultaneously. Therefore, the robots have to go through a role-allocation phase in which they autonomously decide who is doing what, and then execute their respective roles. Moreover, the robots are required to be able to switch from one role to the other (i.e., role switching behaviour) due to the fact that they experience two different types of environment, *Env. A* and *Env. B*. In *Env. A*, *role F* is more important than *role P*. This means that in *Env. A*, a team maximises the fitness if the majority of robots (i.e., more than two robots) visits the foraging site and the minority (i.e., less than three robots) remains in the nest. In *Env. B*, *role P* is more important than *role F*. This means that a team maximises the fitness if the majority of robots (i.e., more than two robots) remains in the nest and the minority (i.e., less than three robots) visits the foraging site. Since a team, throughout its life-span, experiences both types of environment, not all the robots can specialise on a single role. At least one robot has to be able to play both roles and eventually to switch from one role to the other based on the current environmental condition and the roles allocated to the other team mates. How can a robot distinguish between *Env. A* and *Env. B*? The two types of environment can be distinguished by the intensity of grey colouring the floor in the nest site. In *Env. A*, the nest is coloured in dark grey. In *Env. B*, the nest is coloured in bright grey.

During evolution, each team undergoes a set of $E = 4$ evaluation sequences (hereafter, e-sequence). An e-sequence is made of $V = 3$ trials. There are two different types of e-sequence: in *ABA-sequence* the robots experience *Env. A* in trial 1, *Env. B* in trial 2, and *Env. A* in trial 3; in *BAB-sequence* the robots experience *Env. B* in trial 1, *Env. A* in trial 2, and *Env. B* in trial 3. Each group experiences twice each type of e-sequence. At the beginning of trial 1 of each e-sequence, the robots controllers are reset, and each robot is randomly placed within an area corresponding to a sector of the nest. The nest is divided in 6 sectors as illustrated in Fig. 1c. Each robot is randomly placed in one sector randomly oriented in a way that the light can be within an angular distance of ±36° from its facing direction (see Fig. 1c).

Each trial differs from the others in the initialisation of the random number generator, which influences the robots initial position and orientation, all the randomly defined features of the environment, and the noise added to motors and sensors (see [9] for further details on sensors and motor noise). Within a trial, the team life-span is T=900 simulation cycles (with 1 simulation cycle lasting 0.1s). Robots are frozen (i.e., don't move and do not contribute to the team fitness) if they exceed the arena limits (i.e., a circle of 120cm radius, centred in the middle point between the nest and the foraging site). Trials are terminated earlier if all the robots are frozen, or the team exceeds the maximum number of collisions (i.e., 10). In trials following the first one of each e-sequence (trial 2,and 3), the robots are repositioned only if the previous trial has been terminated earlier, or with one or more robot frozen.

Each trial is divided into three phases. During *Phase 1*, which lasts 12s, the green light is on and the red light is off. The robots are required to stay within the nest. During *Phase 2*, which can last from a minimum of 47,5s to a maximum of 52.5s, the red light is on and the green light is off. During *Phase 2*, a team is required to behave according to the rules of the task. That is, in *Env. A*, the majority of robots (i.e., more than two robots) has to visit the foraging site and the minority (i.e., less than three robots) has to remain for the entire length of this phase in the nest. In *Env. B*, the majority of robots has to remain for the entire length of *Phase 2* in the nest and the minority has to visit the foraging site. A robot is considered having visited the foraging site if, during *Phase 2*, it spends more then 100 consecutive time steps at less than 45cm from the light indicating the foraging site. During *Phase 3*, which starts at the end of *Phase 2* and terminates at the end of the trial, the green light is on again and the red light is off. The robots that were foraging during *Phase 2* are required to return in the nest to rejoin their team mates.

We study two slightly different scenarios. In *Scenario I*, the robots can not see each other through the camera. Thus, any robot-robot interaction including those that result in the allocation of roles to robots are based on the activations of the infra-red sensors. In *Scenario II*, the robots can see each other through the camera. Thus, in this scenario there is a wider sensory space in which robot-robot interaction can be generated. This can facilitate the dynamic allocation of roles to agents. However, due to occlusion—only the closest coloured object is perceived in each camera sector—the lights indicating the nest and the foraging site may not be visible all the time by all robots. The robot are coloured in dark yellow. When a robot is perceived through the camera, the red and the green camera sensors return a value of 0.5 and the blue sensor returns a value of 0.

The fitness of a genotype is its average team evaluation score after it has been assessed for four e-sequences (i.e., for a total of 12 trials). In each trial (v) of

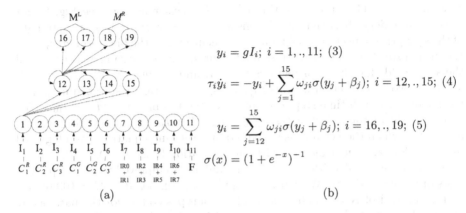

$$y_i = gI_i; \ i = 1, ., 11; \ (3)$$

$$\tau_i \dot{y}_i = -y_i + \sum_{j=1}^{15} \omega_{ji}\sigma(y_j + \beta_j); \ i = 12, ., 15; \ (4)$$

$$y_i = \sum_{j=12}^{15} \omega_{ji}\sigma(y_j + \beta_j); \ i = 16, ., 19; \ (5)$$

$$\sigma(x) = (1 + e^{-x})^{-1}$$

(a)   (b)

Fig. 2. (a) The neural network. (b) The equations used to update the values of sensory, internal, and motor neurons of the robot controller.

each e-sequence (e), the team is rewarded by an evaluation function F_{ev} which is computed in the following:

$$F_{ev} = \left[\left(\frac{\sum_{r=1}^{R} S_r^{ph1}}{R \times T^{ph1}} \times \frac{\sum_{r=1}^{R} S_r^{ph3}}{R \times T^{ph3}} \right) + Q^{ph2} \right] \times PEN; \ \text{with } F_{ev} \in [0,7]; \quad (1)$$

$$Q^{ph2} = \begin{cases} 5 & \text{if robots correctly allocated in } Phase \ 2; \\ 2 \times \frac{N}{R} & \text{if robots incorrectly allocated in } Phase \ 2; \end{cases} \quad (2)$$

with $R = 5$ indicating the number of robots in a swarm; S_r^{ph1} is the number of simulation cycles robot r spends within the nest during *Phase 1*; S_r^{ph3} is the number of simulation cycles robot r spends within the nest during *Phase 3*; N is the number of robots that, during *Phase 2*, play *role F* if the trial is in *Env. A*, or robots that play *role P* if the trial is in *Env. B*; T^{ph1} and T^{ph3} indicate the duration of *Phase 1* and *Phase 3*, respectively. The team collision penalty PEN is inversely proportional to the number of collisions, with $PEN = 1$ with no collisions, and $PEN = 0.4$ with 10 collisions in a trial. The average team evaluation score is $F = \frac{1}{E}\frac{1}{V}\sum_{e=1}^{E}\sum_{v=1}^{V} F_{ev}$, with $E = 4$ and $V = 3$.

4   Controller and the Evolutionary Algorithm

The robot controller is composed of a continuous time recurrent neural network (CTRNN) of 11 sensor neurons, 4 inter-neurons, and 4 motor neurons [2]. The structure of the network is shown in Fig. 2a. The states of the motor neurons are used to control the speed of the left and right wheels as explained later. The values of sensory, internal, and motor neurons are updated using equations 3, 4, and 5. In these equations, using terms derived from an analogy with real neurons, y_i represents the cell potential, τ_i the decay constant, g is a gain factor,

I_i with $i = 1, ., 11$ is the activation of the ith sensor neuron (see Fig. 2a for the correspondence between robot's sensors and sensor neuron), ω_{ji} the strength of the synaptic connection from neuron j to neuron i, β_j the bias term, $\sigma(y_j + \beta_j)$ the firing rate (hereafter, f_i). All sensory neurons share the same bias (β^I), and the same holds for all motor neurons (β^O). τ_i and β_i with $i = 12, ., 15$, β^I, β^O, all the network connection weights ω_{ij}, and g are genetically specified networks' parameters. At each time step, the output of the left motor is $M^L = f_{16} - f_{17}$, and the right motor is $M^R = f_{18} - f_{19}$, with $M_L, M_R \in [-1, 1]$. Cell potentials are set to 0 when the network is initialised or reset, and equation 4 is integrated using the forward Euler method with an integration time step $\Delta T = 0.1$.

An evolutionary algorithm using linear ranking is employed to set the parameters of the networks [6]. We consider populations composed of $M = 100$ teams, each composed of $N = 5$ individuals. The genotypes coding for the parameters of the robots' controllers are vectors comprising 87 real values (76 connections, 4 decay constants, 6 bias terms, and a gain factor) chosen uniformly random from the range [0,1]. In the **clonal approach** teams are genetically homogeneous. Each of the M teams at generation 0 is formed by generating one random genotype and cloning it $N - 1$ times to obtain N identical genotypes. Generations following the first one are produced by a combination of selection with elitism, recombination, and mutation. For each new generation, the highest scoring genotype ("the elite") from the previous generation is retained unchanged, and used to form a new team. Each of the other $M - 1$ new teams are generated by fitness-proportional selection from the 80 best genotypes of the old population. Each new genotype has a 0.3 probability of being created by combining the genetic material of two individual of the old population. During recombination, one crossover point is selected. Mutation entails that a random Gaussian offset is applied to each real-valued vector component encoded in the genotype, with a probability of 0.04. The mean of the Gaussian is 0, and its standard deviation is 0.1. During evolution, all vector component values are constrained to remain within the range [0,1]. In the **aclonal approach** teams are genetically heterogeneous. At generation 0 each of the M teams is formed by generating N random genotypes. For each new generation following the first one, the genotypes of the best team ("the elite") are retained unchanged and copied to the new population. Each of the genotypes of the other teams is formed by first selecting two old teams using roulette wheel selection. Then, two genotypes, each randomly selected among the members of the selected team are recombined with probability 0.3 to reproduce one new genotype. The resulting new genotype is mutated. Mutation and recombination are applied in the same way as for the *clonal approach*. This process is repeated to form $M - 1$ new teams of N genotype each.

5   Results

The aim of this study is to design control systems for swarms of robots that are located in an environment where they have to carry out two simultaneous tasks. One tasks is about moving away from the nest location towards another point indicated by a red light. This task is called *role F*. The other task requires the robot to remain within the nest location indicated by a green light. This task is called *role P*. The task-allocation process is entirely left to the swarm that dynamically allocates roles to robots through physical interactions. Moreover, at least one robot of the swarm is required to change role within consecutive trials to allow the swarm to cope with the environmental variability which dictates that in *Env. A role F* requires more resources than *role P*, whereas in *Env. B role P* requires more resources than *role F* (see Section 3 for details). The experimental design is made of two evolutionary conditions: the *clonal approach* which evaluates homogeneous teams, and the *aclonal approach* which evaluates heterogeneous teams. Moreover, we have two experimental scenarios: *Scenario I* in which the robots can only interact using infra-red sensors, and *Scenario II* in which the robots can perceive each other through the camera and the infra-red sensors. Thus, we have 4 experimental conditions: 1) *clonal approach-Scenario I*; 2) *clonal approach-Scenario II*; 3) *aclonal approach-Scenario I*; 4) *aclonal approach-Scenario II*. 10 evolutionary runs, each using a different random initialisation were carried out for 2500 generations for each of the 4 possible experimental conditions. The effectiveness of the evolved solutions is quantitatively evaluated—using the metric F illustrated in Section 3—through a series of post-evaluation tests. These tests have been applied to the fittest team of each generation for the last 1000 generations of each evolutionary run. The post-evaluation test consists of 40 *ABA-sequence*, and 40 *BAB-sequence* per team (for a total of 240 trials, with $V = 3$ trials times $E = 80$ e-sequences). For each evolutionary run, the team with the highest average re-evaluation fitness score F is considered to be an adequate measure of the success of the run. The graphs in Fig. 3 shows the distributions of the highest average re-evaluation scores achieved by each best team of each run of the *clonal approach* (see Fig. 3, white bars) and *aclonal approach* (see Fig. 3, black bars) in *Scenario I* (see Fig. 3a) and *Scenario II* (see Fig. 3b). Values represent percentage of the optimal evaluation score F on 240 trials. In both task scenarios, all the 10 runs of the *aclonal approach* generated very successful best teams with a fitness higher than 85% of the optimum. In *Scenario I*, 6 best teams, and in *Scenario II*, 8 best teams have a fitness higher than 95% of the optimum score (see Fig. 3a, and 3b, black bars). The best teams generated with the *clonal approach* show less convincing performances in both task scenarios, with 3 best teams scoring more than 85% of the fitness optimum, in *Scenario I*, and 6 best teams in *Scenario II*. In *Scenario II* the gap between *aclonal approach* and *clonal approach* measured in term of

Table 1. Table showing median, mean and standard deviation of the scores of the best evolved teams for each run of each of the four experimental conditions

Scenario I				Scenario II			
	median	mean	s.d.		median	mean	s.d.
clonal approach	5.57	5.07	1.22	*clonal approach*	6.67	5.93	1.38
aclonal approach	6.71	6.60	0.32	*aclonal approach*	6.89	6.81	0.2

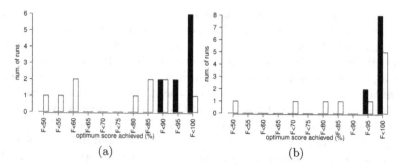

Fig. 3. Histogram showing the distributions of the highest average re-evaluation scores achieved by each best team of each run of the *clonal approach* (white bars) and *aclonal approach* (black bars) in (a) *Scenario I* and (b) *Scenario II*. Values represent percentage of the optimal evaluation score F on 240 trials.

performances of the best teams is less pronounced (see Fig.3b). However, for both task scenarios, the *aclonal approach* out-performs the *clonal approach*. Table 5 shows a comparisons of mean, standard deviation, and median scores of both approaches in both task scenarios. Each measure shows the *aclonal approach* out-performing the *clonal approach*, and the difference between the two set of results is statistically significant (Mann-Whitney U test, $p < 0.01$ for *Scenario I*, and $p < 0.05$ for *Scenario II*). From a statistical point of view, there is enough evidence to prefer one approach over the other for the evolution of multi-robot teams engaged in these dynamic role-allocation and role-switching scenarios. In the next section, we will show the results of further analysis aimed to unveil the behavioural strategies used by the best teams of each evolutionary approach to solve the task in each scenario.

5.1   Behavioural Strategies

In this section, we describe the behavioural strategies of best evolved teams. In particular, we are interested in the capability of the best evolved teams to redistribute agents to task in response to changes in operating conditions. What kind of strategies do the best teams use to keep the majority of the robots engaged on *role F* in *Env. A*, and on *role P* in *Env. B*? To answer this question we have measured for each best team of the *aclonal approach* the frequency with which each robot of the team plays each role in *Env. A* and in *Env. B*. It turned out that, for both scenarios, all the best evolved teams of the *aclonal approach* with an average evaluation score higher than 85% of the fitness optimum employ very similar strategies. The graphs in Fig. 4 illustrate for each scenario the results of the behavioural analysis for the very best team generated aclonally (see Fig. 4a for *Scenario I*, and Fig. 4b for *Scenario II*, with black bars referring to *role F*, and white bars to *role P*). These graphs show that the majority of the robots of this team shows almost no behavioural plasticity. For example, in Fig. 4a robot 1 and 4 play *role F* 100% of the re-evaluation time in *Env. A* and in *Env. B*

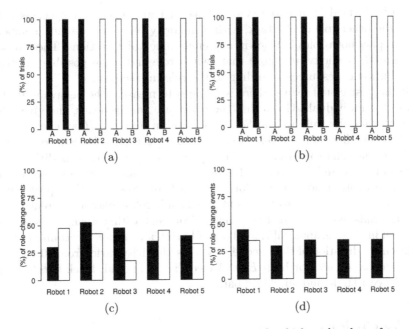

Fig. 4. The graph at the top show the frequency with which each robot of two best aclonal groups plays each role in *Env. A* (A) and *Env. B* (B). (a) refers to *Scenario I*, and (b) refers to *Scenario II*. In (a) and (b) black bars refers to *role F*, white bars refer to *role P*. The graphs at the bottom show the frequency of *role-change events* for each robot of two best evolved clonal groups for (c) *Scenario I*, and (b) *Scenario II*. In (c) and (d) black bars refer to *role-change events* in *ABA-sequence*; white bars refer to *role-change events* in *BAB-sequence*.

(see Fig. 4a, black bars), while robot 3 and 5 play *role P* 100% of the re-evaluation time in both environments (see Fig. 4a, white bars). Only robot 2 is capable of playing both roles, but not in both environment. Robot 2 always play *role F* in *Env. A*, and *role P* in *Env. B*. Similar results are shown in Fig. 4b, which refers to the performances of the very best team of the *aclonal approach* in *Scenario II*. Robot 1 and 3 play *role F* 100% of the re-evaluation time in *Env. A* and in *Env. B* (see Fig. 4b, black bars), while robot 2 and 5 play *role P* 100% of the re-evaluation time in both environments (see Fig. 4b, white bars). Robot 4 always play *role F* in *Env. A*, and *role P* in *Env. B*.

We also looked at the number of identical genes and at the genetic distance between all the possible combination of two robots for the most successful team generated aclonally in *Scenario I*, and the most successful team generated aclonally in *Scenario II* (data not shown). We noticed that some of the robots that play same role share more genes (more than 50% of the genetic material is identical) and they have smaller genetic distance. However, there are also a case in which robots that play same role are not more genetically related than those play different roles (e.g., robot 3 and 5 in teams evolved for *Scenario I*). The results of this analysis show that the best teams generated with the *aclonal approach* tend to be formed by a majority of robots that are genetically specialised for one or the other role of the task, with a minority (one or two team members) that can play both roles but not in any environmental conditions. These robots with limited plasticity are those that switch role and make possible the team redistribution of resources according to the rules of the task.

For the best clonal teams, there is no need to measure the behavioural plasticity of the individuals. Given that the teams are homogeneous, and given that they perform quite well both in *Scenario I*, and in *Scenario II*, we can conclude that the controllers of these very best teams possess the mechanisms requited to play both roles and to switch roles in response to the operating conditions. The graphs in Fig. 4c and 4d show the frequency of *role-change events* for each robot of two best evolved clonal teams for *Scenario I*, and *Scenario II*, respectively. A *role-change events* refers to the event in which a robot play a different role in different trials associated to the same environment within a single e-sequence. As shown in these graphs, all the robots of both teams retain a certain plasticity for the entire duration of the e-sequences. This plasticity varies from 10% to 50% of *role-change events*, for both *Env. A* (see Fig. 4c and 4d, black bars) and for *Env. B* (see Fig. 4c and 4d, white bars).

6   Conclusions

In swarm robotics, plasticity in task-allocation has been successfully obtained by calling upon the principle of individual variability in responding to environmental signals (e.g., [10,14]). In this paper, we looked at task-allocation from the evolutionary swarm robotics perspective. We studied the performances of teams generated with two different evolutionary design approaches. Results indicate that the *aclonal approach* (i.e., heterogeneous teams) outperforms the *clonal approach* (i.e., homogeneous teams) in designing controllers for robot teams engaged in scenarios requiring dynamic task-allocation and task-switching behaviour. The clear advantage of the *aclonal approach* over the *clonal approach* has to be considered in view of evidence showing that the genetic variability is mainly used to generate teams of genetically specialised agents (i.e., agents that can only play a single role), which benefit from the "partial" plasticity of few individuals that can play all the roles of the task but not in all operating conditions. Data not shown indicated that in our scenarios teams generated aclonally are extremely scalable. Incrementing the cardinality of a team by adding those members with partial plasticity (e.g., robot n. 2 in Fig. 4a) has almost no disruptive effects on the performance of the team. However, failure of single team members can be highly disruptive because none of members is capable of playing all the roles in all the operating conditions. Clonal teams do not handle scalability as easily as aclonal teams, but they are capable of successfully facing individual failure by redistributing the agents to tasks. In conclusion, we believe that the *aclonal approach* approach can be a valuable tool to design neural mechanisms that underpin behavioural responses similar to those observed in insects societies. Genetic diversity can facilitate the evolutionary design process of teams that adaptively act in scenarios requiring task-allocation and task-switching behaviour. A similar claim has been made in [15]. However, the emergence of generalist individuals has to be encouraged through the introduction of selective pressures which favour the teams with highly plastic individuals over alternative solutions. Future research work should concentrate on the evolutionary conditions that facilitate the emergence of generalist solutions in the *aclonal approach.*

References

1. Ampatzis, C., Tuci, E., Trianni, V., Christensen, A., Dorigo, M.: Evolving self-assembly in autonomous homogeneous robots. Artificial Life 15(4), 465–484 (2009)
2. Beer, R.D., Gallagher, J.C.: Evolving dynamic neural networks for adaptive behavior. Adaptive Behavior 1(1), 91–122 (1992)
3. Bonabeau, E., Theraulaz, G., Deneubourg, J.L.: Quantitative study of the fixed threshold model for the regulation of division of labour in insects societies. Pro. R. Soc. B 263(1376), 1565–1569 (1996)
4. Dorigo, M., Şahin, E.: Guest editorial. Special issue: Swarm robotics. Aut. Rob. 17(2-3), 111–113 (2004)
5. Duarte, A., Weissing, F., Penn, I., Keller, L.: An evolutionary perspective on self-organised division of labour in social insects. Annu. Rev. Ecol. Evol. Syst. 42, 91–110 (2011)
6. Goldberg, D.E.: Genetic Algorithms in Search, Optimization and Machine Learning. Addison-Wesley, Reading (1989)
7. Gordon, D.: Dynamics of task-switching in harvester ants. Animal Behaviour 38, 194–204 (1989)
8. Gordon, D.: The organisation of work in social insects. Nature 380, 121–124 (1996)
9. Jakobi, N.: Evolutionary robotics and the radical envelope of noise hypothesis. Adaptive Behavior 6, 325–368 (1997)
10. Labella, T., Dorigo, M., Deneubourg, J.L.: Division of labour in a group of robots inspired by ants' foraging behavior. ACM Trans. Aut. Adap. Sys. 1(1), 4–25 (2006)
11. Nolfi, S., Gigliotta, O.: Evorobot*. In: Nolfi, S., Mirolli, M. (eds.) Evolution of Communication and Language in Embodied Agents, pp. 327–332. Springer, Heidelberg (2010)
12. Oldroyd, B., Fewell, J.: Genetic diversity promotes homeostasis in insect colonies. Trends Ecol. E 22(8), 408–413 (2007)
13. Page, R.: The evolution of insects societies. Endeavour 21(7), 114–120 (1997)
14. Pini, G., Brutschy, A., Frison, M., Roli, A., Dorigo, M., Birattari, M.: Task partitioning in swarms of robots: An adaptive method for strategy selection. Swarm Intelligence 5(3-4), 283–304 (2011)
15. Quinn, M.: A comparison of approaches to the evolution of homogeneous multi-robot teams. In: Proc. Int. Conf. Evolutionary Computation (CEC), vol. 1, pp. 128–135 (2001)
16. Quinn, M., Smith, L., Mayley, G., Husbands, P.: Evolving controllers for a homogeneous system of physical robots: Structured cooperation with minimal sensors. Phil. Trans. R. Soc. A 361, 2321–2344 (2003)
17. Trianni, V., Nolfi, S.: Engineering the evolution of self-organizing behaviors in swarm robotics: A case study. Artificial Life 17(3), 183–202 (2011)
18. Tuci, E., Ampatzis, C., Vicentini, F., Dorigo, M.: Evolving homogeneous neuro-controllers for a group of heterogeneous robots. Artificial Life 14(2) (2008)

Influencing a Flock via Ad Hoc Teamwork

Katie Genter and Peter Stone

The University of Texas at Austin, Austin, TX, USA
{katie,pstone}@cs.utexas.edu

Abstract. Flocking is an emergent behavior in which each individual agent follows a simple behavior rule that leads to a group behavior that appears cohesive and coordinated. In our work, we consider how to influence a flock using a set of ad hoc agents. Ad hoc agents are added to the flock and are able to influence the flock to adopt a desired behavior by acting as part of the flock. Specifically, we first examine how the ad hoc agents can behave to quickly orient a flock towards a target heading when given knowledge of, but no direct control over, the behavior of the flock. Then we consider how the ad hoc agents can behave to herd the flock through turns quickly but with minimal agents being separated from the flock as a result of these turns. We introduce an algorithm which the ad hoc agents can use to influence the flock. We also present detailed experimental results for our algorithm, concluding that in this setting, short-term lookahead planning improves significantly upon baseline methods and can be used to herd a flock through turns quickly while maintaining the composition of the flock.

1   Introduction

Consider a flock of migrating birds that is flying directly towards a dangerous area, such as an airport or a wind farm. It will be best for both the flock and the humans in the area if the path of the migratory birds is altered slightly such that the flock will avoid the dangerous area and reach their destination only slightly later than originally expected. However, there is no way to directly control the birds' flight. Rather, we must alter the environment so as to induce the flock to alter their path as desired.

The above scenario is a motivating example for our work on influencing a flock using ad hoc teamwork. We assume that each bird in the flock dynamically adjusts its heading based on that of its immediate neighbors. We assume further that we control one or more *ad hoc* agents — perhaps in the form of robotic birds or ultralight aircraft[1] — that are perceived by the rest of the flock as one of their own. It is through these *ad hoc* agents that we alter the birds' environment so as to induce them to alter their path. We are interested in *how* best to do so.

Flocking is an emergent behavior found in different species in nature including flocks of birds, schools of fish, and swarms of insects. In each of these cases, the animals follow a simple local behavior rule that results in a group behavior that

---

[1] www.operationmigration.org

M. Dorigo et al. (Eds.): ANTS 2014, LNCS 8667, pp. 110–121, 2014.

appears well organized and stable. Research on flocking behavior has appeared in various disciplines such as physics [15], graphics [11], biology [3], and distributed control theory [7,8,13] but the research has focused mainly on characterizing the emergent behavior.

In this work, we are given a team of flocking agents following a known, well-defined rule characterizing their flocking behavior, and we wish to examine how the ad hoc agents should behave. Specifically, this paper addresses two questions: *How should ad hoc agents behave so as to (1) orient the rest of the flock towards a target heading as quickly as possible and (2) herd the rest of the flock through turns quickly but without compromising the composition of the flock?*

The remainder of this paper is organized as follows. Section 2 situates our research in the literature. Section 3 introduces our problem and necessary terminology. The main contribution of this paper is the 1-step lookahead algorithm for influencing a flock to travel in a particular direction; this algorithm is presented in Section 4. We present the results of running experiments using this algorithm in the MASON simulator [10] in Section 5 and then Section 6 concludes.

2 Related Work

Reynolds introduced the original flocking model that we use in this work [11]. His work focused on creating a computer model of flocking that looked and behaved like a real flock of birds. Reynolds' model consists of three simple steering behaviors that determine how each agent maneuvers based on the behavior of the agents around it (henceforth called *neighbors*): *Separation* steers the agent such that it avoids crowding its neighbors, *Alignment* steers the agent towards the average heading of its neighbors, and *Cohesion* steers the agent towards the average position of its neighbors. Vicsek *et al.* considered just the Alignment aspect of Reynolds' model in physics work that studied the emergence of self-ordered motion in flocking [15]. Some related research has also considered how different information provided to the flocking agents affects their behavior. Turgut *et al.* considered how noise in heading measurements, the number of neighbors, and the range of communication affect the self-organization of flocking robots [14]. However, none of these lines of research considered how to influence the flock to adopt a particular behavior by introducing additional agents into the flock.

Jadbabaie *et al.* considered the impact of adding a controllable agent to a flock [8]. They used the Alignment aspect of Reynolds' model and showed that a flock with a controllable agent will always converge to the controllable agent's heading. Su *et al.* also presented work that is concerned with using a controllable agent to make the flock converge [13]. [2] used the same model as [14] and extended it to include informed agents that guide the flock by their preference for a particular direction. Our work is different from these three lines of research in that while they influence the flock to converge to a target heading eventually, we influence the flock to converge quickly.

Couzin *et al.* considered how grouping animals make informed unanimous decision [3]. They showed that only a very small proportion of informed agents is required, and that the larger the group the smaller the proportion of informed individuals needed to orient the group. Cucker and Huepe proposed two Laplacian-based models for a consensus term that balances the trade-off between an informed individuals preference to go in a particular direction and the desire for social interaction [4]. Ferrante *et al.* utilized communication for coordinating movement of a flock towards a common goal [5]. Specifically, informed robots communicated the goal direction while uniformed robots communicated the average of messages received from their neighbors. Yu *et al.* proposed an implicit leadership algorithm that allows all agents to follow a single rule and reach a common group decision without any complex coordination methods [16]. However, none of these lines of research consider how to control some agents from the perspective of knowing and planning for how the other agents will react. Instead, each agent behaves in a fixed way that is pre-decided or based on its type.

Han *et al.* studied how one agent can quickly influence the direction in which an entire flock of agents is moving [7]. In their work each agent follows a simple control rule based on its neighbors, but they only consider one ad hoc agent with unlimited, non-constant velocity. This allows their ad hoc agent to move to any position in the environment within one time step, whereas in our work we assume the agents have bounded velocity.

In our previous work, we considered the problem of leading a flock of agents to a desired orientation using ad hoc agents [6]. In that work we set bounds on the extent to which both stationary and non-stationary ad hoc agents could influence an otherwise stationary team to orient to a desired orientation. The work presented in this paper is substantially different in that we consider a completely non-stationary flock and we present a more advanced algorithm for the ad hoc agents.

Overall, to the best of our knowledge, the work presented in this paper is the first that uses knowledge of how other agents will react to design controllable agents with bounded velocities to influence a flock in motion to converge quickly to a desired heading.

3   Background and Problem Definition

In this section we introduce the concept of ad hoc teamwork and define our problem.

3.1   Ad Hoc Teamwork

Ad hoc teamwork is a relatively new multiagent systems research area [1,9,12] that examines how an agent ought to act when placed on a team with other agents such that there was no prior opportunity to coordinate behaviors. As agents and robots are used with increasing frequency in various cooperative domains, designing agents capable of reasoning about ad hoc teamwork is becoming increasingly important. Ad hoc agents can cooperate within a team without using explicit communication or previously coordinating behaviors among team-mates. One aspect of ad hoc teamwork involves *leading* teammates. Consider a case in which we want to influence a given team of agents to alter their actions in order to maximize the team utility. One way of doing so is by adding one or more agents to the team in order to *lead* them to perform the desired actions.

3.2 Problem Definition

In this work we use a simplified version of Reynolds' Boid algorithm for flocking [11]. Specifically, similarly to other studies such as [8,15], we only consider the Alignment aspect of Reynolds' model. We assume that each agent calculates its orientation for the next time step to be the average heading of its *neighbors*. Throughout this paper, an agent's *neighbors* are the agents located within some set radius of the agent. An agent is not considered to be a neigh-bor of itself, so an agent's current heading is not considered when calculating its orientation for the next time step. In order to calculate its ori-entation for the next time step, each agent com-putes the vector sum of the velocity vectors of each of its neighbors and adopts a scaled version of the resulting vector as its new orientation. Figure 1 shows an example of how an agent's new velocity vector is calculated. At each time step, each agent moves one step in the direction of its current vector and then calculates its new

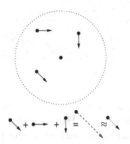

Fig. 1. Calculation of an agent's new velocity vector. The black dot without an arrow repre-sents the agent in question, the dots with arrows represent the agent's neighbors and their ve-locity vectors, and the dotted circle represents the boundary of the agent's neighborhood.

heading based on those of its neighbors, keeping a constant speed.

Over time, agents behaving as described above will naturally gather into one or more groups, and these groups will each travel in some direction. However, in this work we add a small number of *ad hoc agents* to the flock. These ad hoc agents attempt to influence the flock to travel in a pre-defined direction — we refer to this direction as θ^*. This paper addresses two questions: how to *orient* the flock to a target heading and how to *herd* a flock through turns. Hence, throughout this paper we consider two specific cases. In the **Orient** case, the ad hoc agents attempt to influence the flock to travel towards θ^*. In the **Herd** case, the ad hoc agents attempt to influence the flock to travel as a cohesive unit through multiple turns — this can be thought of as influencing the flock towards a frequently changing θ^*.

Note that the challenge of designing ad hoc agent behaviors in a dynamic flocking system is difficult because the action space is continuous. Hence, in our work we make the simplifying assumption of only considering a limited number (*numAngles*) of discrete angle choices for each ad hoc agent.

3.3   Simulation Environment

We situate our research on flocking using ad hoc teamwork within the MASON simulator [10]. Pictures of the Flockers domain are shown in Figure 2. Each agent points and moves in the direction of its current velocity vector.

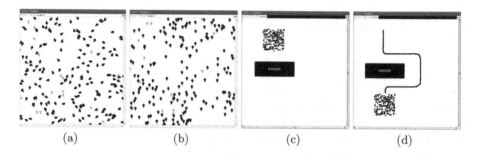

(a)   (b)   (c)   (d)

Fig. 2. Images of (a) the start of an **Orient** trial, (b) the end of an **Orient** trial, (c) the start of a **Herd** trial, and (d) the end of a **Herd** trial (the black line shows the approximate path that the flock travelled to reach their current location). The grey agents are ad hoc agents while the black agents are other members of the flock.

Videos showing the simulator in action in both cases are available on our web page[2]. Our experimental setup is described in more detail in Section 5.2.

4   1-Step Lookahead Behavior

As specified in Section 3, the variable under our control is the heading of each ad hoc agent at every time step of the simulation.

In this section we present Algorithm 1, a 1-step lookahead algorithm for determining the individual behavior of each ad hoc agent. This behavior considers *all* of the influences on neighbors of the ad hoc agent, such that the ad hoc agent can determine the best orientation to adopt based on this information. The 1-step lookahead behavior is a greedy, myopic approach for determining the best individual behavior for each ad hoc agent, where 'best' is defined as the behavior that will exert the most influence on the next time step.

Note that if we only considered the current orientations of the neighbors (instead of the influences on these neighbors) when determining the next orientation for the ad hoc agent to adopt, we would only be estimating the state of each neighbor and hence the resulting orientation adopted by the ad hoc agent would not be 'best'.

The variables used throughout Algorithm 1 are defined in Table 1. Two functions are used in Algorithm 1: *neighbor.vel* returns the velocity vector of neighbor while *neighbor.neighbors* returns a set containing the neighbors of neighbor.

---

[2] http://ants14-flocking.blogspot.com/

Note that Algorithm 1 is called on each ad hoc agent at each time step, and that the neighbors of the ad hoc agent at that time step are provided as a parameter to the algorithm. The output from the algorithm is the orientation that, if adopted by this ad hoc agent, is predicted to influence its neighbors to face closer to θ^* than any of the other $numAngles$ discrete ad hoc orientations considered.

Algorithm 1. bestOrient = 1StepLookahead(neighOfAH)

1: bestOrient ← $(0,0)$
2: bestDiff ← ∞
3: **for** each ad hoc agent orientation vector ahOrient **do**
4:     nOrients ← $\emptyset$
5:     **for** n ∈ neighOfAH **do**
6:         nOrient ← $(0,0)$
7:         **for** n' ∈ n.neighbors **do**
8:             **if** n' is an ad hoc agent **then**
9:                 nOrient ← nOrient + ahOrient
10:            **else**
11:                nOrient ← nOrient + n'.vel
12:         nOrient ← $\frac{nOrient}{\lceil n.neighbors \rceil}$
13:         nOrients ← {nOrient} ∪ nOrients
14:     diff ← average difference between the vectors of nOrients and θ^*
15:     **if** diff < bestDiff **then**
16:         bestDiff ← diff
17:         bestOrient ← ahOrient
18: **return** bestOrient

Conceptually, Algorithm 1 is concerned with how the neighbors of the ad hoc agent are influenced if the ad hoc agent adopts a particular orientation at this time step. Figure 3 presents a pictorial explanation of the calculation of $nOrient$ (lines 6-12 in Algorithm 1). In the figure, $nOrient$, the predicted next step orientation vector of neighbor n of the ad hoc agent, is calculated to be the average of n's neighbors (both marked n') as shown below the diagram. In the example shown, n is the only neighbor of the ad hoc agent, so $nOrients$ would only contain this one $nOrient$. However, $numAngles$ ad hoc agent orientations would be considered by Algorithm 1, re-

Table 1. Variables used in Algorithm 1

Variable	Definition
bestDiff	the smallest difference found so far between the average orientation vectors of $neighOfAH$ and θ^*
bestOrient	the vector representing the orientation adopted by the ad hoc agent to obtain $bestDiff$
neighOfAH	the neighbors of the ad hoc agent
nOrient	the predicted next step orientation vector of neighbor n of the ad hoc agent if the ad hoc agent adopts $ahOrient$
nOrients	a set of the predicted next step orientation vectors of all of the neighbors of the ad hoc agent, assuming the ad hoc agent adopts $ahOrient$

sulting in $numAngles$ different $nOrient$ vectors competing to be $bestOrient$.

Now let us walk through the algorithm in more detail. Algorithm 1 considers each of the $numAngles$ discrete ad hoc agent orientation vectors. For each orientation vector, we consider how each of the neighbors of the ad hoc agent will be influenced if the ad hoc agent adopts that orientation vector (lines 3-13). Hence, we consider all of the neighbors of each neighbor of the ad hoc agent (lines 7-11)

— if the neighbor of the neighbor of the ad hoc agent is an ad hoc agent, we assume that it has the same orientation as the ad hoc agent (even though, in fact, each ad hoc agent orients itself based on a different set of neighbors, line 9), whereas if it is not an ad hoc agent, we calculate its orientation vector based on its current velocity (line 11). Using this information, we calculate how each neighbor of the ad hoc agent will be influenced by averaging the orientation vectors of the each neighbor's neighbors (lines 12-13). We then pick the ad hoc agent orientation vector that results in the least difference between θ^* and the neighbors' current orientation vectors (lines 14-18).

If we assume that there are *numAgents* of agents in the flock, we can calculate the worst-case complexity of Algorithm 1 as follows. Line 3 executes *numAngles* times, line 5 executes at most *numAgents* times, and line 7 executes at most *numAgents*. Hence, the complexity for Algorithm 1 is $O(numAngles * numAgents^2)$.

Results regarding how Algorithm 1 performs in both the **Orient** case and the **Herd** case can be found in Section 5.

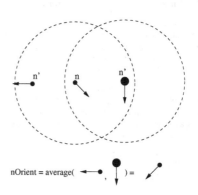

Fig. 3. Diagram illustrating how *nOrient* is calculated in Algorithm 1. Each agent is shown as a dot with an arrow pointing towards its heading. The ad hoc agent is the agent with the larger dot. The dotted circles represent the neighborhood of the agent at the center of the circle.

5   Experiments

In this section we describe our experiments testing the ad hoc agent behavior presented in Section 4 against a baseline method. We describe experiments for both the **Orient** case and the **Herd** case.

5.1   Baseline Ad Hoc Agent Behavior

In this subsection we describe the *Face Desired Orientation* heuristic behavior, which serves as our baseline for comparison. When following this behavior, the ad hoc agents always orient towards θ^*. Note that under this behavior the ad hoc agents do not consider their neighbors or anything about their environment when determining how to behave.

This behavior is modeled after work by Jadbabaie *et al.* [8]. They show that a flock with a controllable agent will eventually converge to the controllable agent's heading. The *Face Desired Orientation* ad hoc agent behavior is essentially the behavior described in their work, except that in our experiments we include multiple controllable agents facing θ^*.

5.2   Experimental Setup

We utilize the MASON simulator [10] for our experiments in this paper. The MASON simulator was introduced in Section 3.3, but in this section we present the details of the environment that are important for completely understanding and replicating our experimental setup.

The baseline experimental settings for variables are given in Table 2 for both the **Orient** case and the **Herd** case. We chose for 10% of the flock to be ad hoc agents as a trade-off between providing enough ad hoc agents to influence the flock and keeping the ad hoc agents few enough to require intelligent behavior in order to influence the flock effectively.

For the **Orient** case, the domain is toroidal. This means that agents that move off one edge of our domain reappear on the opposite edge moving in the same direction. However, for the **Herd** case we removed the toroidal nature of the domain so as to make the domain more realistic. Hence, if agents move off one edge of our domain in the **Herd** case, they will not reappear.

For the **Orient** case, agents are initially randomly placed with random initial headings throughout the domain. For the **Herd** case, agents are initially randomly placed within a square in the top left of the domain, where this square occupies 4% of the domain. Agents are assigned random headings that are within 90 degrees of the initial θ^* for the **Herd** case.

Table 2. Experimental settings for variables in the **Orient** and **Herd** cases. Italicized values were default settings for the simulator.

Variable	Orient Default	Herd Default
toroidal domain	*yes*	no
domain height	*150*	300
domain width	*150*	300
units moved by each agent per time step	*0.7*	0.2
number of agents in flock (*numAgents*)	*200*	*200*
% of flock that are ad hoc agents	10%	10%
neighborhood for each agent (diameter)	*20*	*20*

We only consider *numAngles* discrete angle choices for each ad hoc agent. In all of our experiments, *numAngles* is 50, meaning that the unit circle is equally divided into 50 segments beginning at 0 radians and each of these orientations is considered as a possible orientation for each ad hoc agent. *numAngles*=50 was chosen after some experimentation using the **Orient** case in which *numAngles*=20 resulted in a higher average number of steps for the flock to converge to θ^* and *numAngles*=100 did not require significantly fewer steps for convergence.

In our experiments, we conclude that the flock has converged to θ^* when every agent (that is not an ad hoc agent) is facing within 0.1 radians of θ^*. Other stopping criteria, such as when 90% of the agents are facing within 0.1 radians of θ^*, could have also been used. We tested this alternate stopping criteria in the **Orient** case, but found that using it did not qualitatively alter the results.

In all of our **Orient** experiments, we run 50 trials for each experimental setting. In our **Herd** experiments we run 100 trials for each experimental setting.

In the **Orient** case we use the same 50 random seeds for each set of experiments for the purpose of variance reduction, where in the **Herd** case we use the same 100 random seeds. The random seeds are used to determine the exact placement and orientation of all of the agents at the start of a simulation run.

5.3   Orient Experimental Results

Figure 4 shows the number of time steps needed for the flock to converge to θ^* for the baseline algorithm and the 1-step lookahead algorithm presented in Algorithm 1 using the experimental setup described in Section 5.2 as well as a few variants on this baseline setup. In order to further investigate the dynamics of this domain, in one variant we alter the percentage of the flock that are ad hoc agents while in the other variant we alter the number of agents in the flock. Note that although multiple metrics will be used to judge performance in the **Herd** case, only time to convergence is used in this case since in a toroidal world agents can not become permanently separated from the flock unless they are also travelling towards θ^*.

The results shown in Figure 4 clearly show that the 1-Step Lookahead Behavior performs significantly better than the baseline method in all of our experiments except when the flock size was decreased from 200 agents to 100 agents. In this experiment, although our algorithm did perform better than the baseline, we believe it did not significantly

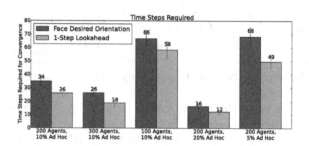

Fig. 4. Results from experiments using the experimental setup described in Section 5.2 as well as four variants on this experimental set-up. The results shown in the figure are averaged over 50 trials and the error bars represent the 95% confidence interval.

improve over the baseline because the agents were too sparse in the environment to have a strong effect on each other.

Altering the percentage of ad hoc agents in the flock clearly alters the amount of agents we can control, which affects the amount of influence we can exert over the flock. Hence, as can be seen in Figure 4, flocks with higher percentages of ad hoc agents will, on average, converge to θ^* in fewer time steps than flocks with lower percentages of ad hoc agents.

5.4   Herd Experimental Results

In our **Herd** experiments, we started all of the agents in a square occupying 4% of the domain in the upper left corner (see Figure 2.c for a picture representing a sample starting configuration). Then the ad hoc agents influenced the flock to travel downward for 300 time steps, then rightward for 300 time steps, then downward for 300 time steps, then leftward for 300 time steps, and finally downward — this path represented the path a flock might need to take to avoid an obstacle in its path.

Different numbers of time steps were used by the ad hoc agents to influence the flock to turn in these four turns. The ad hoc agents were always influencing the flock to orient towards θ^*, so during the turns the value of θ^* was interpolated linearly between the values of θ^* on the surrounding straightaways according to the number of time steps allowed for the turn. Hence, θ^* changed more rapidly when fewer time steps were allowed.

Figure 5 depicts the approximate path along which the flock is influenced to travel, including a depiction of how turns of different lengths affect this path. We maintain approximately the same time to complete all four turns by shortening the straightaway times depending on the amount of time allocated to turning. Flocks that are influenced by the ad hoc agents to turn quicker will inherently have the opportunity to finish their last turn quicker (as can be seen in Figure 5). Hence, *steps-optimal* represents the minimal number of time steps that could be spent by an agent to complete the four required straightaways and turns.

Fig. 5. The approximate path along which the flock is influenced to travel. The dashed line shows the path if turns were instantaneous and the two arcs show the path when 100 or 200 time steps are used to turn. The flock starts in the square.

In the **Herd** experiments, we consider three metrics when determining how much controllability the ad hoc agents exerted on the flock: (1) the average total number of time steps required for the flock to converge to facing downward at the end of the path (*steps-converge*), (2) the difference between *steps-converge* and *steps-optimal* (*diff*), and (3) the average number of agents that become separated from the flock and do not return to the flock before the flock converges to facing downward at the end of the path (*lost*). We also report the number of trials in which at least one agent was separated from the flock and did not return before the flock converged to facing downward at the end of the path, as this makes *lost* easier to interpret.

Table 3 shows results of both the baseline algorithm and the 1-step lookahead algorithm using the experimental setup described above for the **Herd** case. As can be seen in the table, usage of the 1-step lookahead algorithm results in significantly better *steps-converge* and *diff* than the baseline algorithm for each of the turn times tested in the experiment. On average, flocks that are influenced to turn quicker are more likely to have a greater average *diff*. Additionally, note that given this experimental setup, the ad hoc agents would do best to use

Table 3. Results when using the experimental setup described for the **Herd** case. The numbers in parentheses show the 95% confidence interval.

	Steps-Converge	Steps-Optimal	Diff	Lost	Times Lost
10 Steps to Turn - Baseline	1243.0 (4.6)	1205	38.0	17.0	1
30 Steps to Turn - Baseline	1242.3 (2.6)	1215	27.3	17.0	1
50 Steps to Turn - Baseline	1245.8 (2.2)	1225	20.8	0	0
100 Steps to Turn - Baseline	1261.0 (1.6)	1250	11.0	17.0	1
200 Steps to Turn - Baseline	1301.9 (1.0)	1300	1.9	17.0	1
10 Steps to Turn - 1-Step Lookahead	1237.0 (5.4)	1205	32.0	13.5	2
30 Steps to Turn - 1-Step Lookahead	1236.5 (4.6)	1215	21.5	17.0	1
50 Steps to Turn - 1-Step Lookahead	1238.6 (3.0)	1225	13.6	17.0	1
100 Steps to Turn - 1-Step Lookahead	1254.5 (1.3)	1250	4.5	0	0
200 Steps to Turn - 1-Step Lookahead	1300.6 (0.6)	1300	0.6	17.0	1

around 30 time steps to influence the flock through each turn, as *steps-converge* is least when 30 time steps are used for each turn.

Experiments were run in which the percentage of ad hoc agents in the flock was altered to 5% of the flock and 20% of the flock. Results were comparable to those presented in Table 3, but did differ in two significant ways. Specifically, when 20% of the flock consisted of ad hoc agents, no agents were lost during our experiments and turns lasting 10 steps had the least *steps-converge* but were still able to maintain the consistency of the flock. When only 5% of the flock consisted of ad hoc agents, more ad hoc agents were lost on quicker turns and turns lasting about 50 steps were best in terms of *steps-converge*.

6   Conclusion

In this work, we set out to determine how ad hoc agents should behave in order to orient a flock towards a target heading as quickly as possible and to herd a flock around turns quickly but while still maintaining the flock. Our work is situated in a limited ad hoc teamwork domain, so although we have knowledge of the behavior of the flock, we are only able to influence them indirectly via the behavior of the ad hoc agents within the flock. This paper introduces an algorithm that the ad hoc agents can use to influence the flock — a greedy lookahead behavior. We ran extensive experiments using this algorithm in a simulated flocking domain, where we observed that in such a setting, a greedy lookahead behavior is an effective behavior for the ad hoc agents to adopt.

There are plenty of avenues for extensions to this work. We could consider other types of algorithms for the ad hoc agents, such as deeper lookahead searches or algorithms in which the ad hoc agents coordinate their behaviors. Additionally, as this work focused on a limited version of Reynolds' flocking model in which agents calculate their next heading to be the average heading of their neighbors, a promising direction for future work is to extend the algorithms presented in this work to Reynolds' complete flocking model in which agents also consider separation and cohesion when calculating the next heading.

Acknowledgements. This work has taken place in the Learning Agents Research Group (LARG) at UT Austin. LARG research is supported in part by NSF (CNS-1330072, CNS-1305287), ONR (21C184-01), and AFOSR (FA8750-14-1-0070).

References

1. Bowling, M., McCracken, P.: Coordination and adaptation in impromptu teams. In: AAAI, pp. 53–58 (2005)
2. Celikkanat, H., Sahin, E.: Steering self-organized robot flocks through externally guided individuals. Neural Computing & Applications 19(6), 849–865 (2010)
3. Couzin, I.D., Krause, J., Franks, N.R., Levin, S.A.: Effective leadership and decision-making in animal groups on the move. Nature 433(7025), 513–516 (2005)
4. Cucker, F., Huepe, C.: Flocking with informed agents. MathematicS in Action 1(1), 1–25 (2008)
5. Ferrante, E., Turgut, A.E., Mathews, N., Birattari, M., Dorigo, M.: Flocking in stationary and non-stationary environments: A novel communication strategy for heading alignment. In: Schaefer, R., Cotta, C., Kołodziej, J., Rudolph, G. (eds.) PPSN XI. LNCS, vol. 6239, pp. 331–340. Springer, Heidelberg (2010)
6. Genter, K., Agmon, N., Stone, P.: Ad hoc teamwork for leading a flock. In: AAMAS (May 2013)
7. Han, J., Li, M., Guo, L.: Soft control on collective behavior of a group of autonomous agents by a shill agent. Systems Science and Complexity 19, 54–62 (2006)
8. Jadbabaie, A., Lin, J., Morse, A.: Coordination of groups of mobile autonomous agents using nearest neighbor rules. IEEE Transactions on Automatic Control 48(6), 988–1001 (2003)
9. Jones, E., Browning, B., Dias, M.B., Argall, B., Veloso, M.M., Stentz, A.T.: Dynamically formed heterogeneous robot teams performing tightly-coordinated tasks. In: ICRA, pp. 570–575 (2006)
10. Luke, S., Cioffi-Revilla, C., Panait, L., Sullivan, K., Balan, G.: Mason: A multi-agent simulation environment. Simulation: Transactions of the Society for Modeling and Simulation International 81(7), 517–527 (2005)
11. Reynolds, C.W.: Flocks, herds and schools: A distributed behavioral model. SIG-GRAPH 21, 25–34 (1987)
12. Stone, P., Kaminka, G.A., Kraus, S., Rosenschein, J.S.: Ad hoc autonomous agent teams: Collaboration without pre-coordination. In: AAAI (2010)
13. Su, H., Wang, X., Lin, Z.: Flocking of multi-agents with a virtual leader. IEEE Transactions on Automatic Control 54(2), 293–307 (2009)
14. Turgut, A., Celikkanat, H., Gokce, F., Sahin, E.: Self-organized flocking in mobile robot swarms. Swarm Intelligence 2(2-4), 97–120 (2008)
15. Vicsek, T., Czirok, A., Ben-Jacob, E., Cohen, I., Sochet, O.: Novel type of phase transition in a system of self-driven particles. Physical Review Letters 75(6) (1995)
16. Yu, C.H., Werfel, J., Nagpal, R.: Collective decision-making in multi-agent systems by implicit leadership. In: van der Hoek, W., Kaminka, G.A., Lespérance, Y., Luck, M., Sen, S. (eds.) AAMAS, pp. 1189–1196 (2010)

MACOC: A Medoid-Based ACO Clustering Algorithm

Héctor D. Menéndez[1], Fernando E.B. Otero[2], and David Camacho[1]

[1] Departamento de Ingeniería Informática, Universidad Autónoma de Madrid,
Madrid, Spain
{hector.menendez,david.camacho}@uam.es
[2] School of Computing, University of Kent, Chatham Maritime, United Kingdom
F.E.B.Otero@kent.ac.uk

Abstract. The application of ACO-based algorithms in data mining
is growing over the last few years and several supervised and unsupervised learning algorithms have been developed using this bio-inspired approach. Most recent works concerning unsupervised learning have been
focused on clustering, showing great potential of ACO-based techniques.
This work presents an ACO-based clustering algorithm inspired by the
ACO Clustering (ACOC) algorithm. The proposed approach restructures
ACOC from a centroid-based technique to a medoid-based technique,
where the properties of the search space are not necessarily known. Instead, it only relies on the information about the distances amongst
data. The new algorithm, called MACOC, has been compared against
well-known algorithms (K-means and Partition Around Medoids) and
with ACOC. The experiments measure the accuracy of the algorithm for
both synthetic datasets and real-world datasets extracted from the UCI
Machine Learning Repository.

1 Introduction

Ant Colony Optimization (ACO) algorithms have been widely used in several
research fields. One of the most successful application field of ACO algorithms
is data mining [14,17,3,4,18]. Data mining techniques are based on knowledge
extraction or pattern identification inside an information source (dataset). They
usually are focused on supervised or unsupervised learning techniques. Supervised techniques use the information of a class (target) attribute in order to
identify predictive patterns in the data, while unsupervised techniques identify
patterns that can group similar data points into categories. The advantage of
applying ACO algorithm in these problems is that ACO algorithms perform a
global search in the solution space, which in turn has the potential to find more
accurate solutions, and they are less likely to get trap in a local minima.

The work presented in this paper is focused on the application of ACO in
the unsupervised learning task of clustering, where the goal is to group (cluster)
similar data points in the same group and, at the same time, maximise the
difference between different clusters. It has been inspired by a previous algorithm,

M. Dorigo et al. (Eds.): ANTS 2014, LNCS 8667, pp. 122–133, 2014.

called ACO-based Clustering algorithm (ACOC), proposed by Kao and Cheng [10]. ACOC is a centroid-based clustering algorithm, which tries to optimize the centroid (central point) position of each cluster. Following this idea, we focused the proposed algorithm on addressing the centroid-based approaches problems: they need to know the properties of the search space in order to determine the central point, and they are sensitive to noise effects. Inspired by other clustering algorithms [15,16], we reformulated the original ACOC algorithm in a different way to create a medoid-based algorithm. Medoid-based clustering algorithms are usually more robust to noise effects, and do not need the properties of the search space to find a solution—they usually have the distance amongst the data instances, which can be obtained as a Gram matrix of a kernel or a distance measure, and they try to choose those data instances to define the best clusters. These selected instances are called medoids.

In order to check the performance of the proposed algorithm, we have compared it against the original ACOC algorithm using synthetic and real-world datasets and also included the well-known clustering algorithms PAM (Partition Around Medoids) [11] and K-means [13] in the comparisons.

The rest of the paper is structured as follows: Section 2 introduces the related work, Section 3 presents the new algorithm, Section 4 shows the experiments on synthetic and real-world datasets, and, finally, last section explains the conclusions and the future work.

2   Related Work

Ant Colony Optimization (ACO) has become a promising field for data mining problems. ACO algorithms combine the ants foraging behaviour to generate patterns that describe the data according to a supervised or unsupervised learning criteria—depending on the type of algorithm, classification or clustering, respectively.

On the one hand, classification [12] approaches are based on the class information of the data. These classes form part of the data instances and are considered by the algorithm in order to generate a general predictive model that describe (classify) the data. Different classification models have been designed in machine learning. The most common models are based on [12]: decision trees (C4.5), classification rules (RIPPER), artificial neural networks, random decision forest, support vector machines, naïve Bayes, k-nearest neighbours, amongst others. From the ACO point of view, there are several adaptations of these algorithms—e.g., Parpinelli et al. [19] present an ACO to create classification rules; Otero et al. introduce an ACO algorithm for decision tree induction [17]; Blum and Socha introduce a neural network ACO model [3] and Borrotti and Poli focused their work on the naïve Bayes model [4].

There are also approaches that combine ACO with classical classification algorithms in order to improve their results. Some of these techniques, for example, optimize the parameter selection for the classifier (e.g., for SVMs [22]), other are focused on the feature selection process for the data preprocessing phase [6].

On the other hand, clustering [12] is based on a blind search within the data. Clustering techniques try to join similar data points into groups (clusters) according to a cost or objective function, which is usually minimized or maximized, making this clusters different from each other at the same time. There is a large number of clustering approaches, similar to classification, depending of the goal the algorithm should achieve. These techniques are usually divided in three types of clustering [8]: partitional (each instances belongs to a single cluster), overlapping (each instance belongs to one or more clusters) and hierarchical (partitional solutions are nested to generate a tree of clusters). Depending on the clustering algorithm, there are several models focus on the solutions. These models are also divided in parametric [13] or non-parametric model [20], where the former has an statistical estimator that is adapted to the data while the latter separates the data using different topologies or techniques.

The most classical algorithms are K-means [13] and EM [5]. Both K-means and EM are parametrical partitional clustering algorithms, which usually try to optimize an estimator parameters. From a similar perspective, medoid-based algorithms try to find the solution within the data [11]. Medoid-based algorithms, as was mentioned before, do not need the features of the search space in order to find a solution—they can deal with the information extracted from the data distances. This special property makes medoid-based algorithms a good choice for problems where the search space is not well defined, such as time series clustering. There are also bio-inspired algorithms that deal with the clustering problem, several of them focused on genetic algorithms. Hruschka et al. [8] presents a survey of clustering algorithms from different genetic approaches. From other bio-inspired perspectives, ACO algorithms have also produced promising results. Kao and Cheng [10] introduced a centroid-based ACO clustering algorithm; França et al. [7] introduce a bi-clustering algorithm; and Ashok and Messinger focused their work on graph-based clustering [1]; several other approaches are discussed in [9].

3   Medoid-Based ACO Clustering Algorithm (MACOC)

This section presents the Medoid-based ACO Clustering Algorithm (MACOC). The MACOC algorithm is similar to Partition Around Medoids (PAM) algorithm, where the goal of the algorithm is to choose the best M medoids (data instances) based only on distance information. This kind of algorithms usually use a dissimilarity/similarity matrix that measures the distances between the data points. The medoid-based approach is a generalization of the centroid-based approach, but in the medoid case, the properties of the search space are not required—only the distances between the data points.

As an ACO algorithm, MACOC algorithm is based on ACOC algorithm [10]. They have a similar search graph, where the ants try to define the optimal cluster assignment for each of the instances (data points). This graph is based on instances and clusters (Fig. 1). It has an associated NxM matrix, where N is the number of instances and M is the number of clusters (medoids). While in the case of ACOC the construction graph is full-connected, MACOC uses a

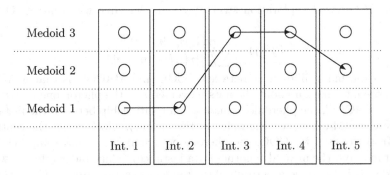

Fig. 1. Representation of the ant travelling around the search graph. Ant visit each instance in order to assign them to a medoid based on the heuristic information and pheromone levels.

graph divided in levels, where each instance defines a level (Fig. 1). The nodes are visited following the sequence of instance, therefore the graph is not full-connected, reducing significantly the memory usage and the complexity of the solution space. It should be noted that this does not incorporate any bias in the search, since the order of the instances is not relevant.

The algorithm is based on several ants looking for the best path in the construction graph. Each ant (k) has the following features:

- set of chosen medoids M^k (which are randomly selected).
- weight matrix W^k (based on the distance between the instances and the ant's medoids). This is similar to ACOC W^k matrix.

The ant has two possible search strategies, exploration and exploitation, similar to the ACOC algorithm. MACOC chooses the strategy for the cluster assignation j according to the ACOC equation:

$$j = \begin{cases} argmax_{u \in N_i}\{[\tau(i,j)][\eta(i,j)]^{\beta}\} & \text{, if } q \leq q_0 \\ S & \text{, otherwise} \end{cases} \quad (1)$$

where N_i is the set of nodes associated to instance i (see Fig. 1), j is the chosen cluster, $\tau(i,j)$ is the pheromone value between i and j, q_0 is the user-defined exploitation probability, q is a random number for strategy selection, $\eta(i,j)$ is the heuristic information between i and j, and S is the ACO-based search strategy. The heuristic information between an instance i and a candidate medoid j is defined by the formula:

$$\eta(i,j) = \frac{1}{d(i,j)} \quad , \quad (2)$$

and the ACO-based exploration strategy S—in the same way than ACOC, is defined by:

$$S = P(i,j) = \frac{[\tau(i,j)] \cdot [\eta(i,j)]^\beta}{\sum_{l=1}^{m}[\tau(i,l)] \cdot [\eta(i,l)]^\beta} \ . \tag{3}$$

One of the main differences between ACOC and MACOC is that MACOC keeps more information about the ants movements in the pheromone matrix. In the case of ACOC, the pheromone matrix is a relationship between the instance and the centroid-label (i.e., the index of the cluster), which is not the centroid itself. In the case of MACOC, the pheromone matrix is a relationship between the instance and the medoid (another data instance), which means that if/when the medoid-label changes as a result of the random selection process, the previous pheromone value is still available. In other words, in the ACOC algorithm, if the centroid value that was previously used as the centroid c_1 is used as the centroid c_2, the previous pheromone values are lost, since they are associated with the label (position) c_1; in the MACOC algorithm, if the medoid instance that was previously used as medoid m_1 is used as the medoid m_2, the previous pheromone values are still used, since the pheromone is associated with the data instance and not with the medoid label.

The MACOC algorithm is structured in the same way than ACOC and can be described as follows:

1. Initialize the pheromone matrix (τ_0), which is global for all ants
2. Initialize ants: choose n random medoids for M^k (n is the number of clusters) and set the matrix W^k to 0. For each ant, until all instances have been visited:
   (a) Select the next data object i
   (b) Select a cluster j: i) Choose a strategy; ii) Calculate neighbouring nodes probability and iii) Visit the node
   (c) Update W^k
3. Choose the best solution:
   (a) Calculate the objective function for each ant:

$$J^k = \sum_{i=1}^{n} \sum_{j=1}^{m} w_{ij}^k \cdot d(x_i, m_j^k) \ , \tag{4}$$

   where $w_{ij}^k \in W^k$ and d is a distance function
   (b) Rank the ants solutions
   (c) Choose the best ant (iteration-best solution)
   (d) Compare it with the best-so-far solution and update this value with the maximum between them
4. Update the pheromone trails (global updating rule): only the r best ants are able to add pheromones:

$$\tau_{ij}(t+1) = (1-\rho)\tau_{ij}(t) + \sum_{h=1}^{r} w_{ij}^h \cdot \Delta\tau_{ij}^h \ , \tag{5}$$

where ρ is the pheromone evaporation rate, $(0 < \rho < 1)$, $w_{ij}^h \in W^h$, t the iteration number, r is the number of elitism ants and $\Delta\tau_{ij}^h = 1/J^k$ is the quality of the solution created by ant h

5. Check termination condition:
   (a) If the number of iterations is greater than the total iterations: re-centrali-
       se the instances assigning each data point to its closest medoid and finish
   (b) Otherwise, go to step 2

4   Experiments

This section presents the experiments which have been carried out to measure
the quality of the proposed MACOC algorithm. The comparisons have been
carried out against K-means, PAM and ACOC algorithms.

4.1   Datasets Description

For the synthetic experiments we have created the following datasets:

- *Synthetic Data 1*: This dataset is formed by 9 two-dimensional gaussian
  models and in this case, there are 3 gaussians which are closer than the rest.
- *Synthetic Data 2*: This second dataset is also formed by 9 two-dimensional
  gaussian models, however, in this case, there are noisy data in the background.

For the real-world experiments, we have chosen four datasets extracted from
UCI Machine Learning Repository [2], which are commonly used as benchmark
for classification and clustering:

- *Iris*: Contains 50 instances distributed over 3 classes, with 4 attributes each.
- *Wine*: Contains 178 instances distributed over 3 classes, with 13 attributes
  each.
- *Vertebral Column* (Ver. Col.): Contains 310 instances distributed over 3
  classes, with 6 attributes each.
- *Breast Tissue* (Bre. Tis.): Contains 106 instances distributes over 6 classes,
  with 10 attributes each.

4.2   Experimental Setup and Evaluation Methods

We selected three algorithms to measure the MACOC quality, namely K-means,
PAM and ACOC.

K-means [13] is an iterative algorithm based on centroids, which are randomly
selected at the beginning. The goal of the algorithm is to find the best centroid
positions. It is executed in two steps: in the first step, it assigns the data to the
closest centroid (cluster); and in the second, it calculates the new position of the
centroid as a centroid of the data which has been assigned to it.

PAM [11] is similar to K-means, but it used medoids instead of centroids. PAM
can works with a dissimilarity/similarity matrix, which is used to calculate the
cost of each medoid belonging to a cluster.

ACOC [10] is the algorithm which our algorithm is inspired. It work with centroids and ants. The main different, apart of the algorithm centroid nature, is that ACOC uses a pheromone matrix from the data instances to the centroid-labels, while our algorithm use a pheromone matrix between all the data to remember the previous medoid assignation. The parameters of ACOC and MA-COC algorithms have been set in a similar way to the original work [10]: the number on ants is 10, the number of elitism is 1, the exploration probability is 0.0001, the initial pheromone values follow an uniform distribution [0.7, 0.8], $\beta = 2.0$, $\rho = 0.1$, and the maximum number of iterations is 1000. The only difference is that the MACOC initial pheromone values have been set as $\frac{1}{n}$ (where n is the number of clusters).

All the experiments have been carried out 50 times—except for K-means, which was carried out 100 times since this algorithm tends to converge to local minima—using the Euclidean distance as the metric, defined by:

$$d(x_i, x_j) = ||x_i - x_j|| = \sqrt{\sum_q (x_i^q - x_j^q)^2} \ , \tag{6}$$

where x_i, x_j represent two data instances and q represents each attribute of the data instance. Additionally, all algorithms need the number of cluster as an initial parameter.

The evaluation of the experiments has been focused on two different ideas: the synthetic dataset has been evaluated according to the cluster discrimination and the performance of the algorithm to discriminate the original clusters in the noisy case; the real-world datasets have been evaluated using the accuracy.

4.3   Synthetic Experiments

The first synthetic dataset is generally easy for all the algorithms (Fig. 2 and 1). The discrimination of the clusters is clearer in this case, resulting in a clear separation of the clusters. The only algorithm that has several problems in identifying the clusters is K-means (see Fig. 2 and Table 1)—probably a result of an early convergence to local minimal solution. PAM provides a good solution of the cluster discrimination and also provides a stable solution (its standard deviation is 0). It means that the algorithm is able to find the medoids with no problems. In the case of ACOC, the solution discriminates the cluster kernels, however, the boundaries are not well-defined (see Fig. 2). MACOC obtains good results for both cluster identification and boundary definition, and also accuracy results (see Table 1).

The second dataset introduces noise and the noise significantly modifies the behaviour of the algorithms (Fig. 3 and Table 1). K-means is not able to identify the cluster kernels and it joins several clusters together, generating a cluster with noisy data. ACOC is also able to find the kernels and discriminate them, however, the boundaries are not well-defined and several instances overlap with other clusters. Finally, PAM and MACOC achieve similar results. In the case of PAM, there are some boundary problems in the central clusters while in the

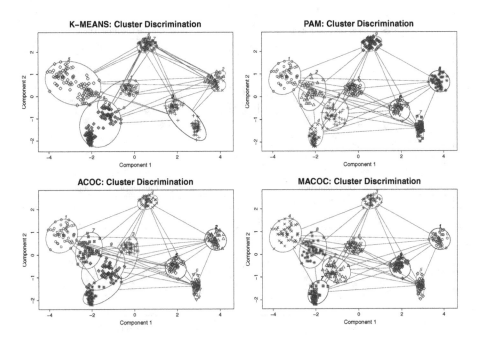

Fig. 2. Results for the synthetic 9 gaussian distribution

case of MACOC the boundaries are clearer, except for one instance (see Fig. 3, at the right of the MACOC image).

These results suggest the following conclusions regarding the comparison with ACOC: while ACOC has boundary problems, which are increased when there is noisy information, MACOC obtains good results for cluster boundary definition and it is more robust to the presence of noise.

4.4   Real-World Experiments

Table 2 shows the results of the algorithms applied to real-world datasets extracted from UCI Machine Learning repository [2].

In the Iris case, K-means and PAM obtains similar results according to the median. K-means obtains the worst minimum accuracy results (58%) and it is the less robust algorithm (its standard deviation is 0.1313). PAM is the most robust algorithm in this case (0 standard deviation), while ACOC and MACOC obtain similar robustness results. The highest minimum value is achieved by both ACOC and PAM. The highest maximum, mean and median values are achieved by MACOC (95.33%, 90.67% and 90.65%, respectively). While MACOC shows better results than ACOC, these results can not be considered different because the null hypothesis can not be refused according to Wilcoxon Test [21].

The application of the algorithms on Wine dataset has shown that K-means also obtains the worst results according to the accuracy and robustness. PAM obtains the highest minimum value and it is again the most robust algorithm.

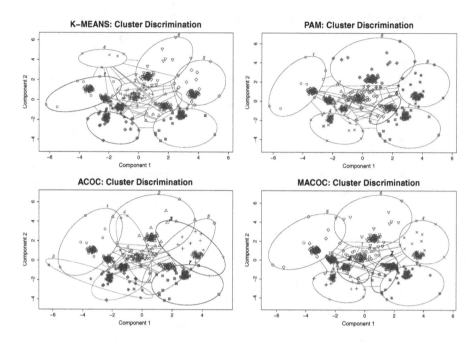

Fig. 3. Results for synthetic 9 gaussian distribution

According to the maximum, mean and median values, MACOC achieves the best results (72.47%, 71.47% and 71.91%, respectively). In this case the null hypothesis is refused with a good significance level, therefore, MACOC results are statistically significantly better than ACOC.

Vertebral column (Ver. Col.) dataset shows different results than the rest of the datasets. In this case, the best algorithm is K-means, which achieves the maximum value for all the metrics. MACOC is the second algorithm according to median, mean and max (52.58%, 53.26% and 65.48%). Again, the Wilcoxon test shows that the null hypothesis can be refused with a high significance level (3e-05), therefore, MACOC results are statistically significantly better than ACOC.

Finally, Breast Tissue dataset shows that MACOC achieves the best results according to mean, median and max (35.55%, 34.91% and 40.57%). In this case, ACOC and PAM achieves similar results, specially according to the median (33.96 %). The null hypothesis can be refused with a significance level of 0.05 (in this case, Wilcoxon test is 0.023), therefore, MACOC results are statistically significantly better than ACOC.

These results show that MACOC improves the performance over ACOC, given that that the solutions obtained by MACOC and ACOC are usually statistically different according to Wilcoxon test in favour of MACOC. Overall, this results are promising regarding the use of medoids instead of centroid, since this is the

Table 1. Results of the application of the algorithms to the synthetic datasets. The p-values for the Wilcoxon test applied to ACOC and MACOC results are: Synthetic 1 (2.394e-13) and Synthetic 2 (7.734e-10)—statistical significant improvements are indicated by a ▲ symbol.

MACOC	Min	Max	Median	Mean		SD
Synthetic 1	99.11%	**100.0%**	**99.78%**	99.75%	▲	± 0.0028
Synthetic 2	96.73%	**100.0%**	**98.18%**	**98.75%**	▲	± 0.0121
ACOC	Min	Max	Median	Mean		SD
Synthetic 1	92.67%	**100.0%**	98.89%	98.57%		± 0.0128
Synthetic 2	82.91%	92.27%	95.27%	94.40%		± 0.0314
K-means	Min	Max	Median	Mean		SD
Synthetic 1	56.67%	99.78%	83.42%	79.47%		± 0.1152
Synthetic 2	65.72%	98.00%	76.73%	79.92%		± 0.0792
PAM	Min	Max	Median	Mean		SD
Synthetic 1	**99.78%**	99.78%	**99.78%**	**99.78%**		± 0.0000
Synthetic 2	**98.00%**	98.00%	98.00%	98.00%		± 0.0000

main difference between MACOC and ACOC. In a similar way that median is more stable than mean, medoids are usually more stable to outliers in the data than centroids. Future studies will be focused on how the algorithm responds to worse conditions, such as large data, more outliers and extremely noisy data.

5 Conclusions and Future Work

This work presented a new ACO-based clustering algorithm, called MACOC, which is focused on a medoid-based approach. MACOC is an adaptation from a previously proposed centroid-based ACOC algorithm to a medoid-based approach. From the ACO perspective, the new algorithm has also improved the use of the pheromone, extending the pheromone matrix to keep the information from different candidate medoid instances across iterations.

The application of the algorithm to synthetic and real-world datasets has shown that MACOC is more robust to noisy information and it defines better cluster boundaries than ACOC. It also showed that MACOC has good general results compared with well-known clustering algorithms.

The future work will be focused on improving the results of MACOC by incorporating a medoid recalculation process. It would be interesting to explore the addition of the medoid selection to the construction graph, allowing ants to share information of the best performing medoid instances.

Table 2. Results of the application of the algorithms to the different datasets extracted from the UCI database. The p-values for the Wilcoxon test applied to ACOC and MACOC solutions are: Iris (0.4439), Wine (4.117e-07), Ver. Col. (3e-05), Bre. Tis. (0.02349)—statistical significant improvements are indicated by a ▲ symbol.

MACOC	Min	Max	Median	Mean		SD
Iris	87.33%	**95.33%**	**90.65%**	**90.67%**		± 0.0187
Wine	69.10%	**72.47%**	**71.91%**	71.47%	▲	± 0.0075
Ver. Col.	46.13%	**65.48%**	52.58%	53.26%	▲	± 0.0488
Bre. Tis.	28.30%	**40.57%**	**34.91%**	**35.55%**	▲	± 0.0328
ACOC	Min	Max	Median	Mean		SD
Iris	**89.33%**	93.33%	90.00%	90.13%		± 0.0080
Wine	70.22%	71.35%	70.79%	70.78%		± 0.0026
Ver. Col.	47.74%	54.19%	49.35%	49.43%		± 0.0095
Bre. Tis.	31.13%	**40.57%**	33.96%	34.47%		± 0.0229
K-means	Min	Max	Median	Mean		SD
Iris	58.00%	89.33%	89.33%	82.46%		± 0.1313
Wine	56.74%	69.33%	70.22%	67.26%		± 0.0632
Ver. Col.	**56.13%**	**65.48%**	**56.13%**	**57.81%**		± 0.0339
Bre. Tis.	33.02%	33.96%	33.21%	33.04%		± 0.0025
PAM	Min	Max	Median	Mean		SD
Iris	**89.33%**	89.33%	89.33%	89.33%		± 0.0000
Wine	**70.79%**	70.79%	70.79%	70.79%		± 0.0000
Ver. Col.	48.71%	48.71%	48.71%	48.71%		± 0.0000
Bre. Tis.	**33.96%**	33.96%	33.96%	33.96%		± 0.0000

Acknowledgments. This work has been partly supported by: Spanish Ministry of Science and Education under project TIN2010-19872 and Savier – an Airbus Defense & Space project (FUAM-076914 and FUAM-076915).

References

1. Ashok, L., Messinger, D.W.: A spectral image clustering algorithm based on ant colony optimization 8390, 83901P–83901P-10 (2012),
   http://dx.doi.org/10.1117/12.919082
2. Bache, K., Lichman, M.: UCI machine learning repository (2013),
   http://archive.ics.uci.edu/ml
3. Blum, C., Socha, K.: Training feed-forward neural networks with ant colony optimization: An application to pattern classification. In: Proceedings of HIS 2005, pp. 233–238. IEEE Computer Society, Washington, DC (2005),
   http://dx.doi.org/10.1109/ICHIS.2005.104

4. Borrotti, M., Poli, I.: Naïve bayes ant colony optimization for experimental design. In: Kruse, R., Berthold, M., Moewes, C., Gil, M.A., Grzegorzewski, P., Hryniewicz, O. (eds.) Synergies of Soft Computing and Statistics. AISC, vol. 190, pp. 489–497. Springer, Heidelberg (2013),
   http://dx.doi.org/10.1007/978-3-642-33042-1_52
5. Dempster, A.P., Laird, N.M., Rubin, D.B.: Maximum Likelihood from Incomplete Data via the EM Algorithm. Journal of the Royal Statistical Society. Series B (Methodological) 39(1), 1–38 (1977),
   http://web.mit.edu/6.435/www/Dempster77.pdf
6. Ding, S.: Feature selection based f-score and aco algorithm in support vector machine. In: Second International Symposium on Knowledge Acquisition and Modeling, KAM 2009, vol. 1, pp. 19–23 (2009)
7. de França, F.O., Coelho, G.P., Von Zuben, F.J.: bicACO: An Ant Colony Inspired Biclustering Algorithm. In: Dorigo, M., Birattari, M., Blum, C., Clerc, M., Stützle, T., Winfield, A.F.T. (eds.) ANTS 2008. LNCS, vol. 5217, pp. 401–402. Springer, Heidelberg (2008), http://dx.doi.org/10.1007/978-3-540-87527-7_45
8. Hruschka, E., Campello, R., Freitas, A., de Carvalho, A.: A survey of evolutionary algorithms for clustering. IEEE Transactions on Systems, Man, and Cybernetics, Part C: Applications and Reviews 39(2), 133–155 (2009)
9. Jafar, O.M., Sivakumar, R.: Ant-based clustering algorithms: A brief survey. International Journal of Computer Theory and Engineering 2, 787–796 (2010)
10. Kao, Y., Cheng, K.: An aco-based clustering algorithm. In: Dorigo, M., Gambardella, L.M., Birattari, M., Martinoli, A., Poli, R., Stützle, T. (eds.) ANTS 2006. LNCS, vol. 4150, pp. 340–347. Springer, Heidelberg (2006),
   http://dx.doi.org/10.1007/11839088_31
11. Kaufman, L., Rousseeuw, P.: Clustering by Means of Medoids. Reports of the Faculty of Mathematics and Informatics (1987),
   http://books.google.co.uk/books?id=HK-4GwAACAAJ
12. Larose, D.T.: Discovering Knowledge in Data. John Wiley & Sons (2005)
13. Macqueen, J.B.: Some methods of classification and analysis of multivariate observations. In: Proceedings of the Fifth Berkeley Symposium on Mathematical Statistics and Probability, pp. 281–297 (1967)
14. Martens, D., Baesens, B., Fawcett, T.: Editorial survey: swarm intelligence for data mining. Machine Learning 82(1), 1–42 (2011)
15. Menéndez, H.D., Barrero, D.F., Camacho, D.: A genetic graph-based approach for partitional clustering. Int. J. Neural Syst. 24(3) (2014)
16. Orgaz, G.B., Menéndez, H.D., Camacho, D.: Adaptive k-means algorithm for overlapped graph clustering. Int. J. Neural Syst. 22(5) (2012)
17. Otero, F., Freitas, A., Johnson, C.: Inducing decision trees with an ant colony optimization algorithm. Applied Soft Computing 12(11), 3615–3626 (2012)
18. Otero, F., Freitas, A., Johnson, C.: A New Sequential Covering Strategy for Inducing Classification Rules With Ant Colony Algorithms. IEEE Transactions on Evolutionary Computation 17(1), 64–76 (2013)
19. Parpinelli, R., Lopes, H., Freitas, A.: Data mining with an ant colony optimization algorithm. IEEE Trans. on Evolutionary Computation 6(4), 321–332 (2002)
20. Schaeffer, S.: Graph clustering. Computer Science Review 1(1), 27–64 (2007)
21. Wilcoxon, F.: Individual comparisons by ranking methods. Biometrics 1(6), 80–83 (1945)
22. Zhang, X., Chen, X., He, Z.: An aco-based algorithm for parameter optimization of support vector machines. Expert Syst. Appl. 37(9), 6618–6628 (2010),
   http://dx.doi.org/10.1016/j.eswa.2010.03.067

Particle Swarm Convergence: Standardized Analysis and Topological Influence

Christopher W. Cleghorn and Andries P. Engelbrecht

Department of Computer Science
University of Pretoria
{ccleghorn,engel}@cs.up.ac.za

Abstract. This paper has two primary aims. Firstly, to empirically verify the use of a specially designed objective function for particle swarm optimization (PSO) convergence analysis. Secondly, to investigate the impact of PSO's social topology on the parameter region needed to ensure convergent particle behavior. At present there exists a large number of theoretical PSO studies, however, all stochastic PSO models contain the stagnation assumption, which implicitly removes the social topology from the model, making this empirical study necessary. It was found that using a specially designed objective function for convergence analysis is both a simple and valid method for convergence analysis. It was also found that the derived region needed to ensure convergent particle behavior remains valid regardless of the selected social topology.

1 Introduction

Particle swarm optimization (PSO) is a stochastic population-based search algorithm that has been effectively utilized to solve numerous real world optimization problems [1]. Despite PSO's widespread use, there still exists a number of important aspects of PSO's behavior that are not completely understood. PSO has also undergone numerous theoretical investigations [2–12]. There is, however, one very evident omission in all of these studies, namely, the impact that the social topology has on a stochastic PSO's ability to converge.

As with most theoretical studies, a number of simplifying assumptions are needed in order to be able to reasonably derive a result. The last remaining assumption that is still present in all theoretical work on the stochastic PSO is the stagnation assumption. The stagnation assumption assumes that the personal and neighborhood best positions remain fixed for each particle. Under the stagnation assumption the notion of a social topology does not exist, as the neighborhood best positions are static. As a result, choosing PSO parameters under arbitrary topologies is non trivial, as the exact impact on convergence is unknown.

This paper has two primary aims. Firstly, to empirically verify the use of a specially designed objective function for convergence analysis. Secondly, to investigate the impact of the PSO's social topology on the parameter region needed to ensure convergent particle behavior.

M. Dorigo et al. (Eds.): ANTS 2014, LNCS 8667, pp. 134–145, 2014.

A brief description of the PSO algorithm is given in section 2. A discussion of the derived parameter regions for particle convergence is given in section 3. The experimental set up and results are given in sections 4 and 5 respectively. Section 6 presents a summary of the findings of this paper, as well as a discussion of topics for future research.

2   Particle Swarm Optimization

Particle swarm optimization (PSO) was originally developed by Kennedy and Eberhart [13] to simulate the complex movement of birds in a flock. The standard variant of PSO this paper focuses on includes the inertia coefficient proposed by Shi and Eberhart [14].

The PSO algorithm is defined as follows: Let $f : \mathbb{R}^k \to \mathbb{R}$ be the objective function that the PSO aims to find an optimum for (if it exists). For the sake of simplicity, a minimization problem is assumed from this point onwards. Let $\Omega(t)$ be a set of N particles in $\mathbb{R}^k$ at a discrete time step t. Then $\Omega(t)$ is said to be the particle swarm at time t. The position $\mathbf{x}_i$ of particle i, is updated using

$$\boldsymbol{x}_i(t+1) = \boldsymbol{x}_i(t) + \boldsymbol{v}_i(t+1), \tag{1}$$

where the velocity update, $\boldsymbol{v}_i(t+1)$, is defined as

$$\begin{aligned} \boldsymbol{v}_i(t+1) = w\boldsymbol{v}_i(t) &+ c_1\boldsymbol{r}_1(t)(\boldsymbol{y}_i(t) - \boldsymbol{x}_i(t)) \\ &+ c_2\boldsymbol{r}_2(t)(\hat{\boldsymbol{y}}_i(t) - \boldsymbol{x}_i(t)), \end{aligned} \tag{2}$$

where $\boldsymbol{r}_1(t), \boldsymbol{r}_2(t) \sim U(0,1)^k$ for all t. The position $\boldsymbol{y}_i(t)$ represents the "best" position that particle i has visited, where "best" means the location where the particle has obtained the lowest objective function evaluation. The position $\hat{\boldsymbol{y}}_i(t)$ represents the "best" position that the particles in the neighborhood of the i-th particle have visited. The coefficients c_1, c_2, and w are the cognitive, social, and inertia weights respectively.

The driving feature of the PSO is social interaction, specifically the way in which knowledge about the search space is shared amongst the particles in the swarm. In general, the social topology of a swarm can be viewed as a graph, where nodes represent particles, and the edges are the allowable direct communication routes. The social topology chosen has a direct impact on the behaviour of the swarm as a whole [15–17]. Some of the most frequently used social topologies are discussed below:

- **Star:** The star topology is one where all the particles in the swarm are interconnected as illustrated in figure 1a. The original implementation of the PSO algorithm utilized the star topology [13]. A PSO utilizing the star topology is commonly referred to as the Gbest PSO.
- **Ring:** The ring topology is one where each particle is in a neighborhood with only two other particles, with the resulting structure forming a ring as illustrated in figure 1b. The ring topology can be generalized to a network structure where larger neighborhoods are used. The resulting algorithm is referred to as the Lbest PSO.

- **von Neumann:** The von Neumann topology is one where the particles are arranged in a grid-like structure. The 2-D variant is illustrated in figure 1c, and the 3-D variant is illustrated in figure 1d.

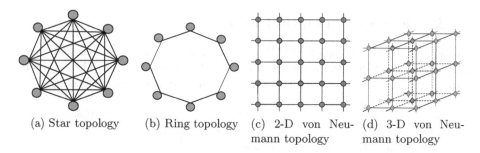

(a) Star topology   (b) Ring topology   (c) 2-D von Neu-   (d) 3-D von Neu-
                                          mann topology   mann topology

Fig. 1. Common social topologies

The PSO algorithm is summarized in algorithm 1.

3   Theoretical Particle Swarm Optimization Background

This section briefly presents each theoretically derived region that is sufficient for particle convergence, along with the corresponding assumptions utilized in the region's derivation.

The primary assumptions that occur in the theoretical PSO research are as follows:

Deterministic Assumption: It is assumed that $\boldsymbol{\theta}_1 = \boldsymbol{\theta}_1(t) = c_1 \boldsymbol{r}_1(t)$, and $\boldsymbol{\theta}_2 = \boldsymbol{\theta}_2(t) = c_2 \boldsymbol{r}_2(t)$, for all t.

Stagnation Assumption: It is assumed that $\boldsymbol{y}_i(t) = \boldsymbol{y}_i$, and $\hat{\boldsymbol{y}}_i(t) = \hat{\boldsymbol{y}}_i$, for all t.

Weak Chaotic Assumption: It is assumed that both $\boldsymbol{y}_i(t)$ and $\hat{\boldsymbol{y}}_i(t)$ will occupy an arbitrarily large finite number of unique positions (distinct positions), ψ_i and $\hat{\psi}_i$, respectively.

Under the deterministic and weak chaotic assumption Cleghorn and Engelbrecht [8], derived the following region for particle convergence:

$$c_1 + c_2 < 2(1 + w), \quad c_1 > 0, \quad c_2 > 0, \quad -1 < w < 1. \tag{3}$$

which generalized the work of Van den Bergh and Engelbrecht [6, 18], and that of Trelea [7]. Equation (3) is illustrated in figure 2, as the triangle AFB.

Kadirkamanathan et al [9], only under the stagnation assumption, derived the following region for particle convergence:

$$\begin{cases} c_1 + c_2 < 2(1 + w) & w \in (-1, 0] \\ c_1 + c_2 < \frac{2(1-w)^2}{1+w} & w \in (0, 1). \end{cases} \tag{4}$$

Algorithm 1. PSO algorithm

Create and initialize a k-dimensional swarm, $\Omega(0)$, of N particles uniformly within a predefined hypercube.

Let f be the objective function.

Let y_i represent the personal best position of particle i, initialized to $x_i(0)$.

Let $\hat{y}_i$ represent the neighborhood best position of particle i, initialized to $x_i(0)$.

Initialize $v_i(0)$ to $\mathbf{0}$.

repeat

    for all particles $i = 1, \cdots, N$ **do**

        if $f(x_i) < f(y_i)$ **then**

            $y_i = x_i$

        end if

        for all particles $\hat{i}$ with particle i in their neighborhood **do**

            if $f(y_i) < f(\hat{y}_i)$ **then**

                $\hat{y}_{\hat{i}} = y_i$

            end if

        end for

    end for

    for all particles $i = 1, \cdots, N$ **do**

        update the velocity of particle i using equation (2)

        update the position of particle i using equation (1)

    end for

until stopping condition is met

Gazi [10] expanded the derived region of equation (4), also under the stagnation assumption only, resulting in the following region:

$$\begin{cases} c_1 + c_2 < \frac{24(1+w)}{7} & w \in (-1, 0] \\ c_1 + c_2 < \frac{24(1-w)^2}{7(1+w)} & w \in (0, 1). \end{cases} \tag{5}$$

The regions corresponding to equations (4) and (5) are illustrated in figure 2 as triangle like regions ADB and AEB respectively. Unfortunately, both equations (4) and (5) are very conservative regions, as they were derived utilizing the Lyapunov condition [19].

Lastly, Poli [11] under the stagnation assumption only, but without the use of the Lyapunov condition, derived the following region:

$$c_1 + c_2 < \frac{24\left(1 - w^2\right)}{7 - 5w}. \tag{6}$$

The region defined by equation (6) is illustrated in figure 2 as the curved line segment AB.

With all of the available convergence regions, the choice of which region to use in practice, is difficult, as each region's derivation relies on at least one simplifying assumption. As a result a study was done by Cleghorn and Engelbrecht [20], which showed with the support of empirical evidence that the region of equation (6) derived by Poli matched almost perfectly with the convergence behavior of

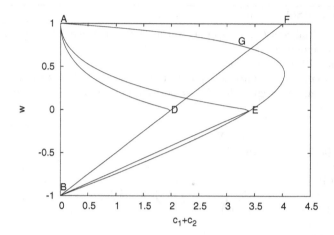

Fig. 2. Theoretically derived regions sufficient for particle convergence

a non simplified Gbest PSO, making the region defined by equation (6) the best choice in practice when utilizing the star topology.

4   Experimental Setup

The experiment conducted in this paper has two primary aims: Firstly, to justify the use of a specifically designed objective function for the convergent parameter region analysis; Secondly, to verify that the theoretically derived region of Poli [11] remains valid under multiple social topologies.

There is an inherent difficulty in empirically analyzing the convergence behavior of PSO particles, specifically with regards to understanding the influence of the underlying objective function's landscape on the PSO algorithm. It is proposed that the following objective function can be used as the reference function for convergent region analysis:

$$CF(\boldsymbol{x}) \in U(-1000, 1000). \tag{7}$$

The objective function in equation (7) is constructed on initialization, and remains static from that point onwards. What the objective function in equation (7) provides is an environment that is rife with discontinuities (actually, it is discontinuous almost everywhere), resulting in a search space the PSO algorithm will battle to become fully stagnate in.

The measure of convergence used in this paper is:

$$\Delta(t+1) = \frac{1}{k} \sum_{i=1}^{i=k} \|\boldsymbol{x}_i(t+1) - \boldsymbol{x}_i(t)\|_2. \tag{8}$$

The experiment utilizes the following static parameters: Population size of 64, 5000 iterations, and a 50-dimensional search space. A population size of 64 is utilized to allow for all the social topologies tested to be complete structures. Particle positions are initialized within $(-100, 100)^k$ and velocities are initialized to $\mathbf{0}$.

The experiment is conducted over the following parameter region:

$$w \in [-1.1, 1.1] \text{ and } c_1 + c_2 \in (0, 4.3], \tag{9}$$

where $c_1 = c_2$, with a sample point every 0.1 along w and $c_1 + c_2$. The experiment is performed for each of the following neighborhood topologies: Star, ring, 2-D and 3-D von Neumann. The experiment is conducted using CF and 11 base objective functions from the CEC 2014 problem set [21]. The functions are as follows: Ackley, High Conditioned Elliptic, Bent Cigar, Discus, Rosenbrock, Griewank, Rastrigin, HappyCat, HGBat, Katsuura, and Expanded Griewank plus Rosenbrock. The region of equation (9) contains exactly 504 points that satisfy equation (6). A total of 989 sample points are used per objective function and topology pair, resulting in 47472 sample points per run. The results reported in Section 5 are the averages over 35 independent runs for each sample point.

In order to allow for a sensible comparison of convergence properties the convergence measure value is bounded as follows:

$$\Delta_{max} = \frac{1}{10}\sqrt{Dimension\,(DomainUpperBound - DomainLowerBound)^2}, \tag{10}$$

which is one tenth the maximum distance of two points in the initialized search space. For this paper, $\Delta_{max} = 1414.214$.

5   Experimental Results and Discussion

In this section a table per topology is presented containing the following measurements per objective function:

- **Measurement A:** The number of PSO parameter configurations that resulted in a final convergence measure value less than or equal to the final convergence measure if the CF objective function was used.
- **Measurement B:** The number of PSO parameter configurations that resulted in a final convergence measure value greater than the final convergence measure if the CF objective function was used.
- **Measurement C:** The number of PSO parameter configurations that resulted in a final convergence measure greater than or equal to Δ_{max}.
- **Measurement D:** The number of PSO parameter configurations that resulted in a final convergence measure less than Δ_{max}.
- **Measurement E:** The number of PSO parameter configurations that satisfied equation (6) and resulted in a final convergence measure less than Δ_{max}.
- **Measurement F:** The number of PSO parameter configurations that satisfied equation (6) and did not result in a final convergence measure less than Δ_{max}.

- **Measurement G:** The average convergence measure value across all parameter configurations, with all elements bounded at Δ_{max}.

Measurements A and B provide a concise way of seeing per objective function how much better or worse the CF objective function performs as a reference convergence analysis function. An ideal convergence analysis function, is one that in general will yield the highest resulting convergence measure for all possible parameter configurations. The higher the resulting convergence measure value is, the harder it was for the PSO to have converged under a given objective function. Measurements C and D give a clear picture of how effectively the underlying objective function highlights possible divergent particle behavior. Given the tested region of equation (9), there are a total of 504 parameter configurations that satisfy equation (6), leaving 485 parameter configurations that should produce divergent behavior. Ideally an objective function utilized for convergence analysis should result in a value for measurement C as close to 485 as possible, and a value for measurement D as close to 504 as possible. Measurements E and F are an extension of measurements C and D, in that an objective function should have at most 504 parameter configurations that both satisfy equation (6) and have a convergence measure value not exceeding Δ_{max}. An objective function with a measurement E value smaller than 504 is more conservative in assigning the label of a convergent particle. A slightly conservative assignment is a positive feature of an objective function being used for convergence analysis, as falsely classifying a parameter configuration as convergent is far from ideal. Measurement G provides an overall view of how difficult the used objective function has made it for the PSO algorithm to converge.

A snapshot of all parameter configurations' resulting convergence measure values are presented for four cases:

- **Case A:** For each parameter configuration the maximum convergence measure value across all 11 objective functions and topologies is reported.
- **Case B:** For each parameter configuration the maximum convergence measure value across all topologies using only the CF objective function.

In order to deduce the convergence region from the empirical data of all 11 base functions and all topologies, the largest recorded convergence measure value of each parameter configuration is reported in case A. Case B is presented to illustrate the similarity between the mapped out convergence region of the PSO algorithm using the CF objective function to the mapped out convergence region of the PSO algorithm in case A, which is constructed using the complete pool of gathered data of the 11 objective functions.

- **Case C:** For each parameter configuration the maximum convergence measure value across all 11 objective functions of the topology that had the greatest Euclidean distance from the optimal region of case A.

Case C is presented to illustrate the maximum deviation between the convergent parameter region under multiple topologies. If the convergent parameter regions between cases A and C are identical, then the topological choice has no influence of the convergent parameter regions.

- **Case D:**For each parameter configuration the maximum convergence measure value across all topologies using only an objective function which has the most similar resulting measurements to case B.

Case D is presented to illustrate that the mapped out convergence region of cases A to D are not identical to the convergence regions of any arbitrary objective function. In particular, cases A to D should result in a subset of the region produced by an arbitrary objective function.

Measurements A and B in table 1 show that the Gbest PSO applied to the CF objective function resulted in a higher convergence measure evaluation than 9 of the 11 other objective functions for nearly all parameter configurations. For the two remaining objective functions, Katsuura is the only objection function close to the CF objective function in terms of measurement A. However, Katsuura has an average convergence measure of 49.672 less than CF has, making CF the better objective function for convergence analysis. The CF objective function also obtained the largest number of parameters configurations that resulted in a convergence measure breach of Δ_{max}, and the highest average convergence measure evaluations. These measurements indicate the effectiveness of CF as an objective function for convergence analysis. The CF objective function under the star topology provides an environment that is much harder for PSO particles to converge in than using any of the other objective functions.

Table 1. Convergence properties per objective function under the Star topology

Measurement / Function	A	B	C	D	E	F	G
CF	–	–	467	522	504	0	683.437
Ackley	879	110	464	525	502	2	676.293
High Conditioned Elliptic	989	0	400	589	504	0	573.601
Bent Cigar	989	0	412	577	504	0	598.593
Discus	989	0	409	580	504	0	592.545
Rosenbrock	988	1	424	565	504	0	622.009
Griewank	989	0	412	577	504	0	596.772
Rastrigin	989	0	411	578	504	0	596.909
HappyCat	989	0	411	578	504	0	595.375
HGBat	989	0	412	577	504	0	595.366
Katsuura	507	482	416	573	504	0	623.765
Expanded Griewank plus Rosenbrock	989	0	416	573	504	0	603.981

Measurements A and B in table 2 show that the Lbest PSO applied to the CF objective function resulted in a higher convergence measure evaluation than 9 of the 11 other objective functions for nearly all parameter configurations. Once again, Katsuura provided the second best results with reference to measurement A, while CF provided the best results for all other measurements. Though inferior, Ackley resulted in values for C, D and G very close to that obtained by the CF objective function. However, CF provided far better results in terms of measurement A, making CF the best choice as an objective function for convergence analysis. The difference in the effect that the ring and star topologies had on convergence is very small, as illustrated by the small changes in the average convergence measure values between tables 1 and 2.

Table 2. Convergence properties per objective function under the Ring topology

Measurement / Function	A	B	C	D	E	F	G
CF	–	–	473	516	503	1	690.797
Ackley	912	77	469	520	504	0	682.57
High Conditioned Elliptic	989	0	400	589	504	0	574.659
Bent Cigar	989	0	415	574	504	0	602.194
Discus	989	0	414	575	504	0	597.668
Rosenbrock	989	0	417	572	504	0	613.094
Griewank	989	0	412	577	504	0	603.304
Rastrigin	989	0	412	577	504	0	601.111
HappyCat	989	0	415	574	504	0	603.536
HGBat	989	0	414	575	504	0	601.403
Katsuura	509	480	413	576	504	0	623.71
Expanded Griewank plus Rosenbrock	989	0	416	573	504	0	609.277

Measurements A through G in tables 3 and 4 show for both the 2-D and 3-D von Neumann topologies that the results remain almost identical to that of the ring and star topologies. This provides evidence that the topology has a negligible impact on the effectiveness on CF as an objective function for convergence analysis. The similarity between tables 1 to 4 under all measurements indicates how minimally the topology influences the parameter region needed for particle convergence.

Table 3. Convergence properties per objective function under the 2-D von Neumann topology

Measurement / Function	A	B	C	D	E	F	G
CF	–	–	480	509	500	4	704.946
Ackley	915	74	475	514	501	3	692.301
High Conditioned Elliptic	989	0	402	587	504	0	577.036
Bent Cigar	989	0	413	576	504	0	600.998
Discus	989	0	414	575	504	0	598.234
Rosenbrock	988	1	415	574	504	0	616.365
Griewank	989	0	414	575	504	0	600.839
Rastrigin	989	0	412	577	504	0	597.999
HappyCat	989	0	414	575	504	0	600.869
HGBat	989	0	413	576	504	0	599.576
Katsuura	525	464	415	574	504	0	622.108
Expanded Griewank plus Rosenbrock	988	1	416	573	504	0	608.545

For case A, the convergent region as illustrated in figure 3a matches the derived region of equation (6) almost perfectly, as does the region seen in figure 3b for case B. While there exists a slight difference between figures 3a and 3b in terms of convergence measure values, the overall convergent region is nearly identical between the two. With this similarity observed between figures 3a and 3b in mind, it is clear that just using the CF function for convergence analysis is sufficient. The similarity between figures 3a and 3b is not observed for the other objective functions. For example, the Katsuura function, when used with PSO,

Table 4. Convergence properties per objective function under the 3-D von Neumann topology

Measurement / Function	A	B	C	D	E	F	G
CF	–	–	479	510	500	4	704.173
Ackley	925	64	473	516	503	1	691.705
High Conditioned Elliptic	989	0	401	588	504	0	576.575
Bent Cigar	989	0	415	574	504	0	601.344
Discus	989	0	416	573	504	0	600.027
Rosenbrock	989	0	416	573	504	0	615.662
Griewank	989	0	417	572	504	0	602.236
Rastrigin	989	0	413	576	504	0	601.2
HappyCat	988	1	415	574	504	0	603.712
HGBat	988	1	415	574	504	0	600.504
Katsuura	532	457	417	572	504	0	624.483
Expanded Griewank plus Rosenbrock	988	1	418	571	504	0	610.503

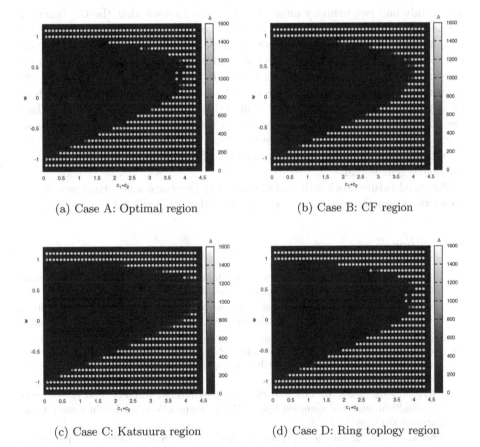

(a) Case A: Optimal region

(b) Case B: CF region

(c) Case C: Katsuura region

(d) Case D: Ring toplogy region

Fig. 3. Convergence snapshots

resulted in properties similar to the PSO using CF in tables 1 through 4. However, Katsuura has a substantially different convergent region to both figures 3a and 3b, as illustrated in figure 3c.

For Case D, the ring topology had the greatest Euclidean distance from the optimal region of case A. The convergent region is illustrated in figure 3d. Despite the ring topology having the greatest Euclidean distance from the optimal region of case A, figure 3d appears identical to the region of figure 3a, as the difference in convergence measure values are very small. The close similarity between figures 3a and 3d is a clear indication that the topology used within the PSO algorithm has no meaningful impact on the convergent region of a PSO.

6   Conclusion

This study had two primary aims: The first was to show that the CF function, defined in equation (7), is an effective objective function to utilize for convergent region analysis. The second objective was to perform an experiment to show that the social topology used by the PSO algorithm has no meaningful effect on the convergent region. It was found that the CF function was able to capture the convergent behavior of the PSO, as the found convergent regions matched both the theoretically derived region of Poli [11] and the "optimal" region. Where the "optimal" region was constructed using the maximum convergence measure value across all topologies and objective functions used (excluding CF). It was also found that the social topology used by PSO had no meaningful impact on the convergent region.

Potential future work will include utilizing the empirical techniques of this paper to obtain the convergence regions for other PSO variants.

References

1. Poli, R.: Analysis of the publications on the applications of particle swarm optimisation. Journal of Artificial Evolution and Applications 2008, 1–10 (2008)
2. Ozcan, E., Mohan, C.: Analysis of a simple particle swarm optimization system. Intelligent Engineering Systems through Artificial Neural Networks 8, 253–258 (1998)
3. Ozcan, E., Mohan, C.: Particle swarm optimization: Surfing the waves. In: Proceedings of the IEEE Congress on Evolutionary Computation, vol. 3. IEEE Press, Piscataway (1999)
4. Clerc, M., Kennedy, J.: The particle swarm-explosion, stability, and convergence in a multidimensional complex space. IEEE Transactions on Evolutionary Computation 6(1), 58–73 (2002)
5. Zheng, Y., Ma, L., Zhang, L., Qian, J.: On the convergence analysis and parameter selection in particle swarm optimization. In: Proceedings of the International Conference on Machine Learning and Cybernetics, Xi'an, China, vol. 3, pp. 1802–1907 (2003)
6. Van den Bergh, F., Engelbrecht, A.: A study of particle swarm optimization particle trajectories. Information Sciences 176(8), 937–971 (2006)

7. Trelea, I.: The particle swarm optimization algorithm: Convergence analysis and parameter selection. Information Processing Letters 85(6), 317–325 (2003)
8. Cleghorn, C., Engelbrecht, A.: A generalized theoretical deterministic particle swarm model. Swarm Intelligence Journal, 1–25 (2014)
9. Kadirkamanathan, V., Selvarajah, K., Fleming, P.: Stability analysis of the particle dynamics in particle swarm optimizer. IEEE Transactions on Evolutionary Computation 10(3), 245–255 (2006)
10. Gazi, V.: Stochastic stability analysis of the particle dynamics in the PSO algorithm. In: Proceedings of the IEEE International Symposium on Intelligent Control, pp. 708–713. IEEE Press, Dubrovnik (2012)
11. Poli, R.: Mean and variance of the sampling distribution of particle swarm optimizers during stagnation. IEEE Transactions on Evolutionary Computation 13(4), 712–721 (2009)
12. Campana, E., Fasano, G., Pinto, A.: Dynamic analysis for the selection of parameters and initial population, in particle swarm optimization. Journal of Global Optimization 48, 347–397 (2010)
13. Kennedy, J., Eberhart, R.: Particle swarm optimization, pp. 1942–1948. IEEE Press, Piscataway (1995)
14. Shi, Y., Eberhart, R.: A modified particle swarm optimizer. In: Proceedings of the IEEE Congress on Evolutionary Computation, pp. 69–73. IEEE Press, Piscataway (1998)
15. Kennedy, J.: Small worlds and mega-minds: effects of neighborhood topology on particle swarm performance. In: Proceedings of the IEEE Congress on Evolutionary Computation, vol. 3, pp. 1931–1938. IEEE Press, Piscataway (1999)
16. Kennedy, J., Mendes, R.: Population structure and particle performance. In: Proceedings of the IEEE Congress on Evolutionary Computation, pp. 1671–1676. IEEE Press, Piscataway (2002)
17. Engelbrecht, A.: Particle swarm optimization: Global best or local best. In: 1st BRICS Countries Congress on Computational Intelligence. IEEE Press, Piscataway (2013)
18. Van den Bergh, F.: An analysis of particle swarm optimizers. PhD thesis, Department of Computer Science, University of Pretoria, Pretoria, South Africa (2002)
19. Kisacanin, B., Agarwal, G.: Linear Control Systems: With Solved Problems and Matlab Examples. Springer, New York (2001)
20. Cleghorn, C., Engelbrecht, A.: Particle swarm convergence: An empirical investigation. In: Proceedings of the Congress on Evolutionary Computation, pp. 1–7. IEEE Press, Piscataway (accepted at, 2014)
21. Liang, J., Qu, B., Suganthan, P.: Problem definitions and evaluation criteria for the cec 2014 special session and competition on single objective real-parameter numerical optimization. Technical Report 201311, Computational Intelligence Laboratory, Zhengzhou University, Zhengzhou China and Nanyang Technological University, Singapore (2013)

Scheduling a Galvanizing Line by Ant Colony Optimization

Silvino Fernandez, Segundo Álvarez, Diego Díaz, Miguel Iglesias,
and Borja Ena

ArcelorMittal Global R&D Asturias
P.O. Box 90 – 33400, Avilés, Asturias, Spain
{silvino.fernandez,segundo.alvarez-garcia,diego.diaz,
miguel.iglesias,borja.ena}@arcelormittal.com

Abstract. In this paper, we describe the successful use of ACO to schedule a real galvanizing line in a steel making company, and the challenge of putting the algorithm to use in an industrial environment. The sequencing involves several calculations in parallel to figure out the best sequence considering the evolution of each important parameter: width, thickness, thermal cycle, weldability, etc.

For solving this combinatorial (NP-hard) problem, new necessity arose to develop an intelligent algorithm able to optimize the scheduling, avoiding traditional manual calculations. Hence, ACO is proposed to translate the scheduling rules and current criteria into a set of technical constraints and cost functions to assure a good solution in a short calculation time.

1 Introduction

The production of steel is a very complex process, with several functions involved in its transformation from coal and iron ore: iron making (conversion of iron ore into liquid iron), steelmaking (conversion of liquid iron into liquid steel), casting (solidification of liquid steel into semi-products: billets of slabs) and finally rolling, aimed to transform the intermediate products into the format accepted by the client (normally coils of steel , bars, heavy plates, wire rod, etc.) to continue the transformation into cars, bridges, beverage cans, or whatever our clients want to produce.

One of the most important products manufactured in our facilities is galvanized steel. The main use of this steel is as a raw material in other industries, and especially in the automotive industry for building car bodies. The process of galvanizing consists in covering the steel with a zinc layer to protect it against corrosion.

The facility in charge of galvanizing steel is the galvanizing line. This facility receives steel coils as an input, and there they are uncoiled, immersed in a bath of molten zinc at a temperature of around 460° Centigrade and finally rolled into a coil again. Every day dozens of steel coils are processed in each of our galvanizing lines, and their sequencing is critical to avoid incidents in the facility and to reduce the cost of the process.

M. Dorigo et al. (Eds.): ANTS 2014, LNCS 8667, pp. 146–157, 2014.
© Springer International Publishing Switzerland 2014

Fig. 1. Steel coils in one of our yards

Depending on this sorting, it is possible to lose lots of meters of strip due to a lack of quality, or even worse, to have a breakage that would halt the facility for several hours or even one day. Every time the strip breaks in the furnace where they are heated before the zinc bath, the line must be stopped until the furnace cools down, in order to remove the steel strip of the furnace: then the furnace must be heated up again before resuming production.

An additional hard constraint coming from the intended use of the model is that execution time is very limited, since it must be able to provide a schedule within a few minutes in order to be useful at the line. This time limit, together with the complexity of the cost function —as we will see later—, results in a much tighter limit on the number of evaluations than usual in these systems aimed to solve combinatorial problems.

2   Related Work

Swarm Intelligence (SI) [8] is a term introduced by Gerardo Beni and Jini Wang in 1989, in the context of robotics [1]. SI is a methodology inspired by the social behavior of agents collaborating in a decentralized and self-organized system with a common aim. Most intelligent animals live, obey rules and reap benefits of a society of kin. Societies vary in size and complexity, but have a key common property: they provide and maintain a shared culture [8].

A main characteristic of SI systems is that each individual itself does not have enough problem solving ability, but the collaboration and experience exchange among a colony of individuals makes it possible to achieve their objectives. SI systems share some similarities with evolutionary techniques such as Genetic Algorithms (GA), and one of their more important applications is the resolution of combinatorial problems [4].

SI systems are considered probabilistic algorithms, and their main characteristic is that they explore the search space trying to obtain the best possible solution in short time.

Normally they use heuristics to select the following candidate to be analyzed. The main features of swarm intelligence algorithms are [2] are: Flexibility (the colony can respond to internal perturbations and external challenges), Robustness (tasks are completed even if some individuals fail), Decentralization (there is no central control in the colony) and Self-Organization (paths to solutions are emergent rather than predefined).

There are many examples of algorithms developed in the framework of SI, some examples are Ant Colony Optimization, Particle Swarm Optimization [9], Charged System Search [6] , or Bees Algorithms [10].

ACO is among the best known and widest spread SI techniques. It was defined by Dorigo, Di Caro and Gambardella in 1999 [3]. Previously, the first ACO system was developed by Dorigo in his PhD thesis called Ant System in 1992 [7]. Since then, several use cases have been developed where ACO has demonstrated its effectiveness solving problems mainly in the field of logistics and job scheduling.

3   Context of the Problem

In steel production, an important task is the galvanization of steel. In this process, steel is coated with a zinc layer, which has the objective to protect the steel against air and moisture. In fact, the zinc layer is considered to be the most effective and low-cost means to achieve this goal. Galvanization is applied to steel coils, which are a finished steel product which has been wound or coiled after rolling slabs. Slabs are semi-finished steel products, obtained by processing through a continuous caster and cut into various lengths. The slab has a rectangular cross section and is used as a starting material in the production process of flat products, i.e. hot rolled coils or plates.

The process of galvanizing is continuous, that is, there is no separation between the coils and the line never stops. The head of the next coil is welded to the tail of the coil in process. Thus, for the line, the input is an infinite strip which has points where there is a change in some of its characteristics: dimensions (width, thickness), steel grade, section, thickness of the zinc layer, etc.

After the welding area, the strip advances towards the accumulator, a kind of buffer that makes it possible to change the speed of the line depending on the needs of the furnace without running out of material. The furnace is the next stage; here the strip has to reach a target temperature depending on its chemical composition (steel grade), and for this line speed and furnace temperature have to be adjusted according to the thickness and width of the coil. Thicker and wider coils require lower speed or higher furnace temperature to reach the same target temperature. Due to the inertia of the system, changing these parameters takes some time, during which part of the coil may not be processed properly, depending on how different the coils and their targets are.

After the furnace, the strip is entered into the zinc pot, and later by passing through an 'air knives' system, the zinc layer is spread evenly and to the thickness specified by the client; this is critical in order to avoid coil rejections.

All this process is dramatically affected by the sequencing. A big difference in the width between two consecutive coils could generate a breakage of the strip in the welding phase or the furnace, and consequently it could stop the line for several days.

3.1   Origin of the Problem: The Importance of the Sequencing

Sequencing is critical in production lines in the steel industry (as it is in the industry in general). If we have a list of n items to process, the possibilities to arrange these items are huge ($n!$). Depending on the sorting, the sequence could have different impact in the process, and it can cause quality issues in the final product, lower productivity, higher energy consumption or even an incident in the facility.

Due to the technical limitations of the lines and the critical efficiency necessary to offer nowadays for each productive plant, the necessity of having a scheduling model able to solve the limitations of current scheduling algorithms has arisen. The objective is to maximize productivity, making the facility more competitive.

Traditionally, commercial tools based on constraint programming have been used for these proposes. These tools are focused on finding a solution with some properties, mainly fulfilling some relations among variables and respecting a set of constraints. Their main problem is their limitation for using complex cost functions, we will solve these issues with an innovative option (at least in this industrial environment): Ants.

3.2   Proposed Solution: ACO

ACO and Swarm Intelligence offer an innovative approach to solve some of the problems of traditional scheduling tools. The idea is to quantify and then minimize the costs of sequencing. In this calculation, intermediate steps, such as as material losses or line stops, are taken into account and finally translated into meters of strip lost. These calculations are in some cases computationally intensive, resulting in long evaluation times for each generated solution.

One important feature of this approach is the flexibility, because if the schedule is finally modified when they are produced (*e.g.* because the foreman thinks that another sequence is better), it is possible to ask why, and then add new cost function modules for including the reason of this deviation between reality and calculation (or maybe, fine-tune some of the current ones). Consequently, it is very easy to fine tune and to maintain the model with the pass of time.

It is important to remark that the line is continuously in evolution, so the definition of a model able to schedule them should follow the same philosophy, and ACO matches this requirement perfectly.

4   How to Apply ACO to a Scheduling Problem in the Steel Industry

One of the best known uses of ACO is solving hard combinatorial optimization problems; typically the most famous example is the TSP (Travel Salesman Problem). Several works have been developed [5] to assay the effectiveness of its application to this problem. In TSP, the objective is to generate the shortest tour of a number of cities, given the distance matrix relating each pair of cities.

Our sequencing problem fits quite nicely this specification, substituting cities for steel coils and distances for costs. Unfortunately, we cannot disclose the details about how these cost are calculated due to confidentiality issues.

Since some of the orderings are not possible, some of the costs will be infinite (or, viewing the problem as a graph, the corresponding arcs do not exist). This is one of the two main differences with TSP. One main goal is the reduction of these infeasibilities, because every time there is one in a sequence, it is necessary to include a transition coil (a coil with no client) to solve the issue. This transition coil has a high cost (if we compare with a normal production cost) in terms of productivity (we produce something we do not need instead our orderbook) and in terms of cost (yield), because maybe the coil will be sold as second quality or even as scrap (cheaper).

The other difference is the impact of whole-sequence costs on the final cost. Some of the aspects that impact the cost function cannot be calculated from the individual transitions, but rather need the whole sequence in order to be defined.

For instance, if we evaluate the cost of width evolution, a change from 950mm to 1,000mm has different implications if the evolution is previously increasing or decreasing. In this case the cost is proportional to the number of sign changes in the first order differences of the sequence.

Thus, we have two types of costs:

Transition Costs. These costs are equivalent to the distances used in the regular TSP; they are costs that arise from the transition between two coils in the sequence. The loss of material due to quality issues caused by the time it takes to adjust the furnace is a typical example. These costs can be used as heuristics for the construction phase.

Sequence Costs. These costs are not calculated as the cost of transitions, but as a function of the whole sequence (or a sub-sequence). The only way of calculating this kind of costs is the evaluation of the sequence once it is complete at the end of the path. Therefore, they can only be used to drive the solution indirectly through their impact on pheromone.

For over a year, the technical experts of the lines defined the losses associated to the transitions, analyzing all the parameters and characteristics of the coils that have any kind of impact on the resulting meters of strip lost of a sequence. These cost functions depend on the parameters of the coils and their differences in the transitions. The losses are mainly generated by width changes between consecutive coils, thickness changes, thermal losses, etc.

The general idea is to try and minimize the transitions for each parameter relevant to sequencing, looking for a smoother evolution and therefore a more stable production.

Our cost function is thus of the form:

$$C(S) = GlobalCosts(S) + \sum_{i=1}^{n-1} TransitionCost(c_i, c_{i+1}) \qquad (1)$$

Where:

S: Sequence of n coils $(c_1, \ldots, c_n)$
$TransitionCost(c_i, c_{i+1})$: They are the costs of the transition.
$GlobalCosts(S)$: They are the costs of the sequence S.

Initially, before any pheromone has been laid out, we use the transition costs as a heuristic for solution construction. Before launching the algorithm, the transition cost for each combination of nodes $\sigma(n_1, n_2)$ is calculated; this cost can be classified as: infinite[1] (if the transition does not respect some of the constraints of the line), zero (no material lost in the transition) or finite (there are losses associated to the transition, but it respects all the constraints of the line).

Given the existence of zero and infinite costs we cannot apply directly the usual selection criterion of selecting the next node with probability inversely proportional to the arc cost. Instead, we do a two-step selection, choosing first among the three categories (we give a higher probability to zero transitions, then to finite and finally to the infinite ones); in the second step, if zero or infinite cost arcs have been selected, all of them are equiprobable, and for finite cost arcs we apply the usual approach.

We accept infeasible transitions because not doing so would require each ant to build a Hamiltonian path. This is a hard problem in itself, and in our problem there is often no such path. A second option would be to accept them only if no feasible transition is left; this leads to much worse sequences, because it tends to leave out the least connected nodes, accumulating large numbers of infeasible transitions at the end of the sequence. We then take infinite cost transitions, with a low probability, at any time to allow the model to discover infeasible transitions that save more of them later.

These unfeasible solutions are "solved" by the operators of the line by means of inserting transition coils between two coils that are not compatible from the production point of view. This *transition coil* is especially expensive, and this is the reason why it is considered as an infinite cost, and the priority of the algorithm will be always to minimize first these infeasible transitions and only then minimizing the total cost. The way of taking into account these infinite costs in the raking process is to translate them into a *huge* cost, in a different scale to the *normal* cost of a transition due to production losses, and add it multiplied by the numer of infeasibilities to the total cost of the sequence.

The function to deposit the pheromone in the path is the standard one, shown in (2), where τ_{ij} is amount of pheromone between nodes i and j, $BestCost$ the cost of the best solution found up to the moment of the update and ρ the evaporation factor:

$$\tau_{ij}^k \leftarrow \begin{cases} (1-\rho)\tau_{ij}^k + \frac{BestCost}{CostAnt_k} & \text{if arc } ij \in Ant_k \\ (1-\rho)\tau_{ij}^k & \text{otherwise} \end{cases} \tag{2}$$

---

[1] We translate it into a huge penalty added to the cost of the transition in the objective function formula.

We set the evaporation parameter to 0.1 to permit some adaptation to new good solutions. With the advance of the iterations, the heuristic based on transition cost loses weight in favor of the pheromone. For this we again follow the standard practice to combine the heuristic and pheromone information:

$$p_{ij}^k = \frac{[\tau_{ij}]^\alpha [\eta_{ij}]^\beta}{\sum_{l \in \mathcal{N}_i^k} [\tau_{il}]^\alpha [\eta_{il}]^\beta}, \quad \forall j \in \mathcal{N}_i^k \tag{3}$$

Where:

p_{ij}^k: Probability of selecting node j after selecting i for the ant k.
τ_{ij}: Amount of pheromone between nodes i and j.
η_{ij}: Heuristic value previously known, $(\frac{1}{C_{ij}})$.
C_{ij}: Meters of strip lost in transition between nodes i and j
α: Parameter to control the influence of the pheromone.
β: Parameter to control the influence of heuristic based on losses.
$\mathcal{N}_i^k$: Set of accessible nodes for an ant k, when being in a node i, that have not been already selected.

The combination of the parameters α and β controls the influence of the pheromone and transition costs over the decision of choosing a path. In our case, we fixed the parameters $\alpha = 1$ and $\beta = 2$, with of objective of having a good balance between the heuristic and experience (pheromone).

We use elitism, allowing only the top 10% ants in each iteration to deposit pheromone in the matrix.

Due to the constrained time for computation and the necessity of obtaining a proposal for the scheduling in just a few minutes, we did not consider to include local search in the current version of our ACO. Few solutions can be explored in the given time and, due to the global costs, each solution needs to be re-evaluated, so we could not afford the computational cost of local search in this situation.

5   Results

The optimal solution to the problem would be the sequence that calling the cost function obtains the cheapest possible value respecting all the technical constraints at the same time. The idea is intuitive, but exhaustive enumeration is the only way to ensure global optimality. Normally in this kind of problems, especially in an industrial setting, the approach is to obtain a "good enough" solution. A good solution must fulfill some characteristics, and the first is that the solution must be feasible; additionally (and not less obvious) the cost of the solution must be very low (compared with other feasible solutions); and last, but not least, the solution must be calculated in a reasonable period of time.

The first characteristic is unquestionable. With the others, a trade-off must be found. A solution a bit more expensive than the optimum is better than the best if we need 1,000 years to calculate it [11]. For us, it is very important to transmit this idea to the final users, in order to avoid confusions. If the user expects to obtain the very best solution, and two consecutive calculations yield different solutions, it would be confusing; and yet, this does not mean the calculation was wrong.

The best benchmark available to us was to use the actual sequences produced in the line. If our algorithm can take the same set of coils an improve the number of infeasible transitions —hence number of transition coils— and cost within the limited time allotted for online execution, we have something valuable.

The experimental setup has been the following: we have taken as reference one entire month of production. This month has not been selected randomly, but the technical team proposed one period of time with high production volume and a typical product mix to produce, so that the results would be representative of the expected normal function of the model.

The month is translated into 29 days of production, for each experiment (day) there are aroung 60 coils (as average) and we have fixed the first and the last coil of the list to assure that the scheduling for the day matches with the scheduling of the following one. We have run ACO 25 times for each day with 200 iterations and 80 ants. We are aware that this setup is very limited from an academic point of view, but the calculation with these parameters takes about 12 minutes, approximately the maximum admitted in real life by the user to calculate a solution. Using better computers, it is possible to reduce this computation time, but another constraint of the reality is that the model runs normally in the laptop or desktop computers at the plant, which are usually not the latest in technology. We have kept the time limit for the tests to replicate the quality that the solutions will have in operation. Obviously, we have done more tests in our high-performance clusters used for scientific computations in our R&D center, but we prefer to present the results and conclusions useful for real life.

As commented, for evaluating the results it is very important to clearly understand the concept of constraint and cost function used in this problem. Every time a constraint is not respected in a transition between two coils, it is assumed that we need to insert a transition coil in the middle to make the sequence feasible. This transition coil is material produced with no client assigned, so it implies a cost of stock and, depending on several circumstances, either the cost of downgrading or scrapping[2] the whole coil[3]. We assume then, an infinite cost for every violated constraint and hence, it has total priority over the cost. For the same number of violated constraints, the algorithm minimizes the production cost.

In the tests, we evaluate in a separate way the number of violated constraints that are reduced with the algorithm (input vs. output) and in a second analysis the savings in costs. Figure 2 shows the number of violated constraints reduced

---

[2] If possible, the coil will be sold at a discount, because of the lower quality; if no client is found, it will be shredded for scrap to be recycled in the steelshop. In either case, the benefit from the material is much less than the sale of a full-price product.

[3] A typical coil is about 2.5Km long; compare with the effect of a transition cost which is normally limited to just a few, maybe tens of meters.

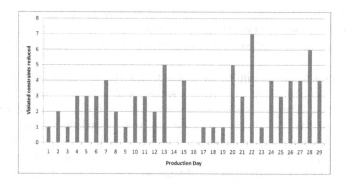

Fig. 2. Number of violated constraints reduced by ACO per day, compared to actual production

per day (difference between the original production sequence and the output of the model).

The results in terms of reduction of violated constraints are very heterogeneous when considered inter-instance[4], but this is a reflection of the different product mix for different days, which results in different levels of complexity. The intra-instance [5] standard deviation is 0 in 80% of the experiments (every one obtained a solution with the minimum possible number of violated constraints), so the model seems to be very stable to calculate the sequence with the minimum violated constraints. In the other 20%, there is no experiment with a standard deviation of more than 1 constraint over the average. There are only two cases where the model does not improve the number of violated constraints, while the rest show important reductions of violated constraints, sometimes up to 6 or 7 constraints reduced.

Figure 3 shows, for each experiment, the average and standard deviation of the percentage of cost savings, referred to the cost of the sequence that was actually produced. Negative values mean that the cost is higher than the original; this occurs, but the number of violated constraints was reduced in those cases (see figure 4), so the overall result still improves the actual sequence.

In terms of production costs, savings are very important. The average of the savings is 52% for the experiments, this would confirm that the model is able to reduce the scheduling cost by a half approximately. As for the standard deviation, it is 25% from a global point of view over the experiments, but it is again a consequence of the different product mixes. The intra-instance deviations are much smaller, as we can see in figure 3.

Finally, figure 4 shows the costs versus the number of constraints reduced in the scheduling of the coils. To make the figure more readable we show here only the average values.

---

[4] Results of ACO for different production days (different instance).
[5] Runs of the ACO algorithm on a sigle day (same instance).

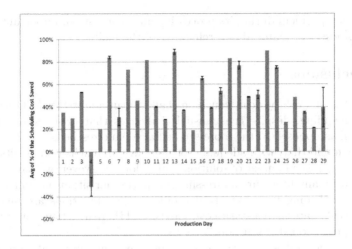

Fig. 3. Average of the % of cost saved for each experiment and their standard deviation

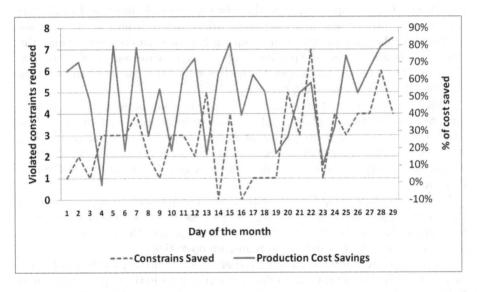

Fig. 4. Number of violated constraints reduced vs cost savings

On the other hand, there are only two cases when the scheduling costs are increased in order to reduce the number of scheduling constraints. As commented, these are especial cases, where the total cost of the sequence is in fact reduced, and hence these negative results are taken as zero for the analysis.

After a first validation by experts, the sequences were validated in tests in the line using the expert knowledge of the foremen. One of the aims of the sequencer is to avoid having breakdowns in the line as well as losses of material, and in this sense both goals have been achieved. The sequences generated by the model do not cause any problem during its processing and it means an important savings in terms of losses of material and reliability of the facility.

6   Conclusions

After months of work, we can conclude that we can adapt the ACO algorithm for line scheduling, despite the unusual challenges that arise from a real life application, as opposed to the more studied academic testbeds.

We have found that, with adequate settings, we can reach good results despite the limitation in the number of solution evaluations. However, we have not been able to identify any literature addressing this particular situation, which is quite normal when dealing with industrial problems; cost functions are much more complex to calculate than a static cost matrix, and there are stringent limitations on running time to get a solution.

We offer a neutral way of calculating complex schedules that improves the traditional manual calculation did in the past. There is hardly any risk. The algorithm takes the input sequence that the expert would have used in the line without any interference, and after some minutes of calculation, the solution is offered to the experts, who can accept, refuse or even compete against the model: the interface offers the possibility of manual re-arrangement of the output sequence. Thus, they can convince themselves about the quality of the solution and the difficulty of improving it.

This technique is completely generic. The separation between the algorithm and the cost functions makes it fully adaptable to another context. The algorithm takes care of arranging a list of items, trying to find a minimum cost, where the context is completely transparent. To adapt the model to a different environment, only the cost functions need to be changed according to the new domain.

Everywhere sequencing has any importance, and its impact can be measured, there is an opportunity of improving the process with a scheduler using these techniques. The results that we have obtained confirm the confidence in the algorithm as well its possibilities in industrial operations.

The results are hardly debatable. Experts have validated them and from a mathematical point of view, they offer an improvement in the sequencing around 50% in average; and even if there is an important inter-instance variability, we have yet to find a specific case where the original sequence is not improved. The model consistently results in an increase in yield and cost-efficiency of the line.

References

1. Beni, G., Wang, J.: Swarm intelligence in cellular robotics systems. In: NATO Advanced Workshop on Robots and Biological Systems (1989)
2. Bonabeau, E.: Swarm intelligence. In: O'Really Emerging Technology Conference (2003)
3. Dorigo, M., Di Caro, G., Gambardella, L.M.: Ant algorithms for discrete optimization. Artificial Life pp. 137–172 (1999)

4. Fernandez Alzueta, S., Diaz, D., Manso Nuño, T., Suarez Rodriguez, M.: Optimization techniques to improve the management of a distribution fleet in the steel industry. In: Proceedings of the SIGCHI Conference on Human Factors in Computing Systems (2010)

5. Gómez, O., Barán, B.: Ant colony optimization and swarm intelligence. In: Proceedings of the 2004 4th International Workshop, ANTS 2004 (2004)

6. Kaveh, A., Talatahari, S.: A novel heuristic optimization method: Charged system search. Acta Mechanica, 267–289 (2010)

7. Marco, D.: Optimization, learning and natural algorithms. Ph.D.Thesis (1992)

8. Mataric, M.: Dedigning emergent behaviors: From local interactions to collective intelligence. In: Proceedings of the SIGCHI Conference on Human Factors in Computing Systems, pp. 526–531 (2000)

9. Parsopoulos, K., Vrahatis, M.: Recent approaches to global optimization problems through particle swarm optimization. Natural Computing, 2–3 (2002)

10. Pham, D.T., Koc, E, Lee, J.Y., Phrueksanant, J.: Using the bees algorithm to schedule jobs for a machine. In: Eighth International Conference on Laser Metrology, CMM and Machine Tool Performance pp. 430–439 (2007)

11. Weise, T.: Global optimization algorithms – theory and application (March 2014), http://www.it-weise.de

SRoCS: Leveraging Stigmergy on a Multi-robot Construction Platform for Unknown Environments

Michael Allwright[1], Navneet Bhalla[2], Haitham El-faham[1], Anthony Antoun[3], Carlo Pinciroli[3], and Marco Dorigo[1,3]

[1] Department of Computer Science, University of Paderborn, Paderborn, Germany
michael.allwright@upb.de, helfaham@mail.upb.de
[2] Sibley School of Mechanical and Aerospace Engineering, Cornell University, Ithaca, New York, USA
navneet.bhalla@cornell.edu
[3] IRIDIA, Université Libre de Bruxelles, Brussels, Belgium
{aantoun,cpinciro,mdorigo}@ulb.ac.be

Abstract. Current implementations of decentralized multi-robot construction systems are limited to construction of rudimentary structures such as walls and clusters, or rely on the use of blueprints for regulation. Building processes that make use of blueprints are unattractive in unknown environments as they can not compensate for heterogeneities, such as irregular terrain. In nature, social insects coordinate the construction of their nests using stigmergy, a mechanism of indirect coordination that is robust and adaptive. In this paper, we propose the design of a multi-robot construction platform called the Swarm Robotics Construction System (SRoCS). The SRoCS platform is designed to leverage stigmergy in order to coordinate multi-robot construction in unknown environments.

1 Introduction

It is possible that a multi-robot construction system will be a practical solution in the future for building basic infrastructure, such as shelter, rail, and power distribution networks on extraterrestrial planets or moons, prior to the arrival of humans [12]. Due to the distances involved, real-time control of the robots or communication supporting the surveying of the remote environment prior to construction are typically not viable options. For this reason, a system that is robust and capable of performing construction in a variety of environments without specific programming is desirable.

Stigmergy is a form of indirect coordination that enables the self-organization observed in social insects such as ants, bees, termites, and wasps. Grassé [8] originally introduced the concept of stigmergy in the context of termite nest construction, where previous work by the termites became a stimulus to perform further work. Although this form of coordination that makes use of stigmergy has been shown to be less efficient than hierarchical coordination, it benefits

M. Dorigo et al. (Eds.): ANTS 2014, LNCS 8667, pp. 158–169, 2014.

from not having a single point of failure, is capable of operating in a variety of environments without specific programming, and requires simpler hardware [7].

A number of multi-robot systems that use exclusively stigmergy to coordinate the building process have been presented in the literature; however, they are only capable of constructing rudimentary structures such as clusters and walls [1, 10, 17, 23–25, 29]. While there are construction systems that make use of stigmergy and are capable of building more complex structures, these systems supplement the use of stigmergy with a blueprint or external infrastructure for positioning and communication [21, 30, 31, 33]. These approaches are not attractive in unknown environments, as the use of a blueprint is a form of specific programming that is unable to compensate for variations in the environment, such as irregular terrain. Furthermore, the use of external infrastructure for positioning and communication is not suitable for rapid deployment in unknown environments.

In order to demonstrate the potential of stigmergy for construction in unknown environments, we propose the design of a multi-robot construction platform called the Swarm Robotics Construction System (SRoCS). The SRoCS platform makes use of stigmergy to coordinate a flexible building process that is capable of adapting to the environment, without relying on external infrastructure for positioning and communication. This is achieved by encoding the construction process as simple rules that use previously completed work, as well as heterogeneities and templates in the environment, to guide the construction process. This approach is inspired by Theraulaz et al. [26, 27] who simulated the construction of wasp nests in a 3D lattice.

Following an overview of the background literature in Section 2, we present in Section 3 the design of our multi-robot construction platform, SRoCS. Experiments using SRoCS are described in Section 4 and the conclusions of this work are then provided in Section 5.

2   Background

In this section, we present some examples where stigmergy is used to coordinate the construction of termite and social wasp nests in nature. Following these examples, we provide an overview of the work done in multi-robot construction in simulation and using real hardware, focusing on the use of stigmergy where present.

2.1   Construction in Nature

Bruinsma [6] used stigmergy to explain the formation of various structural elements in termite nests. For example, Bruinsma described three uses of pheromones by termites that regulate the construction of the royal chamber. First, pheromones are used to form a trail that causes workers to be recruited towards the construction site. Second, a pheromone emitted by the queen termite is used to create a template for the chamber. Third, worker termites add pheromone to the soil pellets during construction. This pheromone attracts workers to place more soil pellets nearby those that have been recently placed.

Karsai et al. [11] demonstrated that social wasps coordinate the construction of brood combs by sensing the local environment using their antennae. Wasps use local information, such as the number of walls in a partially completed comb, to select from various actions such as lengthening a comb or starting a new comb.

2.2 Simulation

Deneubourg et al. [7] presented the first work using stigmergy in simulation using ant-like robots to implement decentralized clustering and sorting algorithms in a 2D lattice. In this system, the robots move around randomly, picking up and putting down objects with probabilities that are a function of the density of similar objects nearby. These actions are coordinated through stigmergy as the previous placement of objects indirectly coordinates further actions taken by other robots. Melhuish et al. [18] showed that this approach was scalable by demonstrating the sorting of up to 20 types of objects in a simulation based on hardware experiments with Holland et al. [10].

Based on a mathematical model developed to explain the emergence of structures observed in termite nests [5], Ladley and Bullock [13, 14] created an agent-based 3D simulation for the formation of chambers and walls, adding in physical and logistical constraints. This work was extended by Linardou [15] who demonstrated the impact of using realistic pheromone dispersion rules. This work showed that stigmergic coordination through interactions of the agents with pheromone gradients and previously completed work was capable of regulating the construction of various termite nest-like structures.

Theraulaz et al. [26, 27] demonstrated the construction of several wasp nest-like structures using algorithms that caused an agent to deposit a brick in a 3D lattice when a condition based on the local configuration was satisfied. These conditions would be in terms of patterns of existing bricks perceived by an agent. This coordination is an example of stigmergy as the patterns of existing bricks are the result of previous actions by other agents. In further work by Bonabeau et al. [4] genetic algorithms were used to search for sets of rules that lead to the construction of structured patterns.

2.3 Multi-robot Construction

Implementations of multi-robot systems have been used to demonstrate construction tasks. In this section, we discuss implementations of decentralized multi-robot construction systems with respect to how and if they use stigmergy in the construction algorithm. Implementations of centralized multi-robot construction systems often depend on external infrastructure for positioning and communication [2, 16, 32–34], which makes them unsuitable for rapid deployment in unknown environments.

Implementations of decentralized multi-robot construction systems are organized with respect to the type of stigmergy used. Stigmergy is classified as being quantitative or qualitative [3]. Quantitative stigmergy is where the likelihood of a response to a stimulus is proportional to the intensity of that stimulus. An example of this type of stigmergy was shown in the work of Bruinsma [6], where the termites would respond to the concentration of pheromones and soil pellets in their immediate environment. Qualitative stigmergy is where the probability of performing a given action is a function of a perceived environmental configuration. For instance in the nest of social wasps, an individual could decide whether or not to add a wall to the brood comb depending on the number of walls already built [11].

Construction Based on Quantitative Stigmergy. Beckers et al. [1] were the first to demonstrate the use of stigmergy in a multi-robot system for distributed clustering. They maintained that the use of stigmergy has a significant advantage over coordination using direct communication, as direct communication would have required the abstraction of the information regarding the type of task, as well as its spatial and temporal locality. Holland and Melhuish [10] extended the work in [1] to the task of clustering and sorting two kinds of Frisbees. In related work, Song et al. [24] used iRobot Creates to cluster square shaped objects using two developed behaviors, *twisting* and *digging* which exploited the geometry of the square tiles to be clustered.

Stewart and Russell [25] constructed a loose wall along a template using a team of robots. The template was formed by a leader robot moving a lamp in a straight line once the current point in the wall had enough material. Soleymani et al. [23] also demonstrated the construction of a wall along a template using soft materials. Napp et al. [19] reasoned that soft materials have advantages over rigid materials, as they conform to the shape of the surface on which they are placed.

Construction Based on Qualitative Stigmergy. Wawerla et al. [29] provided the first application of qualitative stigmergy, demonstrating the construction of a wall from two alternating types of velcro blocks. The wall was built along a laser generated template and the robots would exchange information about the next type of block to be placed.

The TERMES multi-robot construction system by Werfel et al. [21, 31], represents the current state of the art in decentralized multi-robot construction. This system is capable of building staircase-like structures using tiles that the robot can climb on. The system uses an offline compiler to flatten a user-specified cellular 3D structure onto a directed graph whose nodes constitute a height map. The edges of this directed graph specify how the robots can move across the structure. The robots execute an algorithm that selects a subset of these directed edges to traverse the structure. This directed graph is a blueprint containing all the required information to build the structure. In order to avoid deadlock conditions during construction, fixed stigmergic rules are used to regulate the construction order.

2.4   Summary

Decentralized multi-robot construction systems have been shown to be capable of building rudimentary structures like clusters and walls. While more sophisticated structures have been demonstrated using the TERMES system, this system is limited to performing construction in known environments where a blueprint of the structure to be built, is provided by an architect who has prior knowledge of the environment.

In order to enable decentralized construction in unknown environments, we present the design of a multi-robot construction platform called SRoCS. SRoCS aims at leveraging stigmergy to coordinate a flexible building process in a variety of environments.

3   Overview of the Proposed Platform

The design of the SRoCS platform consists of mobile robots and stigmergic building blocks whose prototypes are shown in Fig. 1. The robots are equipped with a specialized manipulator, which has been optimized for assembling the blocks. While disassembly of blocks would support experiments involving the use of temporary scaffolding-like structures, it is not supported in the initial prototype of the SRoCS platform. In addition, the use of a multi-robot simulator is discussed as an alternative to running experiments using real hardware.

(a)   (b)

Fig. 1. Prototypes of (a) the stigmergic building block and (b) the mobile robot

3.1 The Stigmergic Building Blocks

In order to leverage stigmergy in SRoCS, we propose the design of stigmergic building blocks. These blocks aim at emulating the use of pheromones by termites in construction. The blocks contain four multi-color LEDs on each face. The colors of these LEDs can be sensed by the cameras on the robots and updated using the NFC (Near Field Communication) interface between the manipulator and the block.

A prototype of this block is shown in Fig. 1a. We have chosen the geometry to be cubic as it allows the block to be placed into the structure without the need for rotation. Eight spherical magnets in the corners allow the blocks to self-align with each other and allow the blocks to be picked up by a robot. In order to simplify the computer vision required by the robots to see the blocks, localizable 2D barcodes called AprilTags [20] are added to the faces of the block.

Inside the block, a main circuit board hosts a micro-controller, an accelerometer and a Zigbee radio for collecting experimental data and debugging. Depending on the software running on the blocks, additional functionality such as block-to-block communication is also possible.

3.2 The Mobile Robots

The BeBot [9] is selected as the mobile robot in the SRoCS platform due to its small size, modularity, and availability. In the SRoCS platform, the robots move around the environment randomly searching for building blocks that can be used for construction. Proximity sensors around the base of the robots allow for obstacle avoidance with other robots and the structure being built. The camera on the robot allows for the detection of the AprilTag barcodes on the block. The detection of these barcodes allows the robot to localize itself with respect to the building blocks in the environment or to the structure being built.

To pick up and place the building blocks into a structure, the robot is equipped with a specialized manipulator that is shown in Fig. 1b. The manipulator design bears similarities with a fork-lift, with the exception that the block is picked up from the top and held in place using electro-permanent magnets. To detach the block from the manipulator the electro-permanent magnets are activated causing the magnetic field that held the block in place to drop to near zero. We have optimized the manipulator for creating structures of a height of up to three blocks; this provides a good trade-off between flexibility and stability.

The robots are able to communicate indirectly with each other by positioning the building blocks and updating the colors of the LEDs on the blocks. These colors can be assigned various meanings depending on the algorithm in use. For instance, a particular color can be used to indicate a seed block or a block that has already been placed into the structure.

3.3 Simulation Tool

Running experiments with real hardware is time consuming and can be expensive when experimenting with large numbers of robots. It is therefore desirable to use a simulation tool to evaluate the performance of the construction algorithm, before running experiments using the real hardware.

The ARGoS simulator [22] is used as the simulator for this construction platform as it achieves both flexibility and efficiency. Flexibility is necessary as our system requires the simulation of technologies, such as magnetism, that are not commonly found in robot simulation packages. Efficiency is also important as SRoCS is a multi-robot construction platform, and it is desirable that it be possible to run simulations with tens or hundreds of robots.

To simulate the hardware described above, several extensions have been developed for ARGoS. These extensions include a magnetism plugin based on [28] and a new 3D physics plugin based on the open-source physics engine Bullet. These extensions have been shown to simulate the self-alignment behavior of the blocks, as well as the attachment/detachment dynamics of the manipulator.

A prototyping plugin was also developed for the ARGoS simulator that allows for a quick evaluation of designs. This plugin also enables the sensors and actuators required to implement the manipulator, the computer vision, and the communication between the blocks and the robots.

4   Swarm Construction Examples

SRoCS is designed to leverage stigmergy to coordinate construction. Examples are provided to demonstrate the different ways in which stigmergy can be used to coordinate various construction tasks. Figs. 2-4 are visualizations from the ARGoS simulator and are based on the described hardware. These visualizations aim to give examples of the types of experiments that the SRoCS platform has been designed to run.

4.1   Substructure Formation

Blueprints of overall structures to be built are avoided as they are not adaptive to heterogeneities in the environment, such as irregular terrain. It is however useful to encode some substructures as sets of simple rules that use previously completed work to precisely regulate part of the construction.

In the example shown in Fig. 2, the robots are coordinated through the positions of the stigmergic building blocks and the colors of the LEDs on the faces of the blocks, in order to regulate the following construction steps. This approach to leveraging stigmergy is inspired by the work of Theraulaz et al. [26, 27] and is an example of using qualitative stigmergy in the SRoCS platform.

4.2   Construction Using Templates

As discussed in Section 2.1, the formation of the termite royal chamber is in part regulated by the dispersion of a pheromone by the queen. This pheromone stimulates the worker termites to build around her. A similar mechanism can be employed in SRoCS as shown in Fig. 3. In this scenario one or more seed blocks

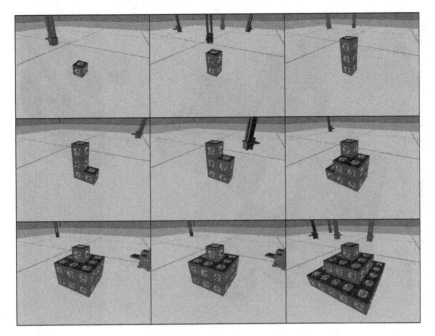

Fig. 2. Construction of a pyramid

are placed in the environment with the LEDs set to a designated color. These seed blocks can form a template in the environment that in conjunction with previously completed work, facilitates the construction of chambers or passage-like structures through stigmergy.

Depending on the implementation, this approach can lead to a stochastic building process. Stochasticity in the building process can be exploited to increase the adaptivity of the system to the environment. It is also possible that stochasticity is fundamental in some cases, such as when the system is required to dynamically explore multiple solutions and adapt the structure being built to the heterogeneities found in the environment.

4.3   Construction Exploiting Environmental Heterogeneities

When designing a system that must be able to build in an unknown environment, heterogeneities need to be taken into account. For example, geographical features in the terrain, such as the presence of a river, must be compensated for in the building process. An example of this is shown in Fig. 4, where the robots are using the previously placed blocks as well as the variations in the simulated terrain to regulate the construction of a wall. This indirect coordination that uses the previously placed blocks as well as variations in the terrain to regulate the construction process is an example of how the SRoCS platform can leverage stigmergy in unknown environments.

Fig. 3. Building a chamber-like structure using a template

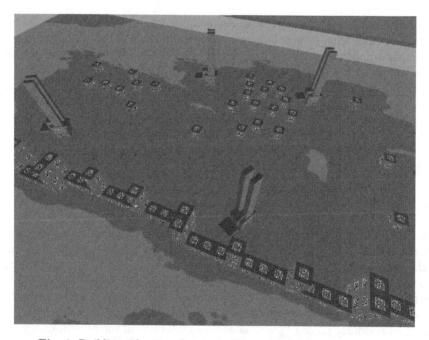

Fig. 4. Building a barrier along a heterogeneity in the environment

5   Conclusions

Current implementations of decentralized multi-robot construction systems are limited to the construction of rudimentary structures such as walls and clusters, or rely on the use of a blueprint or external infrastructure for positioning and communication. In unknown environments, the use of blueprints is unattractive as it cannot adapt to the heterogeneities in the environment, such as irregular terrain. Furthermore, the reliance on external infrastructure is also unattractive, as it is unsuitable for rapid deployment in unknown environments.

In this paper, we have proposed the design of a multi-robot construction platform called SRoCS. In contrast to other multi-robot construction systems, the aim of SRoCS is to provide a flexible building process that is adaptive to heterogeneities and variations in the environment. The coordination of the building process in SRoCS is facilitated through stigmergy, and based on the observations and models of the construction of social wasp and termite nests as described by Karsai et al. [11] and Bruinsma [6] respectively.

Acknowledgements. The research leading to the results presented in this paper has received funding from the European Research Council under the European Union's Seventh Framework Programme (FP7/2007-2013)/ERC grant agreement no. 246939. Marco Dorigo acknowledges support from the Belgian F.R.S.-FNRS of which he is a research director. Navneet Bhalla has been partially supported by a Postdoctoral Fellowship provided by the Natural Sciences and Engineering Research Council of Canada. We would like to thank Eric Klemp and Michael Brand of the Direct Manufacturing Research Center (DMRC) of the University of Paderborn for their ongoing support and advice on manufacturing the prototypes for the stigmergic building blocks and the manipulator. Furthermore, we would also like to thank Prof. Christoph Scheytt and Dr. Uwe von der Ahe from the System and Circuit Technology research group at the Paderborn of University for their support with the BeBot platform.

References

1. Beckers, R., Holland, O.E., Deneubourg, J.L.: From local actions to global tasks: Stigmergy and collective robotics. In: Artificial life IV: Proceedings of the Fourth International Workshop on the Synthesis and Simulation of Living Systems, pp. 181–189. MIT Press, Cambridge (1994)
2. Bolger, A., Faulkner, M., Stein, D., White, L., Rus, D.: Experiments in decentralized robot construction with tool delivery and assembly robots. In: 2010 IEEE/RSJ International Conference on Intelligent Robots and Systems (IROS 2010), pp. 5085–5092. IEEE Press, Piscataway (2010)
3. Bonabeau, E., Dorigo, M., Theraulaz, G.: Swarm Intelligence: From Natural to Artificial Systems. Oxford University Press, New York (1999)
4. Bonabeau, E., Guérin, S., Snyers, D., Kuntz, P., Theraulaz, G.: Three-dimensional architectures grown by simple 'stigmergic' agents. BioSystems 56(1), 13–32 (2000)

5. Bonabeau, E., Theraulaz, G., Deneubourg, J.-L., Franks, N.R., Rafelsberger, O., Joly, J., Blanco, S.: A model for the emergence of pillars, walls and royal chambers in termite nests. Philosophical Transactions of the Royal Society of London B: Biological Sciences 353(1375), 1561–1576 (1998)

6. Bruinsma, O.H.: An Analysis of Building Behaviour of the Termite Macrotermes Subhyalinus (Rambur). Ph.D. thesis, Landbouwhoge School, Wageningen, The Netherlands (1979)

7. Deneubourg, J.-L., Goss, S., Franks, N., Sendova-Franks, A., Detrain, C., Chrétien, L.: The dynamics of collective sorting robot-like ants and ant-like robots. In: Meyer, J.A., Wilson, S. (eds.) Proceedings of the First International Conference on Simulation of Adaptive Behavior on From Animals to Animats, pp. 356–363. MIT Press, Cambridge (1991)

8. Grassé, P.P.: La reconstruction du nid et les coordinations inter-individuelles chez Bellicositermes Natalensis et Cubitermes sp. La théorie de la stigmergie: Essai d'interpretation du comportement de termites constructeurs. Insectes Sociaux 6(1), 41–80 (1959)

9. Herbrechtsmeier, S., Witkowski, U., Rückert, U.: Bebot: A modular mobile miniature robot platform supporting hardware reconfiguration and multi-standard communication. In: Kim, J.-H., et al. (eds.) Progress in Robotics. CCIS, vol. 44, pp. 346–356. Springer, Heidelberg (2009)

10. Holland, O., Melhuish, C.: Stigmergy, self-organization, and sorting in collective robotics. Artificial Life 5(2), 173–202 (1999)

11. Karsai, I., Pénzes, Z.: Comb building in social wasps: Self-organization and stigmergic script. Journal of Theoretical Biology 161(4), 505–525 (1993)

12. Khoshnevis, B.: Automated construction by contour crafting – related robotics and information technologies. Automation in Construction 13(1), 5–19 (2004)

13. Ladley, D., Bullock, S.: Logistic constraints on 3D termite construction. In: Dorigo, M., Birattari, M., Blum, C., Gambardella, L.M., Mondada, F., Stützle, T. (eds.) ANTS 2004. LNCS, vol. 3172, pp. 178–189. Springer, Heidelberg (2004)

14. Ladley, D., Bullock, S.: The role of logistic constraints in termite construction of chambers and tunnels. Journal of Theoretical Biology 234(4), 551–564 (2005)

15. Linardou, O.: Towards Homeostatic Architecture: Simulation of the Generative Process of a Termite Mound Construction. Master's thesis, University College London, London, United Kingdom (2008)

16. Lindsey, Q., Mellinger, D., Kumar, V.: Construction with quadrotor teams. Autonomous Robots 33(3), 323–336 (2012)

17. Martinoli, A., Mondada, F.: Probabilistic modelling of a bio-inspired collective experiment with real robots. In: Distributed Autonomous Robotic Systems, vol. 3, pp. 289–298. Springer, Heidelberg (1998)

18. Melhuish, C., Wilson, M., Sendova-Franks, A.: Patch sorting: Multi-object clustering using minimalist robots. In: Kelemen, J., Sosík, P. (eds.) ECAL 2001. LNCS (LNAI), vol. 2159, pp. 543–552. Springer, Heidelberg (2001)

19. Napp, N., Rappoli, O.R., Wu, J.M., Nagpal, R.: Materials and mechanisms for amorphous robotic construction. In: 2012 IEEE/RSJ International Conference on Intelligent Robots and Systems (IROS 2012), pp. 4879–4885. IEEE Press, Piscataway (2012)

20. Olson, E.: AprilTag: A robust and flexible visual fiducial system. In: 2011 IEEE International Conference on Robotics and Automation (ICRA 2011), pp. 3400–3407. IEEE Computer Society Press, Los Alamitos (2011)

21. Petersen, K., Nagpal, R., Werfel, J.: TERMES: An autonomous robotic system for three-dimensional collective construction. In: Durrant-Whyte, H.F., et al. (eds.) Robotics: Science and Systems VII, pp. 257–264. MIT Press, Cambridge (2011)
22. Pinciroli, C., Trianni, V., O'Grady, R., Pini, G., Brutschy, A., Brambilla, M., Mathews, N., Ferrante, E., Di Caro, G., Ducatelle, F., Birattari, M., Gambardella, L.M., Dorigo, M.: ARGoS: a modular, parallel, multi-engine simulator for multi-robot systems. Swarm Intelligence 6(4), 271–295 (2012)
23. Soleymani, T., Trianni, V., Bonani, M., Mondada, F., Dorigo, M.: Autonomous construction with compliant building material. In: Intelligent Autonomous Systems (IAS 2014). AISC. Springer, Berlin (in press, 2014)
24. Song, Y., Kim, J.H., Shell, D.A.: Self-organized clustering of square objects by multiple robots. In: Dorigo, M., Birattari, M., Blum, C., Christensen, A.L., Engelbrecht, A.P., Groß, R., Stützle, T. (eds.) ANTS 2012. LNCS, vol. 7461, pp. 308–315. Springer, Heidelberg (2012)
25. Stewart, R.L., Russell, R.A.: A distributed feedback mechanism to regulate wall construction by a robotic swarm. Adaptive Behavior 14(1), 21–51 (2006)
26. Theraulaz, G., Bonabeau, E.: Coordination in distributed building. Science 269(5224), 686–688 (1995)
27. Theraulaz, G., Bonabeau, E.: Modelling the collective building of complex architectures in social insects with lattice swarms. Journal of Theoretical Biology 177(4), 381–400 (1995)
28. Thomaszewski, B., Gumann, A., Pabst, S., Straßer, W.: Magnets in motion. ACM Transactions on Graphics 27(5) 162, 162:1–162:9 (2008)
29. Wawerla, J., Sukhatme, G.S., Matarić, M.J.: Collective construction with multiple robots. In: 2002 IEEE/RSJ International Conference on Intelligent Robots and System (IROS 2002), vol. 3, pp. 2696–2701. IEEE Press, Piscataway (2002)
30. Werfel, J., Bar-Yam, Y., Rus, D., Nagpal, R.: Distributed construction by mobile robots with enhanced building blocks. In: 2006 IEEE International Conference on Robotics and Automation (ICRA 2006), pp. 2787–2794. IEEE Computer Society Press, Los Alamitos (2006)
31. Werfel, J., Petersen, K., Nagpal, R.: Designing collective behavior in a termite-inspired robot construction team. Science 343(6172), 754–758 (2014)
32. Willmann, J., Augugliaro, F., Cadalbert, T., D'Andrea, R., Gramazio, F., Kohler, M.: Aerial robotic construction towards a new field of architectural research. International Journal of Architectural Computing 10(3), 439–460 (2012)
33. Wismer, S., Hitz, G., Bonani, M., Gribovskiy, A., Magnenat, S.: Autonomous construction of a roofed structure: Synthesizing planning and stigmergy on a mobile robot. In: 2012 IEEE/RSJ International Conference on Intelligent Robots and Systems (IROS 2012), pp. 5436–5437. IEEE Press, Piscataway (2012)
34. Worcester, J., Rogoff, J., Hsieh, M.A.: Constrained task partitioning for distributed assembly. In: 2011 IEEE/RSJ International Conference on Intelligent Robots and Systems (IROS 2011), pp. 4790–4796. IEEE Press, Piscataway (2011)

Swarm in a Fly Bottle: Feedback-Based Analysis of Self-organizing Temporary Lock-ins

Heiko Hamann[1] and Gabriele Valentini[2]

[1] Department of Computer Science, University of Paderborn, Paderborn, Germany
[2] IRIDIA, Université Libre de Bruxelles, Brussels, Belgium
heiko.hamann@uni-paderborn.de, gvalenti@ulb.ac.be

Abstract. Self-organizing systems that show processes of pattern formation rely on positive feedback. Especially in swarm systems, positive feedback builds up in a transient phase until maximal positive feedback is reached and the system converges. We investigate alignment in locusts as an example of swarm systems showing time-variant positive feedback. We identify an influencing bias in the spatial distribution of agents compared to a well-mixed distribution and two features, percentage of aligned swarm members and neighborhood size, that allow to model the time variance of feedbacks. We report an urn model that is capable of qualitatively representing all these relevant features. The increase of neighborhood sizes over time enables the swarm to lock in a highly aligned state but also allows for infrequent switching between lock-in states.

1 Introduction

Many systems showing pattern formation, such as animal coloration [1], embryogenesis [2], and grazing systems [3], are examples of self-organizing systems. In addition to multiple interactions of sub-components, general features of self-organizing systems are the interplay between positive feedback (also amplification or activation) and random fluctuations as well as that between positive and negative feedback (also inhibition) [4]. Typically the system is initialized to an unordered state (not showing any patterns). Fluctuations generate deviations which are amplified by positive feedback until a spatiotemporal pattern forms. Negative feedback might prevent the system from reaching extreme states (e.g., 100% ordered, extinction). Following this stochastic process, a random dynamical attractor forms and characterizes the dynamics of the system [5].

Swarm systems are an example of self-organizing systems. A frequent setting in the case of swarms is that several stable ordered states exist (multistability) that are symmetrical to each other—a typical situation in a collective decision-making system. A swarm is a distributed agent system where each agent autonomously decides on its actions. With the global knowledge of an external observer, we can classify at least a subset of these actions as positive or negative feedback events that drive the system, respectively, towards or away from a too ordered state [6–8]. By counting these events we are able to calculate the ratio of positive feedback events $f^+ = \frac{F^+}{F^+ + F^-}$, for the number of positive

M. Dorigo et al. (Eds.): ANTS 2014, LNCS 8667, pp. 170–181, 2014.

feedback events $0 \leq F^+ \leq N$ for swarm size N (F^- accordingly). If $f^+ > 0.5$ we say positive feedback is predominant. For several swarm systems, such as density classification, aggregation controlled by BEECLUST, and alignment in locust swarms, negative feedback is initially predominant while positive feedback builds up only over time and independently of the order of the current system state [6–8]. Consequently, there exists a second feature and/or a mechanism besides the order of the system that controls the increase of positive feedback intensity. This feature is very likely a spatial feature which can be determined by the method of elimination due to the simplicity of the investigated systems (agents have only 2 properties: direction of motion and position).

   Our main objective is to determine the above mentioned second feature and to define an appropriate model that covers the interplay between positive and negative feedback as well as the increase of positive feedback over time. We continue the work reported in earlier publications [6, 8] by focusing on questions raised therein. All of the following experiments are based on a swarm model inspired by swarm alignment of locusts. These swarms switch between different aligned states even after having reached high degrees of alignment. Such a special property is also subject to the following investigations.

1.1   Locust Scenario

The desert locust, *Schistocerca gregaria*, shows collective motion in the growth stage of a wingless nymph often called 'marching bands' [9]. The collective motion is expressed in the directional alignment of a majority of locusts, it is density-dependent, and individuals seem to change their direction as a response to neighbors [9]. In experiments, the complexity of the collective motion is reduced to a pseudo-1-d setting by using a ring-shaped arena. Microscopic [10] and macroscopic models [11, 6–8] of this behavior have been reported. Here we use the microscopic model of self-propelled particles by Czirók et al. [10] as our reference model (henceforth 'Czirók model'). The system is defined in 1-d space. A particle i has coordinate $x_i \in [0, C)$ and discrete, dimensionless velocity $u_i \in [-1, 1]$. We refer to particles with velocity $u_i < 0$ as 'left-goers' (respectively, 'right-goers' for $u_i > 0$). The dynamics of a particle is defined by $x_i(t+1) = x_i(t) + v_0 u_i(t)$ where v_0 is the nominal particle velocity and $u_i(t+1) = G(\langle u(t) \rangle_i) + \xi_i$ considers its interaction with neighbors (subject to noise ξ_i uniformly distributed over $[-\eta/2, \eta/2]$). The local average velocity $\langle u(t) \rangle_i$ for the ith particle is calculated over all neighbors located in the interval $[x_i - \Delta r, x_i + \Delta r]$ for perception range Δr (see Table 1 for the parameter settings). G describes both propulsion and friction forces

$$G(u) = \begin{cases} (u+1)/2, & \text{for } u > 0 \\ (u-1)/2, & \text{for } u < 0 \end{cases}. \tag{1}$$

The initial condition is a random uniform distribution for both the particles' coordinates $x_i \in [0, C)$ and their velocities $u_i \in [-1, 1]$. In the locust system, the spatial distribution of particles is biased and undergoes a nontrivial evolution. Fig. 2a gives a simplified picture of the spatial correlations generated by the

Czirók model in the form of the pair correlation function. For a given left-goer ratio, we measure the density of left-goers as a function of the distance from a left-goer at times $t_1 = 30$ and $t_2 = 90$. We consider swarms with $N = 41$ particles and system states with 25 left-goers and 16 right-goers only. The shown results are averaged over many independent runs. The two horizontal dashed lines give the expected distribution under the assumption of a uniform distribution of particles. Early in the simulation, at $t_1 = 30$, a left-goer has an increased density for nearby left-goers (within distances of about 2.6) in comparison to an assumed uniform distribution. Accordingly, right-goers have a decreased density for nearby right-goers due to symmetry. Later in the simulation, at $t_2 = 90$, left-goers have an increased density of nearby left-goers for even longer distances of up to about 6.0 and as a consequence a decreased density for the remaining arena (accordingly for right-goers). These spatial correlations in the particle distributions are discussed next and in Sec. 4 we also interpret the temporal evolution of these correlations.

2   Models

We give a model to investigate the influence of biased spatial distributions as indicated above by the pair correlation function. A Markov chain model for two system variables is introduced to model the above mentioned second feature of the system, for which we choose the neighborhood size. We present an urn model that is able to represent the relevant spatial features and a mathematical model of the underlying feedback processes.

2.1   Well-Mixed and Biased Spatial Distributions

We model the collective decision-making process using Markov chains. A simple model for collective decision with only one state variable was reported before [6, 8]. In the locust scenario, we count left-goers L (without loss of generality) and get a set of $N + 1$ states: $\{0, 1, \ldots, N\}$. As simplifying assumptions, we ignore that the system might stay within the current state (i.e., no self-loops) and that we might have changes in the left-goer number of more than one particle within a small time interval. Without loss of generality, we focus exclusively on transitions that are increasing the number of left-goers $P(L \to L+1)$ due to the symmetry $P(L \to L+1) = 1 - P(L \to L-1)$ for $L \in \{1, 2, \ldots, N-1\}$. These transition probabilities are measured using the Czirók model (see Fig. 1a).

An abstract model that only counts left-goers is not representing space, and therefore, implicitly assumes for the agents a well-mixed distribution in the space independent of their internal state (e.g., heading, opinion). However, swarm systems typically rely on spatial features and show non-homogeneous distributions of agents [12]. In the locust scenario, the first priority for the swarm is to achieve alignment which is generally independent of agents' positions. However, locusts seem to depend heavily on spatial features such as the number of neighbors [7]. In the following, we briefly investigate the difference between well-mixed systems and systems whose agents' spatial distributions are biased by agents' headings.

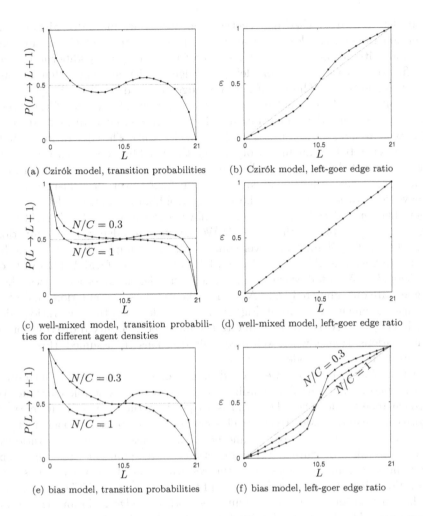

(a) Czirók model, transition probabilities

(b) Czirók model, left-goer edge ratio

(c) well-mixed model, transition probabilities for different agent densities

(d) well-mixed model, left-goer edge ratio

(e) bias model, transition probabilities

(f) bias model, left-goer edge ratio

Fig. 1. Transition probabilities $P(L \rightarrow L+1)$ and left-goer edge ratio ε using the Czirók model and two methods of initially positioning agents following a random uniform distribution or a special biased distribution($N = 21$)

For the following experiment, we initially place the agents by sampling from a uniform distribution and calculate the updates in agent directions u_i according to the Czirók model. We simulate only one time step, and consequently, agents' positions are not correlated due to earlier dynamics. Fig. 1c shows the resulting transition probabilities for two agent densities ($N/C \in \{0.3, 1.0\}$) based on 2×10^5 samples each. Agents' density influences the transition probabilities considerably. In addition, we note qualitative differences in the shapes of the curves compared to Fig. 1a.

Next, we define a measure capable to represent an important spatial feature. The spatial distribution of agents induces a graph. The existence of an edge is simply determined by checking whether two agents are mutually within their perception range Δr. The set of agents with whom an agent shares an edge defines also its neighborhood and the size of this set is its neighborhood size. To define our measure we count 'left-goer edges' which are edges that contain at least one left-goer. The set of left-goer edges E_L is defined by $E_L = \{e | e = (e_1, e_2)$ with e_1 and/or e_2 is left-goer$\}$. The definition of the set of right-goer edges E_r is symmetrical. Our measure is the left-goer-edge ratio ε which is calculated based on the set sizes: $\varepsilon = |E_L|/(|E_L|+|E_R|)$. The edge ratio $\varepsilon \in [0, 1]$ can be interpreted as an indicator of how the neighborhood sizes (i.e., node degrees) are distributed between left-goers and right-goers. If the neighborhood size averaged over all left-goers equals that averaged over all right-goers, then we have edge ratio $\varepsilon = 0.5$. If the average neighborhood size of left-goers is bigger than that of right-goers, then we have edge ratio $\varepsilon > 0.5$. If right-goer neighborhoods are bigger, then $\varepsilon < 0.5$. We measure the edge ratio as a function of time for the Czirók model by averaging over 2×10^5 samples. Fig. 1b shows the presence of bigger neighborhood sizes for left-goers in case of a global majority of left-goers ($L > N/2$, for swarm size N) and of smaller neighborhood sizes when left-goers are outnumbered ($L < N/2$). For the well-mixed simulation we do the same measurements as shown in Fig. 1d. For both densities the edge ratio scales linearly $\varepsilon(L) = L/N$ as expected due to the unbiased well-mixed distribution of agent positions. Hence, when well-mixed distributions are assumed, this spatial features of the Czirók model are ignored.

Next, we consider how the well-mixed simulation can be modified to introduce a spatial bias that results in a non-linear edge ratio $\varepsilon(L)$. A simple constructive approach to influence the edge ratio is to position the agents of the current majority in clusters of two. To create clusters of two, we position two agents from the majority group at the exact same position. Agents are initially positioned according to this procedure and their directions u_i are updated for one time step. Averaging over many samples gives the resulting transition probabilities and edge ratio which are shown in Fig. 1e and f for two densities ($N/C \in \{0.3, 1.0\}$). The biased positioning of agents influences both the edge ratio and transition probabilities. The increased density from 0.3 to 1.0 almost only introduces a downscaling of the edge-bias by a factor of about 0.45.

2.2   Markov Chain Model for Two System Variables

We extend the Markov chain model reported before [6, 8] by considering as second state variable the average neighborhood size $\mathcal{N} \in \{1, \ldots, N\}$ over all agents (with perception range Δr). Fig. 2b shows an example of the resulting chain for swarm size $N = 3$. We get $(N + 1)N$ states for swarm size N. For simplicity, we ignore again that the system might stay within the current state, that a concurrent change of both features might occur, and we also ignore that we might have changes in the left-goer number or neighborhood size of more than one within a small time interval. For any given state $(L, \mathcal{N})$, we measure the probability of observing a transition that increases/decreases the number of left-goers, $P(L \to L \pm 1 | (L, \mathcal{N}))$, and the probability of observing a transition that increases/decreases the neighborhood size, $P(\mathcal{N} \to \mathcal{N} \pm 1 | (L, \mathcal{N}))$. In this way we obtain a periodic Markov chain of period two due to the absence of self-loops in the chain. This kind of Markov chains does not converge to one stationary distribution but jumps between two stationary distributions: one for odd time steps and one for even time steps. We calculate a unique limiting distribution of the process by taking the mean of the two steady-state distributions of the chain. The analysis presented at the end of Sec. 3 is based on the computation of such limiting distributions.

2.3   Urn Model

We define an urn model that represents most of the relevant features of the locust system, especially those that are due to spatial biases. Our aim is to develop a model that is simpler and faster to simulate than the Czirók model but still represents the qualitative key features of this system. The urn model consists of 3 urns: *main*, *edges*, and *resource*. Urn *main* represents the number of left- and right-goers in the swarm and contains a constant number of N marbles. Urn *edges* represents an average neighborhood and contains a variable number of marbles E which represents the neighborhood size and the edge ratio. Urn *resource* provides additional marbles to increase the neighborhood size E and therefore also holds a variable number of marbles. At each round, the drawing process follows four stochastic rules (see Table 2 for used parameters).

Rule 1. We draw E times from *edges* with replacement (if E is even we do $E + 1$ draws to avoid treatment of tie-breakers) and count the left-goers λ and right-goers ρ that we draw. Next, we draw one marble from *main*. If $\lambda > \rho$ and we have drawn a right-goer from *main*, then we put a left-goer back to *main*. If $\lambda < \rho$ and we have drawn a left-goer from *main*, then we put a right-goer back to *main*. Otherwise, that is, $\lambda < \rho$ and we have drawn a right-goer, we do not exchange the marble and put it back in *main* (accordingly for $\lambda > \rho$ and left-goer). This first drawing rule represents the actual decision process of an agent based on counting neighboring agents and a majority rule.

Rule 2. This second drawing rule is executed at each round only with a probability of P_{nsize}. We draw E times from *edges* with replacement (if E is even we do $E + 1$ draws) and count the left-goers λ and right-goers ρ that we draw. Next, we do $\min(\lambda, \rho) + 1$ random experiments: with probability P_{incr} we move a left-goer or a right-goer (with equal probability) from *resource* to *edges* if possible. Finally we do $E(\max(\lambda, \rho)/N - c_{\mathrm{decr}})$ random experiments and move with probability P_{decr} a left-goer or a right-goer (with equal probability) from *edges* to *resource* if possible. This rule models the dynamics of the neighborhood size. Big neighborhoods increase their size faster than small neighborhoods for a balanced distribution of left- and right-goers. For unbalanced distributions, neighborhoods tend to decrease their size.

Rule 3. We draw one marble from *main*. If it is a left-goer, we replace a right-goer in *edges* with a left-goer (if possible) or vice versa in the case of a right-goer (positive feedback). This third drawing rule is executed in each round only with a probability of P_{noise}.

Rule 4. We draw one marble from *main*. If it is a left-goer, we replace a left-goer in *main* with a right-goer (if possible) or vice versa in the case of a right-goer (negative feedback). This fourth drawing rule is executed in each round only with a probability of P_{noise}. These two last rules implement noise. They are executed with the same probability but the positive feedback operates on *edges* and negative feedback operates on *main*.

2.4   Mathematical Model of Feedbacks

We define a mathematical description of the above urn model. A detailed model would not allow for concise equations, therefore, we restrict our attention to the main features. We ignore the dynamics of urn *edges* except for the total number of marbles E which gives the average neighborhoods size $\mathcal{N}$. Our main focus is to model the dynamics of urn *main*. We assume that the ratio of left-goers in the neighborhood (*edges*) is identical to the ratio of left-goers $m = L/N$ in *main* and that $E = \mathcal{N}$ is odd. For $\mathcal{N}$ marbles, of which $m\mathcal{N}$ are left-goers and $(1 - m)\mathcal{N}$ are right-goers, the probability to draw a majority of left-goers is

$$P_{\text{maj}}^{\text{left}} = \sum_{n \in \{\lceil \frac{\mathcal{N}}{2} \rceil, \ldots, \mathcal{N}\}} \binom{\mathcal{N}}{n} m^n (1 - m)^{\mathcal{N}-n} \tag{2}$$

and the probability to draw a majority of right-goers is

$$P_{\text{maj}}^{\text{right}} = \sum_{n \in \{\lceil \frac{\mathcal{N}}{2} \rceil, \ldots, \mathcal{N}\}} \binom{\mathcal{N}}{n} (1 - m)^n m^{\mathcal{N}-n}. \tag{3}$$

Following a heuristic approach, the average change Δm of left-goers within one time step is modeled as

$$\Delta m^h(\mathcal{N}, m) = (1 - m)P_{\text{maj}}^{\text{left}} - mP_{\text{maj}}^{\text{right}} - P_{\text{noise}}(2m - 1), \tag{4}$$

whereas the first two terms model the positive feedback effect implemented by rule 1. An increases (respectively, decrease) in the number of left-goers results from drawing a right-goer while having a majority of left-goers. Besides, the third term models the negative feedback effect of rule 4.

As a second alternative, we model the average change Δm of left-goers with a feedback-based approach as reported in [6, 8]. That is, we neglect the actual processes causing positive and negative feedback in the the urn model and we focus instead on the probability of positive feedback $P_{FB}(\mathcal{N}, m)$. We get

$$\Delta m^{FB}(\mathcal{N}, m) = 1 - 2((1 - P_{FB}(\mathcal{N}, m))P_{\text{maj}}^{\text{left}} + P_{FB}(\mathcal{N}, m)P_{\text{maj}}^{\text{right}}). \tag{5}$$

The probability of positive feedback is calculated by equating and solving the right hand sides of eqs. 4 and 5 which yields

$$P_{FB}(\mathcal{N}, m) = -(2m - 15P_{\text{maj}}^{\text{left}} + 5mP_{\text{maj}}^{\text{left}} + 5mP_{\text{maj}}^{\text{right}} + 4)/(10P_{\text{maj}}^{\text{left}} - 10P_{\text{maj}}^{\text{right}}). \tag{6}$$

With increasing $\mathcal{N}$, we obtain polynomials of increasing degree. The first 3 are:
$P_{FB}(\mathcal{N} = 1, m) = 2/5$, $P_{FB}(\mathcal{N} = 3, m) = 3/4 - 7/(20(-2m^2 + 2m + 1))$ and
$P_{FB}(\mathcal{N} = 5, m) = 3/4 - 7/(20(6m^4 - 12m^3 + 4m^2 + 2m + 1))$. Fig. 2c shows the
behavior of eq. 6 when $\mathcal{N} \in \{1, 3, \dots, 27\}$. A 'negative exponential' increase of
positive feedback intensity with increasing neighborhood size $\mathcal{N}$ is clearly visible.
A similar result was reported in [8, Fig. 8b] for a different swarm experiment
showing temporal dependency. In the locust scenario, the neighborhood size also
increases over time (see Section 3). Hence, the model given by eq. 6 indirectly
confirms the increase of positive feedback in swarm systems as reported in [6, 8].

3   Results

We investigate the Czirók model and the urn model with focus on the key findings that the average neighborhood size and the edge ratio are relevant features
of the locust scenario. In particular, we investigate measured transition probabilities by interpreting both models as Markov chains (Sec. 2.2). An overview
of the complete system dynamics is given by vector fields in Figs. 3a and b for
$N = 41$ (10^6 samples for Czirók model and 5×10^6 for urn model). These plots
are based on the transition probabilities which are put in relation to each other.
Furthermore, the horizontal and the vertical lengths of the arrows were normalized individually to maximize readability (i.e., vector field plots are qualitative,
the quantitative data is given in Figs. 3c-f). In the case of the Czirók model, as a
consequence of the initial random uniform distribution of agents over the whole
ring, the neighborhood size $\mathcal{N}$ is initially small ($2\Delta r N/C \approx 1.2$). Similarly, urn
edges initially holds one left-goer and one right-goer. In the initial unordered
state there are approximately the same number of left-goers and right-goers
($L \approx N - L$). Hence, both systems start in the area at the lower middle of
the vector field. First, the neighborhood size increases. Only later, once a bigger
neighborhood size is formed, the system either increases or decreases in the num-
ber of left-goers L until reaching a stable state, respectively, $(L, \mathcal{N}) \approx (35, 8)$ or
$(L, \mathcal{N}) \approx (6, 8)$.

Projections of the data given in Figs. 3a and b are given in diagrams c-f.
Figs. 3c and d give the transition probabilities for an increase in L for all $\mathcal{N}$
for Czirók and urn model and e and f give the transition probabilities for an
increase in $\mathcal{N}$ for all $\mathcal{N}$. Some curves are noisy because the corresponding configurations occur very rarely. We ignore statistical significance within this qualitative study (big quantitative differences between the two models are obvious).
Positive feedback is found within approximately the same intervals in Figs. 3e
and f (similarly for negative feedback). Noticeable is the extreme positive feedback for $6 < L < 18$ and $23 < L < 35$ for the Czirók model seen in Figs. 3c
and for $12 < L < 29$ in e. Figs. 3g and h give the left-goer edge ratio ε for
different times. The edge ratio of the urn model is more dynamic because the
urn model is always started with one left-goer and one right-goer, that is, an
edge ratio of $\varepsilon = 0.5$. The initial edge ratio for the Czirók model, in turn, is the
direct result of the uniform distribution of agents which gives $\varepsilon(L) = L/N$.

parameter	sym.	value
swarm size	N	$\{17, 21, 25, 33,$
		$41, 49, 57, 61\}$
circumference	C	70 (21)
nominal speed	v	0.1
perception range	Δr	1.0
noise	η	2.5

parameter	symbol	value
prob. neighbh. rule (rule 2)	P_{nsize}	0.2
prob. neighbh. size increase	P_{incr}	0.18
prob. neighbh. size decrease	P_{decr}	0.007
offset neighbh. size decrease	c_{decr}	0.15
probability of noise	P_{noise}	0.2

Table 1: Used parameters for the Czirók model.

Table 2: Used parameters for the urn model.

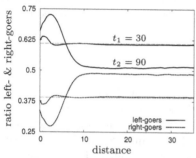

(a) Pair correlation function: measured density of left-/right-goers at distances from a particle of the same kind.

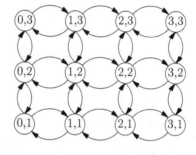

(b) Markov chain for two state variables: number of left-goers L and the average neighborhood size $\mathcal{N}$ (swarm size $N = 3$).

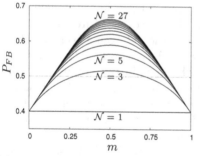

(c) Analytically obtained probabilities of positive feedback (eq. 6) for neighborhood sizes $\mathcal{N} \in \{1, 3, \ldots, 27\}$.

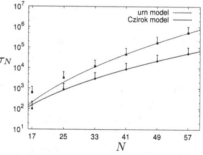

(d) Mean first passage time τ_N over swarm size N for both models fitted to $a \exp(cx)x^b$, error bars give standard deviation.

Fig. 2. Pair correlation, Markov chain, probability of feedback and mean first passage time

Finally, we investigate the scalability of the mean switching time between stable modes of the system. We look at the mean time τ_N necessary to move from an initial set of states with a majority of right-goers to a final set of states where left-goers lead the system. We consider swarm systems whose transient dynamics have already vanished and thus settled down to their limiting behavior. We define the set of initial (respectively, final) states looking at the limiting distribution of the process (computed from the Markov chain defined in Sec. 2.2). We select states with a majority of right-goers (respectively, left-goers) in ascending order

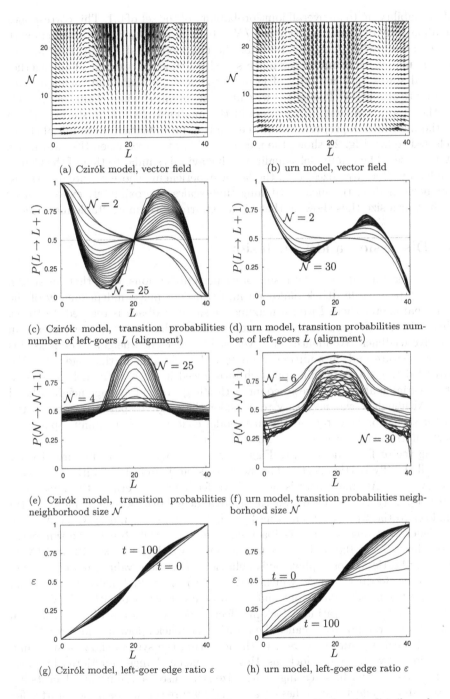

(a) Czirók model, vector field

(b) urn model, vector field

(c) Czirók model, transition probabilities number of left-goers L (alignment)

(d) urn model, transition probabilities number of left-goers L (alignment)

(e) Czirók model, transition probabilities neighborhood size $\mathcal{N}$

(f) urn model, transition probabilities neighborhood size $\mathcal{N}$

(g) Czirók model, left-goer edge ratio ε

(h) urn model, left-goer edge ratio ε

Fig. 3. Vector fields, transition probabilities, and edge ratio for the Czirók and urn model ($N = 41$, 10^6 samples for Czirók model and 5×10^6 for urn model)

of probability up to an overall joint probability of the set of 0.1. This corresponds to a majority of right-goers where $L/N \approx 0.15$ and a majority of left-goers with $L/N \approx 0.85$, while $\mathcal{N} \in [5.8, 14.2]$ for the urn model and $\mathcal{N} \in [4, 10.6]$ for the Czirók model. To compute the mean switching time τ_N, we first lump together all final states with majority of left-goers in a single state and then we make this state absorbing. The mean switching time of our original chain corresponds to the absorption time of a process in the modified chain that starts with an initial distribution proportional to the limiting probabilities of the initial states selected so far. Fig. 2d shows the mean switching time τ_N when the swarm size N increases. The urn model's qualitative behavior is similar to that of the Czirók model for all swarm sizes, with the latter experiencing shorter switching times. For both models, the mean switching time scales approximately exponentially with swarm size thus showing that bigger swarms form more stable majorities.

4   Discussion and Conclusion

This paper started from the result of earlier publications [6, 8] that, in swarm systems, positive feedback builds up in a transient phase independently of the order parameter (here L) until maximal positive feedback is reached. In turns, this indicates the existence of a second feature that controls the increase in positive feedback. We identify the average size of agents' neighborhoods as this second feature and, in addition, we detect the relevance of the edge ratio. We extended the Markov-chain approach introduced in [6, 8] to model the second state variable and therefore to count both left-goers and average neighborhood size. Although it was necessary to consider the original spatial features of the locust scenario, we extended the urn model concept [6, 8] to mimic spatiality, particularly, neighborhood size and edge ratio.

The vector fields depicted in Figs. 3a and b provide a clear picture of what we call the 'fly-bottle effect'[1]. The swarm system is initialized with $L/N \approx 0.5$ and $\mathcal{N} < 5$. At first, there is no positive feedback concerning the number of left-goers L (Figs. 3c and d) but the average neighborhood size $\mathcal{N}$ increases (in analogy to the fly bottle: 'entering from below', Figs. 3e and f). Once $\mathcal{N} \approx 10$ or bigger, a strong positive feedback emerges that easily breaks the symmetry given by $L/N \approx 0.5$ and drives the system towards $L/N \approx 0.12$ or $L/N \approx 0.88$ (fly-bottle analogy: phototaxis behavior). For these values, however, there is negative feedback on $\mathcal{N}$ which decreases to $\mathcal{N} \approx 8$. Finally, positive and negative feedbacks balance out and the system converges to either $L/N \approx 0.15$ or $L/N \approx 0.85$ generating a lock-in effect ('the fly is trapped'). This lock in effect is only temporary because the positive feedback operating on L for $\mathcal{N} \approx 8$ is much weaker than for $\mathcal{N} > 8$. In the long run, the system shows a switching behavior between the two lock-ins. However, such a change in majority becomes more infrequent with increasing swarm size N as also seen in natural locusts [9]. The fly-bottle effect, that relies on a second feature serving as a kick starter for

---

[1] A fly bottle is a traditional device made of clear glass to passively trap flying insects that enter it from below and cannot escape because of their phototaxis behavior.

the whole system, seems to have a certain generality in swarm systems. Indeed, the increase of positive feedback during a transient phase was also reported for other systems such as density classification and aggregation controlled by BEECLUST [6, 8].

An additional interpretation of the fly-bottle effect with reference to Fig. 2a is that the secondary feature is generating only short-ranged correlations early on but not global correlations. These short-ranged correlations seem to be side-effects, such as small clusters of agents in the investigated locust scenario. The long-range correlations seen later in the system are then an effect of the primary feature (here L) and probably could neither be generated by the one or the other feature alone. Future research work will focus on the questions whether the fly-bottle effect is generally observed in swarm systems and, if so, whether it can be used to design swarm behaviors for artificial swarm systems.

Acknowledgments. This work was partially supported by the European Research Council through the ERC Advanced Grant "E-SWARM: Engineering Swarm Intelligence Systems" (contract 246939).

References

1. Camazine, S., Deneubourg, J.L., Franks, N.R., Sneyd, J., Theraulaz, G., Bonabeau, E.: Self-Organizing Biological Systems. Princeton University Press, NJ (2001)
2. Crick, F.: Diffusion in embryogenesis. Nature 225(5231), 420–422 (1970)
3. Noy-Meir, I.: Stability of grazing systems: an application of predator-prey graphs. The Journal of Ecology, 459–481 (1975)
4. Bonabeau, E., Dorigo, M., Theraulaz, G.: Swarm Intelligence: From Natural to Artificial Systems. Oxford Univ. Press, New York (1999)
5. Arnold, L.: Random Dynamical Systems. Springer (2003)
6. Hamann, H.: Towards swarm calculus: Universal properties of swarm performance and collective decisions. In: Dorigo, M., Birattari, M., Blum, C., Christensen, A.L., Engelbrecht, A.P., Groß, R., Stützle, T. (eds.) ANTS 2012. LNCS, vol. 7461, pp. 168–179. Springer, Heidelberg (2012)
7. Hamann, H.: A reductionist approach to hypothesis-catching for the analysis of self-organizing decision-making systems. In: 7th IEEE Int. Conf. on Self-Adaptive and Self-Organizing Systems (SASO 2013), pp. 227–236. IEEE (2013)
8. Hamann, H.: Towards swarm calculus: Urn models of collective decisions and universal properties of swarm performance. Swarm Intelligence 7(2-3), 145–172 (2013)
9. Buhl, J., Sumpter, D.J.T., Couzin, I.D., Hale, J.J., Despland, E., Miller, E.R., Simpson, S.J.: From disorder to order in marching locusts. Science 312(5778), 1402–1406 (2006)
10. Czirók, A., Barabási, A.L., Vicsek, T.: Collective motion of self-propelled particles: Kinetic phase transition in one dimension. Phys. Rev. Lett. 82(1), 209–212 (1999)
11. Yates, C.A., Erban, R., Escudero, C., Couzin, I.D., Buhl, J., Kevrekidis, I.G., Maini, P.K., Sumpter, D.J.T.: Inherent noise can facilitate coherence in collective swarm motion. Proc. Natl. Acad. Sci. USA 106(14), 5464–5469 (2009)
12. Hamann, H.: Space-Time Continuous Models of Swarm Robotics Systems: Supporting Global-to-Local Programming. Springer, Berlin (2010)

Temporal Task Allocation
in Periodic Environments
An Approach Based on Synchronization

Manuel Castillo-Cagigal[1], Arne Brutschy[2], Alvaro Gutiérrez[1],
and Mauro Birattari[2]

[1] ETSIT, Universidad Politécnica de Madrid, Madrid, Spain
manuel.castillo@upm.es, aguti@etsit.upm.es
[2] IRIDIA, CoDE, Université Libre de Bruxelles, Brussels, Belgium
{arne.brutschy,mbiro}@ulb.ac.be

Abstract. In this paper, we study a robot swarm that has to perform task allocation in an environment that features periodic properties. In this environment, tasks appear in different areas following periodic temporal patterns. The swarm has to reallocate its workforce periodically, performing a *temporal task allocation* that must be synchronized with the environment to be effective.

We tackle temporal task allocation using methods and concepts that we borrow from the signal processing literature. In particular, we propose a distributed temporal task allocation algorithm that synchronizes robots of the swarm with the environment and with each other. In this algorithm, robots use only local information and a simple visual communication protocol based on light blinking. Our results show that a robot swarm that uses the proposed temporal task allocation algorithm performs considerably more tasks than a swarm that uses a greedy algorithm.

1 Introduction

In dynamical environments, real-time resource allocation commonly involves situations in which events occur periodically, with a certain frequency [14]. Periodicity can originate from both natural and artificial phenomena, for example, earth's rotation and revolution, tides, cyclic production processes, and customer demands. In artificial systems, the designer typically wishes to allocate resources so as to increase the system performance and achieve predefined goals [11]. To this end, it is paramount that information on the nature of the periodic events involved is available during the design process [9].

Task allocation as studied in swarm robotics [5] is a class of resource allocation problems: the workforce of the swarm can be seen as the resource to be allocated—see [2] for a recent review of the swarm robotics literature including works on task allocation. In this paper, we study a case in which a robot swarm needs to perform task allocation in an environment that features periodic properties. Specifically, the periodicity of the environment lies in the temporal pattern

M. Dorigo et al. (Eds.): ANTS 2014, LNCS 8667, pp. 182–193, 2014.

in which new tasks appear. To operate effectively, the swarm needs to reallocate its workforce according to the periodicity of the environment. We call *temporal task allocation* a task allocation that takes into account temporal properties of the environment.

To exploit environments with periodic properties, a task allocation algorithm needs to adapt to the periodicity of the environment. In this paper, we propose a novel temporal task allocation algorithm that adapts to the environment. This algorithm is based on concepts that we borrow from the signal processing and collective synchronization literature.

Collective synchronization has been previously observed and studied in biological systems (e.g., [6]). In these systems, the components converge to a common phase and oscillate in unison. Collective synchronization is usually modeled via coupled oscillators [15]. A model that is commonly adopted is the one proposed by Kuramoto [8]. A direct application of Kuramoto's model in swarm robotics is not appropriate because it would require that each robot knows with which phase the others oscillate. Other models exist that do not require that a robots knows the phase of the others. Examples are models based on firefly synchronization and chorusing mechanisms [4,7], which are commonly based on local communication.

In this paper, we propose a temporal task allocation algorithm in which robots synchronize with each other and with the environment. The synchronization with the environment is the novelty of our work.

2   Environment and Robots

We consider a rectangular environment that is divided in three areas: workspace A, workspace B, and a transition area. Fig. 1a shows a schematic representation of the environment. Tasks appear either in workspace A or B, following a temporal pattern. Robots have to travel from workspace to workspace to attend tasks where they appear. The workspaces are separated by the transition area: a robot that moves from one workspace to the other has to cross the transition area. The time spent by the robot i to cross the transition area is called switching cost ξ_s^i. It is measured in time units and is independent for each robot. Due to possible collisions with other robots, ξ_s^i is a random variable.

We use an abstracted representation of tasks: to carry out a task, a robot has to reach the location at which the task appeared and stay there for a certain amount of time. The time ξ_e that a robot spends working on a task is fixed. Tasks have a life time ξ_l after which they expire: if a task remains unattended for longer than ξ_l, it is removed from the environment. At time k, $N^A[k]$ and $N^B[k]$ are the number of tasks present in workspace A and B, respectively. The amount of tasks in each workspace is bounded by the task capacity Γ, which is the same for the two workspaces.

The periodicity of the environment that we consider in this paper lies in the temporal pattern with which tasks appear. During a period of time T^A, new tasks appear in workspace A. After the end of T^A, new tasks appear in workspace B

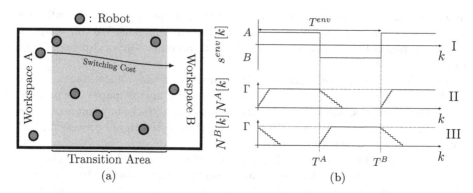

Fig. 1. Environment definition. a) Schematic representation of the arena, with workspaces A and B in white and transition area in gray. b) Environment period and location of the task appearance example: I) signal $s^{env}[k]$ of task appearance with period T^{env}, II) number $N^A[k]$ of tasks in workspace A, III) number $N^B[k]$ of tasks in workspace B.

for a period of time T^B. After the completion of T^B, new tasks appear again in workspace A, and so on. The full cycle has a period $T^{env} = T^A + T^B$. In this paper, we assume $T^A = T^B$. The location of the appearance of tasks in the environment can be described as a square signal denoted by $s^{env}[k]$ that takes a value of A or B. An example of T^{env} and $s^{env}[k]$ is shown in Fig. 1b-I.

Regardless of the workspace, the tasks appear in the environment with a certain incoming task rate λ. If the task capacity Γ of a workspace is reached, additional tasks are dismissed. When tasks no longer appear in a workspace, the number of tasks in this workspace decreases as tasks expire. This effect can be observed in Fig. 1b-II and 1b-III for both workspaces: the number of tasks increases until Γ is reached and decreases after new tasks cease to appear.

The robots move in the arena between workspace A and B in order to attend to the tasks. Robots act independently of each other, but are able to exchange simple messages via short-range line-of-sight communication. The number of robots in a workspace are the workforce allocated to this workspace by the swarm. In order to maximize performance, the swarm needs to allocate its complete workforce to the workspace where tasks are available. To achieve this goal, the robots need to switch between workspaces so that their movement is synchronized with the temporal pattern of task appearance, performing a temporal task allocation.

3   Collective Synchronization Algorithm

In this section, we present the *collective synchronization* (CS) algorithm. The goal of CS is to synchronize the movement of the robots between workspaces with the appearance of tasks in the environment. In CS, each robot i has an internal timer τ^i that governs its transitions between workspaces. This timer increases each time step and resets to zero when it reaches the period T^i of robot i. The robot switches between workspaces depending on τ^i:

$$s^i[k] = \begin{cases} A & \tau^i \leq T^i/2 \\ \\ B & \tau^i > T^i/2 \end{cases} \tag{1}$$

This equation produces a square signal $s^i[k]$ as shown in Fig. 2a-I. The timer τ^i might not be synchronized with the appearance of tasks in the environment. The difference between τ^i and task appearance is $\bar{\tau}^i$, defined in the range $[-T^i/2, T^i/2]$.

CS achieves synchronization in two steps. First, each robot i evaluates the extend to which it is synchronized with the environment. This is measured by the fraction of time during which the robot finds tasks in its current workspace—see Sect 3.1. Second, each robot i modifies its internal timer τ^i and period T^i to synchronize with the environment—see Sect. 3.2 and 3.3. A robot i is synchronized with the environment when $T^i = T^{env}$ and $\bar{\tau}^i = 0$. Additionally, CS features a visual communication protocol to avoid physical interference between robots—again, see Sect. 3.2.

3.1   Assessment of Synchronization

Each robot i assesses its synchronization with the environment by measuring the correlation between its internal timer and the appearance of tasks. Robot i switches between workspaces every $T^i/2$, where T^i is updated by CS and is therefore not constant. Let l^i be a sequential number that identifies switches between workspaces for robot i, and let k_{l^i} be the moment in time at which switch l^i happens. Let $W^i[l^i]$ be the amount of time spent in a workspace by robot i between switch $l^i - 1$ and l^i, as opposed to transitioning between workspaces. See Fig. 2a-II.

Robot i can perform a task when it is in a workspace that contains available tasks. Let $w^i[k]$ a signal that takes the value 1 if robot i is working on a task at instant k and 0 if it is not. See Fig. 2a-III.

In order to assess its synchronization with the environment, robot i should ideally compute the correlation between the signal $s^i[k]$ of its internal timer and the signal $s^{env}[k]$ of the actual appearance of tasks:

$$g^i[k] = \begin{cases} 1 & \text{if } s^{env}[k] = s^i[k] \\ \\ 0 & \text{if } s^{env}[k] \neq s^i[k] \end{cases} \tag{2}$$

$g^i[k]$ can be integrated over a time interval yielding a cross-correlation by which robots can evaluate the similarity of the two signals during the chosen interval. Let $r^i[l^i]$ define the cross-correlation between these signals during $W^i[l^i]$:

$$r^i[l^i] = \frac{1}{W^i[l^i]} \sum_{\kappa=0}^{W^i[l^i]} g^i[k_{l^i} - \kappa] \tag{3}$$

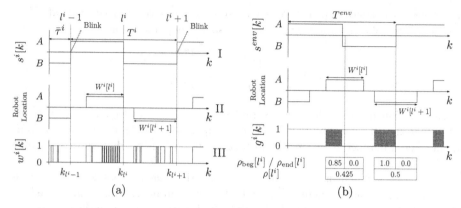

Fig. 2. Example of robot operation and cross-correlation of signals. a) Robot operation: I) signal $s^i[k]$ of the internal timer τ^i of robot i; II) robot location in the environment and amount of time W^i spent in a workspace (as opposed to transitioning between workspaces); III) work signal $w^i[k]$. b) Cross-correlation and partial cross-correlation, from top to bottom: signal $s^{env}[k]$ of task appearance; actual robot location; correlation of $s^{env}[k]$ and $s^i[k]$.

If robot i is perfectly synchronized with the environment, the two signals are identical; that is, $r^i[l^i] = 1$, $\forall l^i$. Typically, the two signals are not identical and $r^i[l^i]$ takes values lower than 1 for any value of l^i.

Unfortunately, $r^i[l^i]$ can not be directly measured because $s^{env}[k]$ is not known by the robots. Nevertheless, we can approximate $r^i[l^i]$ using the work signal $w^i[k]$: if robot i is performing a task, its internal timer and the location of task appearance match. Hence, we assume $w^i[k] \approx g^i[k]$. Let $\rho^i[l^i]$ define the cross-correlation between $s^i[k]$ and $w^i[k]$ during $W^i[l^i]$:

$$\rho^i[l^i] = \frac{1}{W^i[l^i]} \sum_{\kappa=0}^{W^i[l^i]} w^i[k_{l^i} - \kappa] \qquad (4)$$

Contrarily to $r^i[l^i]$, $\rho^i[l^i]$ can be measured by robot i and provides it with an estimated assessment of the synchronization between its internal timer and task appearance.

3.2 Synchronization of the Internal Timer

To achieve internal timer synchronization with $s^{env}[k]$, robot i uses the cross-correlation $\rho^i[l^i]$ defined in (4). We additionally define two partial cross-correlations:

$$\rho^i_{beg}[l^i] = \frac{2}{W^i[l^i]} \sum_{\kappa=W^i[l^i]/2}^{W^i[l^i]} w^i[k_{l^i} - \kappa]$$
$$\rho^i_{end}[l^i] = \frac{2}{W^i[l^i]} \sum_{\kappa=0}^{W^i[l^i]/2} w^i[k_{l^i} - \kappa] \qquad (5)$$

where $\rho_{beg}^i[l^i]$ is computed for the first half of $W^i[l^i]$ and $\rho_{end}^i[l^i]$ for the second. The comparison between these two quantities measures the balance of work between the two halves of $W^i[l^i]$: $\rho_{beg}^i \neq \rho_{end}^i$ indicates that robot i is working more during one half of $W^i[l^i]$ than the other. See Fig. 2b.

Robot i uses the relationship between ρ_{beg}^i and ρ_{end}^i to shift its internal timer as follows:

$$\Delta \tau^i[l^i] = \frac{W^i[l^i]}{2} \left(\rho_{beg}^i[l^i] - \rho_{end}^i[l^i] \right) \tag{6}$$

where $\Delta \tau^i[l^i]$ denotes the timer modifier of robot i at switch l^i, which occurs at the end of $W^i[l^i]$.

Collective Synchronization: Additionally to (6), we propose a mechanism for collective synchronization that is based on short-range line-of-sight communication. Communication is implemented using a simple visual protocol: a robot emits a light blink when its internal timer has finished a full cycle, thereby signaling to other robots that its timer is zero—see Fig. 2a-I. Other robots perceiving this light blink adjust their timers to achieve collective synchronization.

Upon perceiving a light blink, robot i shifts its internal timer as follows:

$$\Delta \tau^i = \begin{cases} 0.1 \left(\beta T^i - \tau^i \right) & \text{if } \tau^i \leq T^i/2 \\ 0.1 \left(T^i - \beta T^i - \tau^i \right) & \text{if } \tau^i > T^i/2 \end{cases} \tag{7}$$

where $\beta \in [0, 0.5)$ is a parameter.

In case $\beta T^i < \tau^i < T^i - \beta T^i$, robot i shifts its timer such that its reset point is closer to one of the emitting robot. This provokes a coupling and clustering of the timers. On the level of the swarm, the cluster of timers tends to synchronize with $s^{env}[k]$ because each timer is modified by (6). On the other hand, if $\tau^i < \beta T^i$ or $\tau^i > T^i - \beta T^i$, robot i shifts its timer so that its reset point is farther from the one of the emitting robot. This avoids that timers are too closely clustered, which would cause all robots to cross the transition area at the same time, thereby creating physical interference. Notice that in (7) there is no reference to an absolute time as robots do not share a common time reference and the visual communication protocol is asynchronous.

3.3   Period Synchronization

To achieve period synchronization of signals $s^i[k]$ and $s^{env}[k]$, robot i uses two statistics of the cross-correlation $\rho^i[l^i]$: the exponential moving average $avg\left(\rho^i[l^i]\right)$ and the variance $var\left(\rho^i[l^i]\right)$. These statistics are updated with the current value of the cross-correlation $\rho^i[l^i]$, using a memory factor η:

$$avg\left(\rho^i[l^i]\right) = \eta \, avg\left(\rho^i[l^i - 1]\right) + (1 - \eta)\rho^i[l^i]$$

$$var\left(\rho^i[l^i]\right) = \eta \, var\left(\rho^i[l^i - 1]\right) + (1 - \eta)\left(\rho^i[l^i] - avg\left(\rho^i[l^i]\right)\right)^2 \tag{8}$$

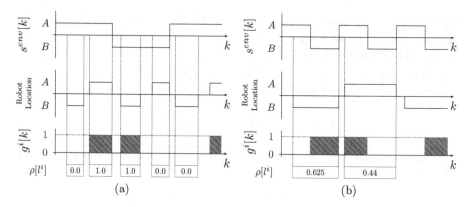

Fig. 3. Cross-correlation examples for a) $T^i < T^{env}$, where $avg\left(\rho^i[l^i]\right) \to 0.5$ and $var\left(\rho^i[l^i]\right) \to 0.25$ and b) $T^i > T^{env}$, where $avg\left(\rho^i[l^i]\right) \to 0.53$ and $var\left(\rho^i[l^i]\right) \to 0.008$. From top to bottom: location of task appearance ($s^{env}[k]$), robot location and cross-correlation.

with $\eta \in [0,1]$ being a configurable parameter. The higher η, the more relevant the current value. $avg\left(\rho^i[l^i]\right)$ measures the difference between T^i and T^{env}: the closer $avg\left(\rho^i[l^i]\right)$ to 1, the smaller the difference is. As $avg\left(\rho^i[l^i]\right)$ only indicates a difference, but not if T^i is shorter or longer, we use $var\left(\rho^i[l^i]\right)$ to measure the length of T^i in relation to T^{env}: if T^i is shorter than T^{env}, $var\left(\rho^i[l^i]\right)$ goes to 1. Figure Fig. 3 illustrates these statistics and their values in two example situations.

Robot i modifies the period of its internal timer $s^i[k]$ as follows:

$$\Delta T^i[l^i] = W^i[l^i]\left(f_{inc}\left(var\left(\rho^i[l^i]\right)\right) - f_{dec}\left(avg\left(\rho^i[l^i]\right)\right)\right) \tag{9}$$

where $f_{inc}\left(var\left(\rho^i[l^i]\right)\right)$ and $f_{dec}\left(avg\left(\rho^i[l^i]\right)\right)$ are positive functions that increase and decrease the period, respectively.

We use the following function $f_{inc}(var\left(\rho^i[l^i]\right))$ for increasing the period:

$$f_{inc}\left(var\left(\rho^i[l^i]\right)\right) = W^i[l^i]\alpha_{var}\left(var\left(\rho^i[l^i]\right)\right)^2 \tag{10}$$

where $\alpha_{var} \in \mathbb{R}$ is a configurable parameter that regulates the influence of $var\left(\rho^i[l^i]\right)$ on the period. We use the following function $f_{dec}\left(avg\left(\rho^i[l^i]\right)\right)$ for decreasing the period in this paper:

$$f_{dec}\left(avg\left(\rho^i[l^i]\right)\right) = W^i[l^i]\alpha_{avg}\left(1 - avg\left(\rho^i[l^i]\right)\right)^2 \tag{11}$$

where $\alpha_{avg} \in \mathbb{R}$ is a configurable parameter that regulates the influence of $avg\left(\rho^i[l^i]\right)$ on the period.

4   Experiments

We conduct the experiments in simulation using ARGoS [13]. ARGoS is a discrete-time physics-based simulation framework whose focus is the simulation of large robot swarms. The arena that we use in the experiments has the same layout as shown in Fig. 1a, with a length of 120 cm, a width of 60 cm, and a workspace width of 30 cm.

For our experiments, we use a swarm of 6 *e-puck* robots [12], which are randomly distributed in the transition area upon the start of the experiments. We use the following sensors of the e-puck: proximity sensors for obstacle avoidance, ground sensors to detect floor color, light sensor for phototaxis and the camera for task detection and visual communication. We use the wheel actuator with a maximum speed of 8 cm/s. Additionally, we use the LED actuator to implement the visual communication protocol.

We represent tasks using a device called *task allocation module* (TAM) [3]. A TAM represents a task to be executed by an e-puck robot at a given location and at a given moment in time. TAMs are programmable booths that signal the availability of a task to the robots through a set of color LEDs. A robot can work on the task that is represented by a TAM by driving into it and waiting inside until ξ_e has elapsed. We placed 10 TAMs in each workspace; hence, the task capacity of each workspace is $\Gamma = 10$. The time that a robot needs to perform a task is $\xi_e = 0.5$ s. The task life time is $\xi_l = 5$ s.

Robots use phototaxis to navigate: a light source identifies the right side of the arena. Robots navigate towards the light to work in workspace B, and do the opposite to work in workspace A. Robots can detect the workspace they are in by reading the color of the floor. When a robot is in a workspace, it perceives the available tasks by the color of the LEDs of the nearby TAMs. Robot i can calculate $W^i[j]$ by measuring the time at which it arrives in a workspace and the time at which the internal timer switches to the other workspace. The robot can sense the work signal $w^i[k]$ by the color of the TAMs in its current workspace.

The parameters of the environment are the incoming task rate $\lambda = 10$ tasks/s and the period $T^{env} = 80$ s. The configurable parameters of CS used for this example are $\eta = 0.65$, $\alpha_{avg} = 0.58$ and $\alpha_{var} = 41.92$. These values have been obtained through a tuning process using I/F-Race [1,10]. The parameter $\beta = 0.0375$ has been obtained by exhaustive search. Initially, the period T^i of each robot i is uniformly sampled from the interval $[40, 240]$, and the initial time difference between the internal timers and the task appearance $\bar{\tau}^i$ is uniformly sampled from the interval $[-T^i/2, T^i/2]$.

Figure 4 shows the development of T^i, $\bar{\tau}^i$ and number of tasks performed over the duration of the experiment. The synchronization of the periods in the swarm is shown in Fig. 4a. Notice that T^{env} is constant during the experiment. We can observe that every T^i converges to T^{env} in the first 2500 s. The timer synchronization is shown in Fig. 4b. We can observe that all $\bar{\tau}^i$ converge to zero. The number of tasks performed by the swarm during the experiment is a cumulative metric shown in Fig. 4c. We define the *performance rate* as the number of tasks performed per second which is the derivate of this metric. In this example, the robots achieve a performance rate of 0.65 tasks/s.

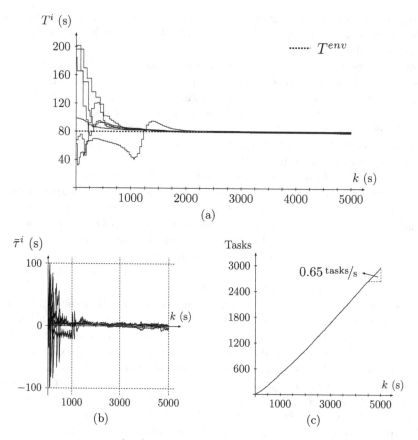

Fig. 4. Development of T^i, $\bar{\tau}^i$ and number of tasks performed over the duration of the experiment. a) Periods T^i of the robots compared to the period T^{env}. b) Time difference $\bar{\tau}^i$ between the internal timers and the task appearance. c) Number of tasks performed and final performance rate.

In order to analyze the performance of CS, we compare it with two other algorithms:

- No-synchronization algorithm (NS): robots using NS have internal timers for switching between areas, but do not attempt to synchronize with the environment or with other robots. This means that $\bar{\tau}^i$ and T^i, which are randomly initialized, remain unchanged throughout the experiment. This represents the initial situation of CS. The comparison of CS with NS allows us to quantify the improvement obtained by synchronizing.
- Greedy algorithm (GR): robots using GR do not switch between workspaces depending on an internal timer, but on task availability. To this end, robots switch workspace with a given probability in case they to not find tasks in their current workspace. The comparison of CS with GR allows us to observe the difference in task performance between a temporal task allocation algorithm and an algorithm that is commonly used in task allocation.

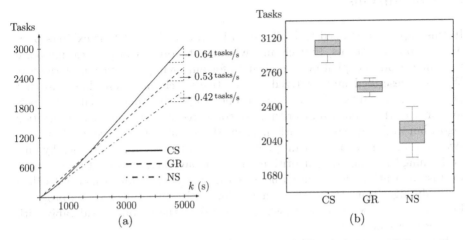

Fig. 5. Comparison of the performance of CS, NS and GR. a) Average of the number of tasks performed during the experiments; and final performance rate. b) Boxplot of the number of task performed by each algorithm during the experiment, based on 15 repetitions per algorithm. The whiskers represent the lowest value still within $1.5\,IQR$ of the lower quartile, and the highest value still within $1.5\,IQR$ of the upper quartile.

The comparison between algorithms is based on the number of tasks performed in experiments of 5000 s. We perform 15 experiments for each algorithm.

The average number of tasks performed by the swarm, calculated every second for each algorithm, is shown in Fig. 5a. Notice that the performance rates of NS and GR remain constant during the experiment because these algorithms do not implement a synchronization process. We can observe that GR has a higher performance rate than NS. This implies that a temporal task allocation algorithm that is not perfectly synchronized performs less tasks than a greedy algorithm. Furthermore, we can observe that the performance rate of CS increases over time due to the synchronization process. At the beginning of the experiments, CS and NS have a similar performance rate. After 800 s, the performance rate of CS increases and eventually the number of tasks performed by CS exceeds the number of tasks performed by NS. Similarly, the number of tasks performed by CS exceeds the number of tasks performed by GR after 1600 s.

Figure 5b shows a boxplot representation of the number of task performed at the end of the experiments. The NS algorithm has the highest dispersion because there is no synchronization; the results strongly depend on the initial conditions. CS and GR have a lower dispersion than NS as they adapt the behavior of the robots to the environment, reducing the dependence on the initial conditions. Notice that the lower whisker of CS is longer than the upper whisker of the GR algorithm. This implies that the previous deductions made on Fig. 5a are valid.

5   Conclusions

In this paper, we studied task allocation in an environment that exhibits periodic temporal patterns. In such an environment, robots can perform temporal task allocation to exploit synchronization for improving their task performance. We have described and analyzed a collective synchronization algorithm that performs temporal task allocation for robot swarms. In order to analyze the performance of the proposed algorithm, we compared it with a no-synchronization algorithm and a greedy algorithm. From the results, we can conclude that a swarm using our algorithm can synchronize with the environment, thereby outperforming the competing algorithms. The comparison also shows that a temporal task allocation algorithm without synchronization performs less tasks than a greedy algorithm. However, an algorithm with synchronization as proposed by us increases the number of tasks performed by the robots considerably with respect to a reactive behavior.

In this paper, we applied the concepts of synchronization and signal processing to task allocation in swarm robotics, with satisfactory results. In the immediate future, we plan to study the proposed approach on a swarm of real robots. The potential of our approach opens several possible directions for future research. One is to study environments that exhibit more complex temporal patterns, for example, environments in which the task appearance is not only a square signal but a signal with multiple frequency components. Another direction is the application of this approach to other resource allocation problems such as energy management. For example, they can be used to organize the consumption of a scarce energy resource by a swarm of robots or other autonomous agents such as electric vehicles.

Acknowledgments. M. Castillo-Cagigal is sponsored by the Spanish Ministry of Education with a PhD grant (FPU-2010). Arne Brutschy and Mauro Birattari acknowledge support from the Belgian F.R.S.–FNRS.

References

1. Balaprakash, P., Birattari, M., Stützle, T.: Improvement strategies for the F-Race algorithm: Sampling design and iterative refinement. In: Bartz-Beielstein, T., Blesa Aguilera, M.J., Blum, C., Naujoks, B., Roli, A., Rudolph, G., Sampels, M. (eds.) HCI/ICCV 2007. LNCS, vol. 4771, pp. 108–122. Springer, Heidelberg (2007)
2. Brambilla, M., Ferrante, E., Birattari, M., Dorigo, M.: Swarm robotics: A review from the swarm engineering perspective. Swarm Intelligence 7(1), 1–41 (2013)
3. Brutschy, A., Garattoni, L., Brambilla, M., Francesca, G., Pini, G., Dorigo, M., Birattari, M.: The TAM: abstracting complex tasks in swarm robotics research. Tech. Rep. TR/IRIDIA/2014-006, IRIDIA, Université Libre de Bruxelles, Belgium (2014)
4. Christensen, A.L., O'Grady, R., Dorigo, M.: From fireflies to fault-tolerant swarm of robots. IEEE Transactions on Evolutionary Computation 13(4), 754–766 (2009)

5. Dorigo, M., Birattari, M., Brambilla, M.: Swarm robotics. Scholarpedia 9(1), 1463 (2014)
6. Glass, L.: Synchronization and rhythmic processes in physiology. Nature 410(6825), 277–284 (2001)
7. Holland, O., Melhuish, C., Hoddell, S.E.J.: Convoying: using chorusing for the formation of travelling groups of minimal agents. Robotics and Autonomous Systems 28, 207–216 (1999)
8. Kuramoto, Y.: Chemical Oscillations, Waves, and Turbulence. Springer, Berlin (1984)
9. Liu, F., Picard, R.W.: Finding periodicity in space and time. In: Proceedings of the 6th International Conference on Computer Vision, pp. 376–383. IEEE Computer Society, Los Alamitos (1998)
10. López-Ibáñez, M., Dubois-Lacoste, J., Stützle, T., Birattari, M.: The irace package, iterated race for automatic algorithm configuration. Tech. Rep. TR/IRIDIA/2011-004, IRIDIA, Université Libre de Bruxelles, Belgium (2011)
11. Martin, H.J.A., de Lope, J., Maravall, D.: Adaptation, anticipation and rationality in natural and artificial systems: computational paradigms mimicking nature. Natural Computing 8(4), 757–775 (2009)
12. Mondada, F., Bonani, M., Raemy, X., Pugh, J., Cianci, C., Klaptocz, A., Magnenat, S., Zufferey, J.C., Floreano, D., Martinoli, A.: The e-puck, a robot designed for education in engineering. In: Gonçalves, P.J.S., et al. (eds.) Proceedings of the 9th Conference on Autonomous Robot Systems and Competitions, pp. 59–65. IPCB: Instituto Politècnico de Castelo Branco, Portugal (2009)
13. Pinciroli, C., Trianni, V., O'Grady, R., Pini, G., Brutschy, A., Brambilla, M., Mathews, N., Ferrante, E., Di Caro, G., Ducatelle, F., Birattari, M., Gambardella, L.M., Dorigo, M.: ARGoS: a modular, parallel, multi-engine simulator for multi-robot systems. Swarm Intelligence 6(4), 271–295 (2012)
14. Rosu, D., Schwan, K., Yalamanchili, S., Jha, R.: On adaptive resource allocation for complex real time applications. In: Proceedings of the 18th Real-Time System Symposium, pp. 320–329. IEEE Computer Society, Los Alamitos (1997)
15. Strogatz, S.H.: From Kuramoto to Crawford: exploring the onset of synchronization in populations of coupled oscillators. Physica D: Nonlinear Phenomena 143(1-4), 1–20 (2000)

Towards a Cognitive Design Pattern for Collective Decision-Making

Andreagiovanni Reina[1], Marco Dorigo[1], and Vito Trianni[2]

[1] IRIDIA, CoDE, Université Libre de Bruxelles, Brussels, Belgium
[2] ISTC, Italian National Research Council, Rome, Italy
{areina,mdorigo}@ulb.ac.be, vito.trianni@istc.cnr.it

Abstract. We introduce the concept of *cognitive design pattern* to provide a design methodology for distributed multi-agent systems. A cognitive design pattern is a reusable solution to tackle problems requiring cognitive abilities (e.g., decision-making, attention, categorisation). It provides theoretical models and design guidelines to define the individual control rules in order to obtain a desired behaviour for the multi-agent system as a whole. In this paper, we propose a cognitive design pattern for collective decision-making inspired by the nest-site selection behaviour of honeybee swarms. We illustrate how to apply the pattern to a case study involving spatial factors: the collective selection of the shortest path between two target areas. We analyse the dynamics of the multi-agent system and we show a very good agreement with the predictions of the macroscopic model.

1 Introduction

Several recent studies describe swarm systems as information-processing systems capable of some cognitive ability, which is strongly determined by the interaction patterns among the system components [4,14]. In this paper, we propose to take a similar perspective in the design of large-scale distributed systems. The studies mentioned above suggest that—to a large extent—the cognitive processing of natural decentralised systems takes place in inter-individual interactions, therefore limiting the need to postulate explicit representations within the single units. By viewing artificial swarm systems as distributed cognitive systems, it will be possible to maximise their information processing capability while keeping a low complexity of the individual units. That is, individual units would contribute to the overall system behaviour without having the global picture about the cognitive process they are collectively producing.

Designing such an information-processing system is clearly a complex endeavour, and in general, modelling, predicting and controlling large-scale distributed systems are complex tasks. Therefore, a successful design methodology for such systems should be grounded on solid theoretical premises. We propose a design methodology that leverages the current understanding of cognitive processing in (natural) distributed systems, and that puts this knowledge in use for the

M. Dorigo et al. (Eds.): ANTS 2014, LNCS 8667, pp. 194–205, 2014.
© Springer International Publishing Switzerland 2014

design of artificial ones. Our proposal is based on the concept of *cognitive design patterns*, that is, reusable solutions to tackle problems requiring cognitive processing (e.g., collective decision-making among multiple alternatives). Similarly to common practice in software engineering [5], these design patterns can be used to guide the design and development of distributed cognitive processes, independently of the particular implementation technique. The idea of using design patterns in distributed systems has already been partially explored, but previous studies are not grounded on the theoretical understanding of collective dynamics [6,1,12]. Our proposal aims at providing general solutions grounded on the principled understanding of the basic mechanisms underlying cognitive processing.

In this paper, we propose a cognitive design pattern for collective decision-making. We refer to decision-making as the process of choosing the best option among a (finite, possibly unknown) number of different alternatives. This process requires the estimation of the quality of the available alternatives (possibly with uncertainty) to select the best one. Recent studies have identified optimal decision strategies in decentralised systems [8,13,11].We propose a cognitive design pattern hinged on these studies, to be applied to artificial distributed systems. Although decision-making has been largely studied in this context [2,7,9,12], general purpose solutions are still missing. Our work aims at filling this gap by providing a design methodology that can be applied to several application domains.

The cognitive design pattern we propose is composed of the following elements: problem, inspiration, solution and case study. The collective decision-making problem and its relevance in artificial distributed systems have been discussed above. The biological inspiration is presented in Section 2, in which we discuss nest-site selection in honeybees and the related theoretical models accounting for the collective decision-making process. From these models, we derive the solution, which is discussed in Section 3 along with the causal relationship between the microscopic and macroscopic levels. To ease the pattern comprehension, we instantiate the design pattern in a case study presented in Section 4. In Section 5, we show the agreement between the collective decision implemented following the design pattern and the macroscopic model predictions. In Section 6, we discuss the proposed methodology and identify directions for further improvements.

2   Biological Inspiration and Theoretical Models

A remarkable example of collective decision-making is given by honeybee swarms during nest site selection. In spring, honeybee swarms reproduce by colony fission: the queen bee and several thousand workers leave the parent hive and create a cluster in the neighbourhood. Several hundred *scout bees* start searching for new potential nest sites, and return to the swarm to advertise through waggle dances what they have discovered. A number of alternatives may be discovered during the selection process, and a consensus decision is necessary to lead the

whole swarm to the best one. Decision-making is based on peer-to-peer interactions among bees: scouts committed to a potential site recruit other scouts.
Additionally, scouts have a certain probability of spontaneously abandoning com-
mitments. As a consequence, a competition between populations committed to
different sites takes place, eventually leading to a quorum of individuals committed for the site that is finally chosen. Recently, cross-inhibition between different
populations has been discovered. This is implemented through a *stop signal* that
scout bees selectively deliver to nest-mates advertising for a different option [13].
A bee receiving several stop signals abandons the recruitment and becomes uncommitted. Thanks to the stop signal, poor-quality sites are quickly abandoned
in favour of better ones. Most importantly, the stop signal allows to break decision deadlocks when same-quality alternatives are available, leading to a random
decision for one of the two. In this way, the system can optimise the decision
making, resulting in a choice that maximises the colony reward.

An analytical model of the nest-site selection process has been developed and
confronted with empirical results, confirming the existence of both positive and
negative feedback loops that determine the collective decision [13]. The model
describes the decision-making process in a binary-choice scenario. The swarm
is composed of N individuals (e.g., the scout bees), which can belong to three
different groups: uncommitted individuals (population U with size N_U), and
individuals committed to one of the alternatives (respectively population A and
B, with sizes N_A and N_B). A continuous-time Markov process describes the way
in which individuals switch between populations. Four types of transitions are
sufficient: *discovery, abandonment, recruitment* and *cross-inhibition*.

Uncommitted individuals spontaneously *discover* and become committed to
the alternative i at the rate γ_i:

$$\begin{aligned} \langle N_U, N_A, N_B \rangle &\xrightarrow{\gamma_A} \langle N_U^-, N_A^+, N_B \rangle \\ \langle N_U, N_A, N_B \rangle &\xrightarrow{\gamma_B} \langle N_U^-, N_A, N_B^+ \rangle \end{aligned} \tag{1}$$

where N_i^+ and N_i^- represent an increment or a decrement in population i. Committed individuals *abandon* the alternative i and thus get uncommitted at the
rate α_i:

$$\begin{aligned} \langle N_U, N_A, N_B \rangle &\xrightarrow{\alpha_A} \langle N_U^+, N_A^-, N_B \rangle \\ \langle N_U, N_A, N_B \rangle &\xrightarrow{\alpha_B} \langle N_U^+, N_A, N_B^- \rangle \end{aligned} \tag{2}$$

Individuals from population i actively *recruit* uncommitted ones at the rate
$\rho_i N_i/N$ proportional to the recruiting population size:

$$\begin{aligned} \langle N_U, N_A, N_B \rangle &\xrightarrow{\rho_A N_A/N} \langle N_U^-, N_A^+, N_B \rangle \\ \langle N_U, N_A, N_B \rangle &\xrightarrow{\rho_B N_B/N} \langle N_U^-, N_A, N_B^+ \rangle \end{aligned} \tag{3}$$

Finally, individuals from population i actively *inhibit* individuals of population
j at the rate $\sigma_i N_i/N$ proportional to the inhibiting population size:

$$\begin{aligned} \langle N_U, N_A, N_B \rangle &\xrightarrow{\sigma_A N_A/N} \langle N_U^+, N_A, N_B^- \rangle \\ \langle N_U, N_A, N_B \rangle &\xrightarrow{\sigma_B N_B/N} \langle N_U^+, N_A^-, N_B \rangle \end{aligned} \tag{4}$$

Here, all transition rates—$\gamma_i, \alpha_i, \rho_i, \sigma_i$—are greater than zero. It is worth noting that this model does not require any explicit comparison of the alternatives' quality by the single individuals. The quality value of the two alternatives—hereafter labelled v_A and v_B—is instead encoded in the transition rates (e.g., through value-dependent discovery or recruitment rates [11]): different-quality alternatives correspond to biased transition rates, while same-quality alternatives to unbiased ones. Overall, the collective decision is based purely on the system dynamics resulting from individual-to-individual interactions.

Starting from this stochastic model, it is possible to obtain the continuous-time master equation that describes how the probability of being in each state evolves over time. In the limit of large N, it is possible to extract a mean-field, population-level model of the system dynamics [13]. This takes the form of two coupled ordinary differential equations that describe the dynamics of the fraction $\Psi_i = N_i/N$ of individuals belonging to population $i \in \{U, A, B\}$:

$$\begin{cases} \dot{\Psi}_A = \Psi_U(\gamma_A + \rho_A\Psi_A) - \Psi_A(\alpha_A + \sigma_B\Psi_B) \\ \dot{\Psi}_B = \Psi_U(\gamma_B + \rho_B\Psi_B) - \Psi_B(\alpha_B + \sigma_A\Psi_A) \end{cases} \tag{5}$$

where $\Psi_U = 1 - \Psi_A - \Psi_B$. An extensive analysis of this model showed that the cross-inhibition rates σ_i crucially determine the dynamics of decision-making (see [11]). In case of same-quality alternatives ($v_A = v_B$), the transition rates for different alternatives have the same value (i.e., $\gamma_i = \gamma$, $\rho_i = \rho$, $\alpha_i = \alpha$ and $\sigma_i = \sigma$). In this case, for low rates of cross-inhibition, the system remains deadlocked at indecision with an equal number of individuals committed to either alternative ($\Psi_A = \Psi_B$). Through linear stability analysis [13], it is possible to identify the cross-inhibition level for which the system breaks the deadlock and converges to the choice of one alternative. The working region is $\{\rho > \alpha, \sigma > \sigma^*\}$, with critical value:

$$\sigma^* = \frac{4\alpha\gamma\rho}{(\rho - \alpha)^2}. \tag{6}$$

Additionally, in case of options of different quality, the cross-inhibition rate determines the minimum quality difference between the two alternatives to break the symmetry and make a systematic choice [11]. Therefore, there exist parameterisations of the system that allow to obtain accurate decisions when the quality of the options differs sufficiently, or random decisions when the values are similar.

3   Design Guidelines

The models discussed above provide a link between the individual-level description—given by the continuous time Markov process—and the population-level dynamics—given by the dynamical system in (5). However, the models alone are not sufficient to guide the implementation of a distributed multi-agent system. In fact, the transition rates depend on several factors: they incorporate global knowledge about the populations size, which may not be available to the agents, and they embed spatial and topological factors that partially determine

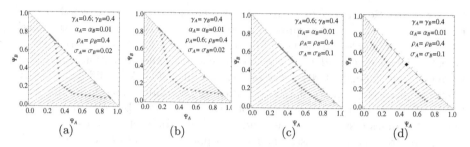

Fig. 1. Macroscopic dynamics for different parameterisations (trajectories and equilibrium points, shown as light green triangles for stable points and dark blue rhombus for unstable saddle points)

the probability of interaction among agents. For design purposes, it is therefore necessary to identify the causal relationship between microscopic transitions and macroscopic dynamics, and to provide a mechanistic description of the working principles.

The analysis of the macroscopic model in (5) reveals that, when the two alternatives have different value (e.g., $v_A > v_B$), an unbalanced agent distribution between the two populations is obtained thanks to a similarly biased commitment rate. This can be obtained through either discovery or recruitment (e.g., $\gamma_A > \gamma_B$ or $\rho_A > \rho_B$). Different discovery rates directly lead to a distribution of agents between the two alternatives that follows the rate ratio (see for instance Figure 1(a)). Similarly, a difference in the recruitment rates results in a unbalanced distribution even when the discovery rates are equal (see Figure 1(b)). This is due to the positive feedback mechanism that can be noted in (3), which states that the recruitment rate is proportional to the recruiting population size: the larger the population, the stronger the recruitment for the same population. Conversely, cross-inhibition provides a negative feedback loop that reduces the size of a population committed to an alternative, as stated in (4). Crucially, the rate of cross-inhibition is proportional to the size of the inhibiting population, and therefore contributes to the creation of an unbalanced distribution of individuals between committed populations, even for unbiased inhibition rate (i.e., $\sigma_A = \sigma_B$, see Figure 1(c)). This is true also for same-quality alternatives (i.e., $v_A = v_B$). In this symmetric case, discovery, abandonment and recruitment are equal and are therefore not sufficient to break the symmetry. However, a sufficient level of cross-inhibitio (i.e., $\sigma > \sigma^*$) makes the equilibrium point unstable, therefore leading to a symmetry breaking, as shown in Figure 1(d).[1]

This mechanistic description clarifies the working regimes and suggests how the transition rates should be chosen to obtain the desired macroscopic behaviour. However, to guide the implementation of a distributed multi-agent

---

[1] A full characterisation of the parameter space is out of the scope of the present paper, and is object of ongoing studies. The interested reader can find a dynamical systems analysis for the parameter σ in [11].

system it is also necessary to define the main features of the individual agent behaviour and of the agent-to-agent interactions in order to obtain the transition rates in the appropriate range. In doing this, spatial and topological factors need to be suitably taken into account. The definition of these guidelines should provide the minimal requirements to obtain the desired system behaviour, and should incorporate the knowledge gained from the theoretical models.

In the particular case of collective decision making, we define general-purpose design guidelines as follows:

(i) *discovery*: an uncommitted agent must commit to the alternative i with probability per unit time $P_{\gamma,i}$ (possibly proportional to the value v_i);

(ii) *abandonment*: an agent committed to the alternative i must become uncommitted with probability $P_{\alpha,i}$ (possibly inversely proportional to v_i);

(iii) *recruitment*: an uncommitted agent interacting with an agent committed to any alternative i must commit to i with probability $P_{\rho,i}$ (possibly proportional to the value v_i);

(iv) *cross-inhibition*: an agent committed to alternative i that interacts with an agent committed to alternative $j \neq i$ must become uncommitted with fixed probability P_σ;

(v) *interaction*: the system must be *well-mixed*, that is, the probability of interaction between any two agents is uniform.

Note that *cross-inhibition* is asymmetric, that is, only one of the two agents (e.g., the one initiating the interaction) can change commitment state, this way respecting the stop signals mechanism. These guidelines are sufficient to generate a collective decision, provided that the agent probabilities of changing the commitment state are in the correct range. In the following section, we follow the design guidelines to implement a multi-agent system.

4   A Simple Spatial Scenario

We introduce a simple, spatial multi-agent scenario to demonstrate the application of the cognitive design pattern for collective decision making. The simplicity of the case study eases the analysis and the comprehension of the implemented mechanisms. At the same time, the studied scenario preserves the relevant ingredients of the collective decision making, therefore can well represent application scenarios where spatiality influences the system dynamics.

We study the collective choice of the shortest path between two alternatives in a 1D space: agents move on a circle and need to collectively select and exploit the shortest path between two *target areas* (see a pictorial representation in Figure 2). Two alternatives are possible: the upper and the lower path, respectively labelled A and B. The angle θ between the target areas defines the decision problem: the best alternative is A for $\theta < \pi$, B for $\theta > \pi$ and any of the two for $\theta = \pi$. To identify and exploit a path, agents need to navigate back and forth between target areas. We assume that agents can move at maximum angular speed $\omega = \pi/18 \ s^{-1}$, and that movements are subject to noise modelled as a Gaussian displacement per arc degree following a $\mathcal{N}(0, \xi)$ distribution, with $\xi = \pi/4500$. During navigation, agents track the angular distance of the two areas through dead reckoning, and use their estimates to attain previously visited areas without exploring or sensing the environment. However, due to the movement noise, position estimates are subject to cumulative errors. As a consequence, agents may end up with incorrect information and may be unable to attain a target area. Finally, agents have a sensing range of $\beta = \pi/36$ within which they can identify target areas and interact with other agents. All agents start with no knowledge about the target areas, and are therefore uncommitted.

Given the above specifications, we have developed the agent behaviours and interactions following the cognitive design pattern, as stated below.

(i) *Discovery*: an agent explores the environment through a correlated random walk with persistence rate $\lambda = 0.8$ [3], and gets committed to a path as soon as it stores the position of the two target areas, allowing to navigate back and forth between them on the discovered path. Here, the probability $P_{\gamma,i}$ is determined by spatial factors (such as position and size of the areas) and by the parameters ω and λ. We obtain $\gamma_i \propto v_i$ because shorter paths are easier to discover through random walk.

(ii) *Abandonment*: an agent abandons its commitment if it fails to attain a target area due to errors in the position estimates. Consequently, it erases the stored locations and resumes exploration. Also in this case, $P_{\alpha,i}$ is not directly under control of the agent's behaviour, but depends on the parameter β and on the movement noise variance ξ. Here, we obtain $\alpha_i \propto 1/v_i$ because lower abandonment rates result from smaller cumulative error on shorter paths: the variance increases proportionally to the path length.

(iii) *Recruitment*: an uncommitted agent that interacts with an agent committed to alternative i gets recruited with fixed probability P_ρ: it receives the location of the target areas and transforms them in its own reference frame.

(iv) *Cross-inhibition*: an agent committed to alternative i that interacts with an agent committed to alternative $j \neq i$ becomes uncommitted with fixed probability P_σ: it erases the stored locations and resumes exploration.

(v) *Interaction*: to provide an equal probability of interaction with agents exploiting different paths, interactions are possible only when agents are within the same target area. Each agent has a maximum of one interaction per time unit. Additionally, agents remain in the target area with

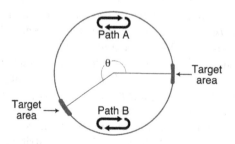

Fig. 2. A graphical representation of the multi-agent scenario. The monodimensional environment is a circle in which the agents move on the circumference line to navigate back and forth between the two target areas.

probability $P_s = 0.9$ per time unit, or until a state change. This helps in creating well-mixed conditions and also increases the interaction rate.

Note that we have not specified a direct way to control the transition rates for discovery and abandonment, while recruitment and cross-inhibition are determined by the control probabilities P_ρ and P_σ. We choose fixed probabilities independently of the possible differences in the path lengths. As discussed in Section 3, this should be sufficient to produce a collective choice, provided that the discovery rates are biased toward the best option. Additionally, a sufficient level of cross-inhibition will contribute to make a collective choice and to break decision deadlocks (see Figure 1). To simplify the system analysis, we fix $P_\rho = P_\sigma = P$, which we refer to as the *interaction probability*. We study the system behaviour varying P and θ, while the other parameters are kept constant.

5   Results

To verify the correctness of the design pattern and to study how the collective behaviour changes as a function of the interaction probability P and of the decision problem given by θ, we check the adherence of the multi-agent system with the macroscopic model. To this purpose, we statistically estimate the transition rates γ_i, α_i, ρ_i and σ_i directly from the multi-agent system. Parameter estimation is performed through survival analysis, which permits to estimate how the probability of an event changes over time directly from the experimental data. Using the Nelson-Aalen estimator [10], we computed the hazard curve for the cumulative number of expected events (e.g., the discovery of alternative A):

$$H(t) = \sum_{t_i \le t} d_i/n_i, \tag{7}$$

where d_i is the number of events recorded at t_i and n_i is the number of events occurring at $t \ge t_i$ (or not occurring at all, e.g., censored cases). In a memoryless process, the instantaneous transition rate is constant over time, therefore events cumulate at a constant rate, which corresponds to the hazard curve being a line. Its slope represents the estimated transition rate (see for instance Figure 3(a)). By employing a survival analysis, we can at the same time estimate all the

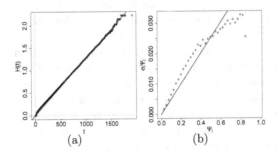

(a)   (b)

Fig. 3. (a) Hazard curve to estimate the transition rate α_A for $\theta = \pi$ and $P = 0.1$. (b) Transition $\sigma_i \Psi_i$ as function of Ψ_i for $i = A, B$, $\theta = \pi$ and $P = 0.1$. Points are unnormalised estimations, and the fitting line slope represents the normalised transition rate.

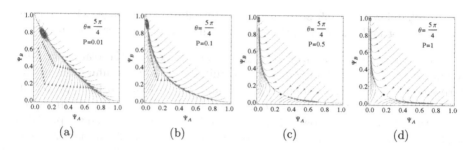

Fig. 4. Comparison between macroscopic dynamics (trajectories and fixed points, shown as light green triangles for stable points and dark blue rhombus for unstable ones) and the multi-agent simulations (final repartition of agents between the two populations, shown as red empty dots) for the asymmetric case $\theta = 5\pi/4$

transition rates and check the Markov assumption by looking at the shape of the hazard curves. Indeed, even if the agent behaviour is purely memoryless, departure from the Markov assumption is possible due to the spatial factors.

We perform $M = 300$ multi-agent simulations with $N = 400$ agents, and we vary both P and θ. The events we consider are the changes in the commitment status of each agent resulting from discovery, abandonment, recruitment and cross-inhibition, for both alternatives A and B. Discovery and abandonment are transitions spontaneously triggered by an agent, therefore it is possible to directly estimate the parameters γ_i and α_i from the corresponding hazard curve (e.g., Figure 3(a)). Conversely, recruitment and cross-inhibition result from the interaction of agents belonging to different populations, and the transition rates are proportional to the size of the recruiting and inhibiting populations. In this case, it is necessary to first estimate the transition rates for each population fraction (i.e., we estimate $\rho\Psi_i$ and $\sigma\Psi_i$), and then normalize for Ψ_i to obtain ρ and σ (see for instance Figure 3(b)). In this study, we limit the number of different events to consider to $N/10$ by approximating the population fractions within fixed-width windows of 0.025.

We first consider the asymmetric case of $\theta = 5\pi/4$. In this case, the decision problem should lead to the systematic choice of the alternative B. In Figure 4, we show the dynamics of the macroscopic model of equation (5) with the parameters estimated from the multi-agent system. We note that for low values of the interaction probability P there exists a single stable fixed point for $\Psi_B > 0.8$, and all trajectories converge to it. The agreement between the multi-agent system and the macroscopic dynamics is remarkable: the final distribution of agents from the simulations perfectly matches the model predictions (see Figure 4(a) and 4(b)). For higher values of P, the macroscopic dynamics show that the system undergoes a bifurcation, and a second stable fixed point appears that corresponds to the choice of the inferior option (see Figure 4(c) and 4(d)). The basin of attraction of the inferior fixed point is however smaller, and the trajectories starting from the origin (i.e., when all agents are uncommitted) always lead to

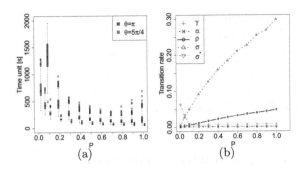

Fig. 5. (a) Convergence time for symmetric and asymmetric cases as a function of the interaction probability P, with decision threshold at $\Psi_i = 0.7$ and $t_{max} = 2000s$. (b) Transition rates as a function of interaction probability P for the symmetric case, with $\theta = \pi$.

the selection of the best option. This is confirmed by the multi-agent simulations, as all the repetitions resulted in a systematic choice of the alternative B. The bifurcation observed in the macroscopic dynamics appears when cross-inhibition is sufficiently strong compared to the other transition rates. On the one hand, this may lead to errors in the decision making if the system happens to be in the basin of attraction of the inferior choice. On the other hand, as also noted in [11], larger cross-inhibition rates lead to increased decision speed (see Figure 5(a)). Similar dynamics can be observed for different values of $\theta < \pi$, for which we observed that smaller differences between the alternatives sometimes lead to the wrong choice, as predicted by the macroscopic model (see supplementary material http://iridia.ulb.ac.be/supp/IridiaSupp2014-005/sm.pdf).

In Figure 6 we show the case for $\theta = \pi$, which corresponds to equal alternatives and—potentially—to a decision deadlock. This is actually the case for very low values of the interaction probability (e.g., $P = 0.01$ shown in Figure 6(a)). In this case, the model predicts a single stable fixed point for $\Psi_A = \Psi_B$, in agreement with the multi-agent simulations that equally remain deadlocked at indecision (see the red dot-cloud around the fixed point in Figure 6(a)). However, a phase transition is observed for increasing interaction probability, corresponding to a higher cross-inhibition rate and therefore to the ability of breaking the symmetry: two stable solutions appear indicating a collective choice for either A or B. The accordance between the multi-agent simulations and the macroscopic dynamics is very good also in this case, as shown in Figure 6(b)-(d). Note also that the macroscopic dynamics are highly symmetric, in accordance with the underlying multi-agent system. Similarly to the asymmetric case, we observe that higher values of P lead to a more definite choice of one or the other option, and that the convergence speed is also increased (see Figure 5(a)).

Figure 5(b) shows how the estimated transition rates vary with respect to the interaction probability P for the symmetric case. While discovery and abandonment remain roughly constant, both ρ and σ increase quasi-linearly with P, indicating that a higher probability of interaction among agents directly translates in increased recruitment and cross-inhibition rates. We note that the estimated cross-inhibition rate is initially below the critical value ($\sigma < \sigma^*$) for small interaction probabilities ($P < 0.07$). These are actually the values at which the multi-agent simulations remain deadlocked at indecision. For larger P, cross-inhibition is sufficiently high and the collective decision is efficiently performed.

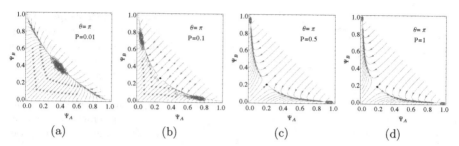

Fig. 6. Macroscopic dynamics vs. multi-agent simulations for the symmetric case $\theta = \pi$

6   Conclusions

Overall, the results we obtained confirm that the multi-agent system that we implemented following the guidelines given by the cognitive design pattern corresponds very well to the reference macroscopic model. This result is non-trivial due to the differences between the two levels: while the model is continuous and deterministic, the multi-agent system, due to the finite swarm size, is discrete and stochastic. We have exploited the understanding of the basic mechanisms underlying the collective decision-making process in order to perform important implementation choices, such as the use of a fixed interaction probability independent of the decision problem to be faced. This minimalistic choice would not have been safe without knowledge about the system dynamics and about the correspondence between microscopic rules and macroscopic behaviour. Similarly, the implementation choices to grant a uniform interaction probability among agents—and therefore the *well-mixed* property—are also a result of the design pattern guidelines, which allowed to pinpoint the important aspects to be considered (e.g., the need to limit peer-to-peer interactions in a location containing a good sample of the population distribution). The parameter estimation we performed and the subsequent analysis for varying P and θ suggest that our implementation results in a well-behaved system not violating any assumption, despite the spatial factors that hinder the adherence to a Markov process.

The main feature of the design pattern we have developed consists in the possibility to perform decisions with minimal complexity at the individual level. Indeed, the only requirement is that agents can interact and recognise that their peers have a different opinion. Quality comparison is not necessary, which allows to implement the system in a large number of possible applications.

The present paper represents the very first step toward the definition and formalisation of cognitive design patterns for swarm systems. Several aspects must be investigated further in order to provide a proper engineering methodology. For what concerns collective decision-making mechanisms, future work will be dedicated to the characterisation of the full parameter space, in order to identify the parameter ranges that result in desired macroscopic dynamics. Additionally, we will characterise the relationship between individual-level parameters (e.g., the interaction probabilities P_ρ and P_σ) and the corresponding transition rates (e.g., ρ and σ). Another issue to be considered is the effect of spatial and topological constraints on the collective decision process. In some preliminary studies, we have observed that spatiality influences the macroscopic dynamics (e.g., violating the well-mixed condition), and therefore needs to be characterised properly in order to provide guidelines re-usable in multiple domains.

Finally, following software engineering common practices [5], to let a solution become a design pattern we need to apply such solution to at least three different problems. In particular, we aim, as future work, to implement this cognitive design pattern in the fields of cognitive radio networks and language games.

References

1. Babaoğlu, O., Canright, G., Deutsch, A., Di Caro, G., Ducatelle, F., Gambardella, L.M., Ganguly, N., Jelasity, M., Montemanni, R., Montresor, A., Urnes, T.: Design patterns from biology for distributed computing. Transactions on Adaptive and Autonomous Systems 1(1), 26–66 (2006)
2. Campo, A., Garnier, S., Dédriche, O., Zekkri, M., Dorigo, M.: Self-Organized Discrimination of Resources. PLoS One 6(5), e19888 (2011)
3. Codling, E.A., Plank, M.J., Benhamou, S.: Random walk models in biology. Journal of the Royal Society, Interface 5(25), 813–834 (2008)
4. Couzin, I.: Collective cognition in animal groups. Trends in Cognitive Sciences 13(1), 36–43 (2009)
5. Gamma, E., Helm, R., Johnson, R., Vlissides, J.: Design Patterns: Elements of Reusable Object-Oriented Software. Addison-Wesley Professional (1995)
6. Gardelli, L., Viroli, M., Omicini, A.: Design patterns for self-organising systems. In: Burkhard, H.-D., Lindemann, G., Verbrugge, R., Varga, L.Z. (eds.) CEEMAS 2007. LNCS (LNAI), vol. 4696, pp. 123–132. Springer, Heidelberg (2007)
7. Hamann, H.: Towards swarm calculus: Urn models of collective decisions and universal properties of swarm performance. Swarm Intelligence 7(2-3), 145–172 (2013)
8. Marshall, J.A.R., Bogacz, R., Dornhaus, A., Planqué, R., Kovacs, T., Franks, N.R.: On optimal decision-making in brains and social insect colonies. Journal of the Royal Society, Interface 6(40), 1065–1074 (2009)
9. Montes, M., Ferrante, E., Scheidler, A., Pinciroli, C., Birattari, M., Dorigo, M.: Majority-rule opinion dynamics with differential latency: A mechanism for self-organized collective decision-making. Swarm Intelligence 5(3-4), 305–327 (2010)
10. Nelson, W.: Hazard plotting for incomplete failure data. Journal of Quality Technology 1, 27–52 (1969)
11. Pais, D., Hogan, P.M., Schlegel, T., Franks, N.R., Leonard, N.E., Marshall, J.A.R.: A mechanism for value-sensitive decision-making. PLoS One 8(9), e73216 (2013)
12. Parker, C.A.C., Zhang, H.: Cooperative decision-making in decentralized multiple-robot systems: the best-of-N problem. IEEE Transactions on Mechatronics 14(2), 240–251 (2009)
13. Seeley, T.D., Visscher, P.K., Schlegel, T., Hogan, P.M., Franks, N.R., Marshall, J.A.R.: Stop signals provide cross inhibition in collective decision-making by honeybee swarms. Science 335(6064), 108–111 (2012)
14. Trianni, V., Tuci, E., Passino, K.M., Marshall, J.A.R.: Swarm Cognition: An interdisciplinary approach to the study of self-organising biological collectives. Swarm Intelligence 5(1), 3–18 (2010)

A Novel Competitive Quantum-Behaviour Evolutionary Multi-Swarm Optimizer Algorithm Based on CUDA Architecture Applied to Constrained Engineering Design

Daniel Leal Souza[1,2], Otávio Noura Teixeira[1,3], Dionne Cavalcante Monteiro[2], Roberto Célio Limão de Oliveira[3], and Marco Antônio Florenzano Mollinetti[1]

[1] Laboratory of Natural Computing (LCN),
Area of Exact and Natural Sciences (ACET), University Centre of Pará (CESUPA),
Belém, PA, Brazil
{onoura,marco.mollinetti}@gmail.com
[2] Laboratory of Applied Artificial Intelligence (LAAI),
Institute of Exact and Natural Sciences (ICEN), Federal University of Pará (UFPA),
Belém, PA, Brazil
{daniel.leal.souza,dionnecm}@gmail.com
[3] Post-Graduate Program in Electrical Engineering (PPGEE),
Institute of Technology (ITEC), Federal University of Pará (UFPA),
Belém, PA, Brazil
limao@ufpa.br

Abstract. This paper presents a new bio-inspired algorithm named *Competitive Quantum-Behaviour Evolutionary Multi-Swarm Optimization* (CQEMSO) based on CUDA parallel architecture applied to solve engineering problems, using the concept of master/slave swarm working under a competitive scheme and being executed over the paradigm of General Purpose Computing on Graphics Processing Units (GPGPU). The efforts on implementing the CQEMSO algorithm are focused at generating a solution which includes greater quality of search and higher speed of convergence by using mechanisms of evolutionary strategies with the procedures of search and optimization found in the classic QPSO. For performance analysis, the proposed solution was submitted to some well-known engineering problems (WBD, DPV) and its results compared to other solutions found on scientific literature.

1 Introduction

As notable examples of modifications of the classical Particle Swarm Optimization (PSO) that has shown superior results in comparison to the classical PSO, the following can be cited: the *Quantum-Behaviour Particle Swarm Optimization* (QPSO), which is one of the most well known optimization algorithm based on quantum properties, proposed by [7] to be a more complex solution than the classic PSO based on Newtonian mechanics; the *Evolutionary Particle Swarm Optimization* (EPSO), developed by [3] where it uses a modified PSO trajectory scheme combined with a robust mechanism of evolutionary strategies; and

M. Dorigo et al. (Eds.): ANTS 2014, LNCS 8667, pp. 206–213, 2014.

lastly, there is the existence of another category of PSO algorithms where it involves the concept of creating multiple populations for the sake of maximizing the search and exchange of information obtained by each particle in order to compare and refine the search based on the results already obtained by the neighbouring clusters [1].

In this paper, a hybrid multi-swarm QPSO which includes mechanisms of Evolutionary Strategies proposed by [3] over a competitive approach of master and slaves swarms implemented under CUDA architecture is presented by the name of Competitive Quantum-Behaviour Evolutionary Multi-Swarm Optimization (CQEMSO).

2 An Evolutionary Competitive Multi-Swarm Approach for QPSO on CUDA Architecture (CQEMSO)

The solution proposed in this paper is a result of the integration between the mechanisms found in Miranda's EPSO [3] (e.g. Evolutionary Strategies), Jun Sun's QPSO [7] and a slave/master multi-swarm approach proposed by [5]. It can be described as an environment where two or more slave swarms compete among themselves and the master swarm in the search of the global best. The optimization process that occurs in the master swarm takes advantage of the best results found by slave swarms and by the master swarm itself in order to increase the search performance. In other words, the master swarm explores the best results obtained by the slave swarm and use it as an exploitative approach.

The use of CUDA architecture for CQEMSO allows the swarms to be executed in parallel and with a low processing time. In comparison to other parallel and distributed computing techniques, CUDA takes advantage of many threads on low cost computers with no need for clusters. Concerning the parallel scheme for slave swarms, CUDA provides a programming environment that allows to execute the Evolutionary Strategies (ES) with a high level of parallelism.

The data structure of CQEMSO in CUDA is configured as it follows [6]: *One thread, one particle*: Each thread is responsible for handling a single particle; *One block, one swarm*: Each block (a vector or a matrix (2D or 3D) of threads) is responsible for handling a single swarm; *One grid, one collection of swarms*: Each grid (a vector or a matrix (2D) of blocks) is responsible for handling the slaves and master swarm.

2.1 Quantum-Behaviour Particle Swarm Optimization with Evolutionary Strategies (QPSO+ES) on Slave Swarms

The Quantum-Behaviour Particle Swarm Optimization with Evolutionary Strategies emerges as a new hybrid metaheuristic where the mechanisms of ES found in EPSO are inserted in the QPSO algorithm with a weighted mean best. The main goal behind the use of ES in QPSO is to include a search engine to assist the optimization process by generating replicas with new values based on the original particle position. Among the main features proposed for the

QPSO+ES algorithm, there are: A modified calculation of Learning Inclination Point, a global best that undergoes perturbation process (adapted from the EPSO's ES) and a contraction-expansion factor (β) submitted to mutation process ($m\beta^*_{(i)}$) applied to the replicas (based on the procedure developed by [3]).

Equations 1, 2 and 3 shows the mutation process applied to the replicas that are related to the following variables: Global best disturbance factor (ω_i) [3], Swarm's global best undergone disturbance process (P^*_g) [3] and mutated contraction-expansion factor ($m\beta^*_{(i)}$).

$$\omega_i = P_g + (1 + \sigma_g N(0, 1)) \tag{1}$$

$$P^*_g = P_g + (1 + \omega_i N(0, 1)) \tag{2}$$

$$m\beta^*_{(i)} = \beta + (1 + \sigma N(0, 1)) \tag{3}$$

The variables described in equations 1, 2 and 3 are: Gaussian distribution of mean value 0 and standard deviation value 1 ($N(0, 1)$); Disturbance variable for global best (σ_g); Mutation parameter for $m\beta^*_{(i)}$ (σ); Swarm's global best (P_g); Stochastic star probability constant (θ).

Along with the mutation process for β factor, the QPSO+ES algorithm introduces some modifications to the calculation of Learning Inclination Point (LIP), where the constant θ acts as a selection parameter between two LIP_i solutions. The logic behind this procedure is based on the stochastic star topology proposed by [4], which involves the probability of the particle's element to use the information available on the global best in the respective iteration. If the value obtained by a uniform random number generator between 0 and 1 ($rand()$) is less than the value of θ, the particle's element will have the global best information available and therefore the calculation of the LIP will include both components (local and global), otherwise, the particle's element will have anything but the local best information available, therefore, the LIP calculation will only use the local best.

Another important change is the use of global best that has been undergone disturbance (P^*_g) applied to the Learning Inclination Point update, which is described in equation 2. This addition provides a new search reference and consequently, improves the optimization process. Equation 4 demonstrates a Learning inclination Point calculation for the replicas ($R(LIP_i)$).

$$R(LIP_i) = \begin{cases} {}^{(R_{(c1)}*P_i)+(R_{(c2)}*P^*_g)}/_{(R_{(c1)}+R_{(c2)})} & \text{if } rand() < \theta \\ {}^{(R_{(c1)}*P_i)}/_{R_{(c1)}} & \text{if } rand() > \theta \end{cases} \tag{4}$$

With the values of $R(LIP_i)$ and $m\beta^*_{(i)}$, the new replica's position ($R(X_i^{(t+1)})$) is calculated by equation 5. The variables $R_{(c1)}$, $R_{(c2)}$ $R_{(u)}$ are uniform random numbers between 0 and 1.

$$R(X_i^{(t+1)}) = \begin{cases} R(LIP_i) + m\beta^*_{(i)} * \left| P_{M(i)} - X_i^{(t)} \right| * \ln(\frac{1}{R_{(u)}}) & \text{if } rand() < 0.5 \\ R(LIP_i) - m\beta^*_{(i)} * \left| P_{M(i)} - X_i^{(t)} \right| * \ln(\frac{1}{R_{(u)}}) & \text{if } rand() > 0.5 \end{cases} \tag{5}$$

It is noteworthy to state that the calculation of Learning Inclination Point and position update applied to original particles in the slave swarms are the same as those found in classical QPSO. Algorithm 1 describes the optimization process in slave swarms.

Update contraction-expansion factor ($\beta = 0.5 - (0.5 * rand())$);
Update α factor ($\alpha = 0.5 - (0.5 * rand())$);
Update weighted mean best ($P_{M(i)}$) with α as described in [10];
Run in Parallel For All Slave Swarms (multi-*Block*)
> Update Learning Inclination Point using classic QPSO equation [10];
> Update original particles' position ($X_i^{(t+1)}$) using classic QPSO equation [10];
> Apply position correction to the replicated particles ("Damping");
> Update fitness for the original particles;
> **foreach** *Particle's replica* **do**
>> Generate $m\beta_{(i)}^*$ (equation 3) and ω_i^* (equation 1) by mutation;
>> Update Learning Inclination Point using equation 4;
>> Update replicated particles' position ($R(X_i^{(t+1)})$) using equation 5;
>> Apply position correction to the replicated particles ("Damping");
>> Update fitness for the replicated particles;
>> Select the best particles for the next iteration (slave swarms);
> **end**
> Update particle's local best (P_i);
> Sync threads (wait for all threads to finish);
> **if** *threadIndex=0* **then**
>> Update global best (P_g) of each slave swarm;
> **end**
end
Sync blocks (wait for all blocks to finish);
Update the best global value found in the slave swarms (P_g^S);

Algorithm 1. CQEMSO under CUDA Architecture - Slave Swarms

2.2 Quantum-Behaviour Particle Swarm Optimization with Evolutionary Strategies (QPSO+ES) on Master Swarm

In CQEMSO, the search procedure applied to the master swarm optimization is based on an exploratory search, where the master swarm uses the best solution obtained by the entire system, either by the master or slave swarms in order to enhance its particles.

In order to adapt QPSO+ES for the master swarm, a comparative process between the original master swarm (P_g^M) and the original slave swarms' best solution (P_g^S) is included to select the best result to be used on the calculation of the LIP. The update equation for the master swarm's LIP described in equation 8 is based on stochastic star topology and can be summarized by the following conditions: If the value obtained by a uniform random number generator between 0 and 1 ($rand()$) is less than the constant θ, both slave and master swarms' global best are compared and the best one is subjected to disturbance process and then inserted to the calculation of LIP_i^M. Otherwise, the master and slave swarm's global best remains unchanged. Equations 6 and 7 shows the perturbation process for slave (P_g^{S*}) and master swarms' global best (P_g^{M*}).

$$P_g^{S*} = P_g^S + (1 + \omega_i N(0,1)) \tag{6}$$

$$P_g^{M*} = P_g^M + (1 + \omega_i N(0,1)) \tag{7}$$

The update process for LIP_i^M (master swarm) is described by equation 8

$$R(LIP_i^M) = \begin{cases} {}^{(R_{(c1)}*P_i)+(R_{(c2)}*P_g^{S*})}\!/_{(R_{(c1)}+R_{(c2)})} & \text{if } rand() < \theta \text{ and } P_g^S < P_g^M \\ {}^{(R_{(c1)}*P_i)+(R_{(c2)}*P_g^{M*})}\!/_{(R_{(c1)}+R_{(c2)})} & \text{if } rand() < \theta \text{ and } P_g^S \geq P_g^M \\ {}^{(R_{(c1)}*P_i)+(R_{(c2)}*P_g^M)}\!/_{(R_{(c1)}+R_{(c2)})} & \text{if } rand() \geq \theta \text{ and } P_g^S < P_g^M \\ {}^{(R_{(c1)}*P_i)}\!/_{R_{(c1)}} & \text{if } rand() \geq \theta \text{ and } P_g^S = P_g^M \\ {}^{(R_{(c1)}*P_i)+(R_{(c2)}*P_g^S)}\!/_{(R_{(c1)}+R_{(c2)})} & \text{if } rand() \geq \theta \text{ and } P_g^S > P_g^M \end{cases} \tag{8}$$

After obtaining the values from the Learning Inclination Point, the next step is to calculate the position of the replicas by using equation 9.

$$R(X_i^{(t+1)}) = \begin{cases} R(LIP_i^M) + m\beta_{(i)}^* * \left| P_{M(i)} - X_i^{(t)} \right| * \ln(\frac{1}{R_{(u)}}) & \text{if } rand() < 0.5 \\ R(LIP_i^M) - m\beta_{(i)}^* * \left| P_{M(i)} - X_i^{(t)} \right| * \ln(\frac{1}{R_{(u)}}) & \text{if } rand() > 0.5 \end{cases} \tag{9}$$

LIP and Position Update for the Original Particles. The position update applied to the original particle in master swarm is similar to that found in the classical QPSO algorithm. However, some changes in the Learning Inclination Point calculation have been made in order to use the best value found throughout the system as a global best reference for the calculation of the LIP_i^M.

The major differences between the method used in particle's replicas and its original counterpart consists on the use of comparative process between the original master swarm (P_g^M) and the original slave swarms best solution (P_g^S) in order to select which global best will be used on the Learning Inclination Point calculation, and also for the use of the original contraction-expansion factor (β). Equation 10 shows the calculation of the LIP_i^M applied to the original particles.

$$LIP_i^M = \begin{cases} {}^{(R_{(c1)}*P_i)+(R_{(c2)}*P_g^S)}\!/_{(R_{(c1)}+R_{(c2)})} & \text{if } P_g^S < P_g^M \\ {}^{(R_{(c1)}*P_i)+(R_{(c2)}*P_g^M)}\!/_{(R_{(c1)}+R_{(c2)})} & \text{if } P_g^S \geq P_g^M \end{cases} \tag{10}$$

After obtaining the values from the Learning Inclination Point, the position of the original particles are calculated by equation 11.

$$X_i^{(t+1)} = \begin{cases} LIP_i^M + \beta * \left| P_{M(i)} - X_i^{(t)} \right| * \ln(\frac{1}{R_{(u)}}) & \text{if } rand() < 0.5 \\ LIP_i^M - \beta * \left| P_{M(i)} - X_i^{(t)} \right| * \ln(\frac{1}{R_{(u)}}) & \text{if } rand() > 0.5 \end{cases} \tag{11}$$

Based on the proposed mechanisms and procedures found in this paper, the CQEMSO algorithm could be treated as an enhanced and evolutionary version of the QPSO algorithm applied to a multi-swarm approach with slave/master topology. As seen under a single swarm perspective it is possible to adapt the QPSO+ES algorithm to be executed under sequential environments. Therefore it can be said that the CQEMSO algorithm can be divided in two distinct algorithms: one for slave swarms and for sequential single swarm implementations (QPSO+ES), and another one applied for the master swarm based on the multi-swarm approach described in this work. Algorithm 2 describes the optimization process in master swarm.

Update master swarm's mean best $(P_{M(i)})$ as described in [10];
Run in Parallel For Master Swarm (single-*Block*)
   Update Learning Inclination Point using equation 10;
   Update position from the original particles using equation 11;
   Apply position correction to the replicated particles ("Damping");
   Update fitness for the original particles;
   foreach *Particle's replica* **do**
      Generate $m\beta_{(i)}^{*}$ (equation 3) and ω_i^{*} (equation 1) by mutation;
      Update Learning Inclination Point using equation 8;
      Update position from the replicated particles using equation 9;
      Apply position correction to the replicated particles ("Damping");
      Update fitness for the replicated particles;
      Select the best particles for the next iteration (master swarm);
   end
   Update particle's local best (P_i);
   Sync threads (wait for all threads to finish);
   if *threadIndex=0* **then**
      Update master swarm's global best (P_g^M);
      if P_g^S *is better than* P_g^M **then**
         Assign value of P_g^S to P_g^M;
      end
   end
end
Sync blocks (wait for all blocks to finish);

Algorithm 2. CQEMSO under CUDA Architecture - Master Swarm

3   Experiments and Results

The CQEMSO algorithm was subjected to tests with two engineering problems widely used in scientific literature: Welded Beam Design (WBD); Design of Pressure Vessel (DPV). Tables 1 and 2 shows the best results obtained in 500 executions. For means of comparison, a multi-swarm approach with slave/master topology version of QPSO called *Classic Competitive Multi Swarm Optimization* (COQMSO) has been implemented.

The parameter values are: Number of iterations = 1000; Particles for each slave swarm = 80; Particles for master swarm = 80; Replicas per particle = 4; Number of slave swarms = 4; Disturbance parameter for global best (σ_g) = 0.005; Mutation parameter for β factor (σ) = 0.22.

Table 1. Comparison of results for the WBD

Variables	CQEMSO	COQMSO	[8]	[9]
$X_1(h)$	0.205727	0.204985	0.171937	0.202369
$X_2(l)$	1.517707	1.523215	4.122129	3.544214
$X_3(t)$	9.036655	9.036094	9.587429	9.048210
$X_4(b)$	0.205729	0.205760	0.183010	0.205723
G_1	-0.106445	-0.257812	-8.067400	-12.839796
G_2	-0.189453	-0.833984	-39.336800	-1.247467
G_3	-0.000003	-0.000774	-0.011070	-0.001498
G_4	-3.607639	-3.607062	-3.467150	-3.429347
G_5	-0.080727	-0.079985	-0.236390	-0.079381
G_6	-0.235540	-0.235540	-16.024300	-0.235536
G_7	-0.001465	-2.382324	-0.046940	-11.681355
Violations	0	0	0	0
Avg. time (s)	1.083012	0.353992	N/A	N/A
Variance	**1.288738e-08**	9.671503e-07	N/A	N/A
Mean	**1.4590260**	1.4607121	N/A	N/A
Fitness	**1.458890**	1.459244	1.664373	1.728024

Table 2. Comparison of results for the DPV

Variables	CQEMSO	COQMSO	[8]	[2]
$X_1(T_s)$	0.778169	0.794552	0.812500	0.812500
$X_2(T_h)$	0.384649	0.401580	0.437500	0.437500
$X_3(R)$	40.319626	41.098831	42.092732	42.097398
$X_4(L)$	199.999954	190.783829	195.678619	176.654050
G_1	-0.000000	-0.001345	-0.000110	-0.000020
G_2	-0.000000	-0.009497	-0.035935	-0.035891
G_3	-0.000000	-7184.000000	-1337.994634	-27.886075
G_4	-40.000046	-49.216171	-63.052220	-63.345953
Violations	0	0	0	0
Avg. time (s)	0.945307	0.295996	N/A	N/A
Variance	**1.574957e+01**	7.628519e+05	N/A	N/A
Mean	**5886.6388926**	6869.2100430	N/A	N/A
Fitness	**5885.334961**	5979.822266	6066.029360	6059.946300

4   Conclusions and Future Works

Based on results obtained from the experiments involving engineering problems, the CQEMSO algorithm showed the best results compared to the values obtained by other implementations. In CQEMSO, by adding the Evolutionary Strategies proposed by [3], every particle tends to a better exploration of the search space by generating mutated copies that represents a large amount of new solutions, thus increasing its search capability. Despite being implemented under CUDA architecture, due to its feasibility of execution for multi-populations algorithms with a small processing time, CQEMSO can be easily ported to other parallel solutions such as Beowulf with MPI.

In regards to the experiments, some conclusions can be outlined: 1) The average time of the CQEMSO is higher than the COQMSO due to the usage of the ES mechanisms; 2) The results obtained by CQEMSO (i.e. mean, variance, best solution) were higher than the ones obtained by COQMSO, reflecting the impact of the Evolutionary Strategies in the optimization process.

For future works, we can highlight the inclusion of mechanisms based on Game Theory developed by [8], a study based on the impact caused by boundary conditions in the optimization process, as well as the inclusion to new approaches applied to the calculation of the Learning Inclination Point.

Acknowledgments. This work is supported financially by the Research Support Foundation of Pará (FAPESPA) and Federal University of Pará (UFPA).

References

1. El-Abd, M., Kamel, M.: A taxonomy of cooperative particle swarm optimizers. International Journal of Computational Intelligence Research, 137–144 (2008)
2. He, Q., Wang, L.: An effective co-evolutionary particle swarm optimization for constrained engineering design problems. Engineering Applications of Artificial Intelligence, 89–99 (2007)
3. Miranda, V., Fonseca, N.: EPSO - evolutionary particle swarm optimization, a new algorithm with applications in power systems. In: Transmission and Distribution Conference and Exhibition 2002: Asia Pacific. IEEE/PES, vol. 2, pp. 745–750. IEEE Press (2002)
4. Miranda, V., Keko, H., Duque, A.J.: Stochastic star communication topology in evolutionary particle swarm optimization(EPSO). IJCIR - International Journal of Computational Intelligence Research 4(2) (2007)
5. Niu, B., Zhu, Y., He, X.: Multi-population cooperative particle swarm optimization. In: Capcarrère, M.S., Freitas, A.A., Bentley, P.J., Johnson, C.G., Timmis, J. (eds.) ECAL 2005. LNCS (LNAI), vol. 3630, pp. 874–883. Springer, Heidelberg (2005)
6. Souza, D.L., Teixeira, O.N., Monteiro, D.C., de Oliveira, R.C.L.: A new cooperative evolutionary multi-swarm optimizer algorithm based on CUDA architecture applied to engineering optimization. In: Hatzilygeroudis, I., Palade, V. (eds.) Combinations of Intelligent Methods and Applications, vol. 23, pp. 95–115. Springer (2013)
7. Sun, J., Feng, B., Xu, W.: Particle swarm optimization with particles having quantum behavior. In: Congress on Evolutionary Computation (CEC 2004), vol. 1, pp. 325–331 (2004)
8. Teixeira, O.N., Lobato, W.A.L., Yanaguibashi, H.S., Cavalcante, R.V., Silva, D.J.A., de Oliveira, R.C.L.: Algoritmo Genético com Interação Social na Resolução de Problemas de Otimização Global com Restrições, ch. 10, 1st edn., pp. 197–223. Editora OMNIPAX (2011)
9. Wang, Y., Feng, X.Y., Huang, Y.X., Pu, D.B., Zhou, W.G., Liang, Y.C., Zhou, C.G.: A novel quantum swarm evolutionary algorithm and its applications. Neurocomputing 70, 633–640 (2007)
10. Xi, M., Sun, J., Xu, W.: An improved quantum-behaved particle swarm optimization algorithm with weighted mean best position. Applied Mathematics and Computation 205(2), 751–759 (2008)

Cooperative Object Recognition: Behaviours of a Artificially Evolved Swarm

David King and Philip Breedon

School of Architecture Design and the Built Environment,
Nottingham Trent University, Nottingham, UK

Abstract. Having simple agents capable of cooperatively distinguishing one shape from another is an interesting problem, which will have
benefits in the future. An abstracted model of this was tested with neighbouring agents who change state according to their immediate surroundings, changing the surroundings themselves. Using a genetic algorithm
to determine the agent behaviours at each state it was possible to train
a swarm to remove only one out of two unknown types of shape in an
enclosed space. Estimates of the difficulty the genetic algorithm would
have in finding a suitable solution had a significant correlation with the
measured difficulty in the eleven tested scenarios, with known solutions.

1 Introduction

Swarm robotics, which takes inspiration from the behaviours of social insects,
provide methods of cooperation for groups of agents [8]. Groups of homogeneous
regular shaped robots can connect into lattice formations, providing useful capabilities. Two-dimensional lattice structures can be formed by triangular, square
and hexagonal robots [1][4][5][9]. Three-dimensional lattice structures are formed
by cubical and spherical robots [2][7][10].

An advantage of lattice robots, especially with smaller and simpler robots, is
their ability to become a unit of measurement. This method is used for duplicat-
ing inert shapes built from blocks the same size as the agents [3]. To determine
the shape these cubic robots surround the object and pass a message from agent
to agent in a loop. This information is then used to replicate the shape from the
robots themselves.

In the case of identification, without duplication, the whole shape does not
require mapping, only an identifying feature needs to be found. This method of
identifying only key features was shown to be possible on a hexagonal lattice
where agents cooperate to distinguish two shapes without completely surrounding the shapes [6].

This paper proposes using a genetic algorithm (GA) to train the swarm to
remove only one out of two types of shape in an enclosed arena. Scenarios, with
known solutions, are tested to determine how difficult it is for a GA to find
suitable behaviours for a swarm of homogeneous agents to complete the task.

M. Dorigo et al. (Eds.): ANTS 2014, LNCS 8667, pp. 214–221, 2014.

2   The Simulation Method

A hexagonal lattice forms an arena where each hexagonal cell in the arena can represent: an object cell, which are grouped to form object shapes; a single agent, termed a hBot; a border, which the hBots cannot pass; or an empty space, Fig.1.

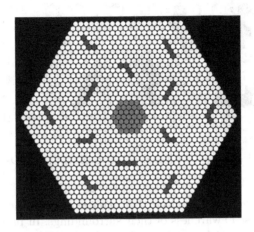

Fig. 1. The arena with six each of two types of object shape and 37 hBots in the centre

2.1   Object Shapes

Neighbouring object cells form object shapes, which are described by data-chains, without notation of rotation or location. If each cell that neighbours an object shape has a value determined by how many object cells it touches, the data-chain is an array of these values (in clockwise order), represented by the sequence which is first in lexicographical order. All the object shapes with four object cells (ID0 - ID9) are shown, Fig. 2, with the numbers in each of their surrounding cells indicating how many object cells they neighbour. For the purposes of these experiments only object shapes with the values, 1, 2 and 3 will be considered, negating object shape ID6, future implementation will allow for higher values.

2.2   hBot Agents

The hBots used in the cooperative object recognition task are homogeneous, anonymous, have no common coordinate system and are not aware of their position relative to the arenas coordinate system. The hBots are modelled with local sensor and communication, both with a range of one cell. This capability allows the hBots to determine their current state from their immediate surroundings and to communicate this state to any hBots that they neighbour, providing a feedback loop between neighbouring hBots. As the hBots share more information

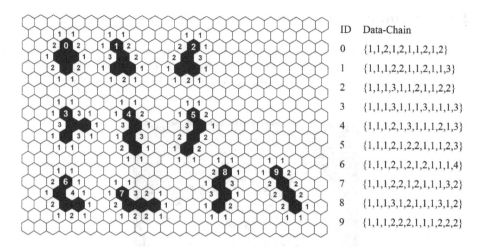

Fig. 2. All ten object shapes with four object cells and their IDs and data-chains

with their neighbours the states that they achieve are at higher state-levels, representing an increased awareness of their surroundings. In practice a hBot that neighbours two hBots with a state-level equal to or higher than it's own state, excluding state-level 0, will change its current state to the relevant state at the next state-level. In the case that a hBot is at state-level 0, a hBot changes states dependant on the number of object cells that it is neighbouring. The states and state-levels are summarised as follows:

- A hBot at state-level 0 is not in contact with an object shape and therefore has no knowledge about the object.
- A hBot at state-level 1 knows the number of object cells it is neighbouring: one, two, or three which is equivalent to knowing a single value in a data-chain. This knowledge is represented by states 1, 2 and 3 respectively.
- A hBot at state-level 2 knows as much as three individual agents at state-level 1, as it knows its own state-level 1 state and the states of its neighbours. This is equivalent to knowing three consecutive values in a data-chain and includes states 4 - 21.
- A hBot at state-level 3 knows as much as five individual agents at state-level 1, or three agents at state-level 2. This is equivalent to knowing five consecutive values in a data-chain and includes states 22 - 264.

Distinguishing between Object Shapes. A scenario is defined by the two types of object shapes that the hBots must cooperate to distinguish between. In a scenario one of the object shapes is always valid and the other object shape is invalid. Object shapes are placed such that a hBot will only ever be in contact with a single shape at any moment. In a given scenario the hBots are required to identify and destroy all six valid object shapes whilst leaving the six invalid object shapes intact. To achieve this the hBots have four different behaviours:

0. Attempt to move a single cell in a random direction, if the cell is occupied remain stationary.
1. Destroy whole object shape which is currently being neighboured.
2. 10% chance of returning to state 0, allowing movement from neighboured object shape.
3. 1% chance of returning to state 0, allowing movement from neighboured object shape.

Behaviour 0 is always used when the hBot is in state 0, state-level 0. If the hBots have perfect information about the two object shapes and therefore the states they can achieve whilst interacting with the object shapes, the behaviours that they should exhibit for each of the states can be determined using the following rules:

- If state is achievable for valid object shapes only: Behaviour 1
- If state is achievable for invalid object shapes only: Behaviour 2
- If state is common to both object shapes and state-level 3: Behaviour 2
- If state is common to both object shapes and state-level 1 or 2: Behaviour 3

The Simulation of hBot Actions. Once per time-step, each of the three steps listed is performed by all the hBots before moving to the next step.

1. Perform one of the four behaviours as determined by the current hBot state.
2. Sense the surrounding six cells of the hBot.
3. Update the hBot's current state.

3   Genetic Algorithm

In a scenario where the hBots do not have prior information about the two object shapes they would be required to learn to determine suitable behaviours in order to distinguish between the objects. To achieve this task a GA was chosen. In this GA each member of the population (size 30) is swarm of 37 homogeneous hBots with identical behaviours. The behaviours for each of the states were determined by the genome (length L, 41). Although there are 264 states over the three state-levels only the states achievable with the four cell object shapes are considered. An integer representation which had a restricted set of $\{1,2,3\}$ representing the three behaviours, excluding behaviour 0 which remains constant at state 0. Recombination is 2-point crossover. The mutation probability is $1/L$ with equal probabilities of the two other integer values occurring. A tournament selection with group size 4 was used for the parent selection using a generation model. The population was initialised with random seeding and was terminated after 30 generations. Each member of the population attempted the given scenario 15 times. The maximum number of time-steps allowed for each member to complete each of its 15 repeated runs was 7000.

3.1 The Fitness Value

The overall fitness, normalised between 0 and 1, for a candidate solution was found with the following formula:

$$fitness = \frac{154C + 28(V_n - I_n) + \frac{(I_{first}+I_{last})-(V_{first}+V_{last})}{1000} + 336}{672} \qquad (1)$$

Where:
V_n is the median number of valid object shapes removed.
I_n is the median number of invalid object shapes removed.
V_{first} is the mean time-steps to remove the first valid object shape.
V_{last} is the mean time-steps to remove the last valid object shape.
I_{first} is the mean time-steps to remove the first invalid object shape.
I_{last} is the mean time-steps to remove the first invalid object shape.

$$C = \begin{cases} 1 & if \ V_n > 0 \ and \ I_n = 0 \\ -1 & if \ V_n = 0 \ and \ I_n > 0 \\ 0 & otherwise \end{cases} \qquad (2)$$

The formula was determined to insure that the calculated fitness values give the following relationships:

- Removing only the correct object shapes is more important than the difference in the number of each type of object shape removed.
- The number of each type of object shape removed is more important than the number of time-steps it took to remove them.
- The number of time-steps taken to remove the object shape is the least important factor.

3.2 Selected Scenarios

Of the different possible pairings of the object shapes with four object cells, eleven scenarios were selected to test the GA. These eleven scenarios were selected to give a suitable range of task difficulty as determined by the number of time-steps taken to complete the task when the behaviours are pre-determined. The chosen scenarios are listed in Table 1 along with the average and maximum number of time-steps taken to complete the scenario, after fifty repeated tests.

Table 1. The number of time-steps required to complete each of the eleven scenarios when the behaviours are pre-determined

Object Shape												
	Valid	0	0	0	1	1	3	5	5	5	9	9
	Invalid	1	4	9	0	9	5	0	9	4	0	4
Number of Time-Steps	max	7877	9982	6678	3712	5744	6344	2724	9015	3445	5969	8623
	average	3980	3645	2392	1400	1699	3331	1355	3728	1501	1993	2686

4   Results

Each of the eleven selected scenarios was repeated three times. The following notation is used to describe a scenario:

- F5I9A: Find and destroy valid object shape ID5 whilst ignoring invalid object shape ID9 experiment run A.

A suitable solution to the scenario is one where, over the average of the fifteen repeated candidate member tests, the candidate solution removed all of the six valid object shapes whilst not removing the invalid object shapes within the allotted 7000 time-steps. This requires a candidate solution fitness value of above 0.9792. For all but one of the scenarios a suitable solution was found for each of the three repeated runs of the GA. The exception to this is F5I4B, where all of both types of object shape were removed, however, the valid shapes were removed earlier in subsequent generations and the invalid ones later.

4.1   Measuring the Genetic Algorithm Scenario Difficulty

The boxplots indicate the relative difficulty of finding a suitable solution using a GA for each scenario. If on average a population's fitness increases rapidly early it suggests an easier scenario to solve, for example F1I9. Where a boxplot that increases gradually and later suggest a harder scenario to solve, for example F3I5, both in Fig. 3.

An estimation of the difficulty that the GA had in determining a suitable solution was quantified by calculating the average number of successful candidate solutions over the three GA runs. This gives an indication of how quickly the solutions were found and the consistency over the repeated test runs.

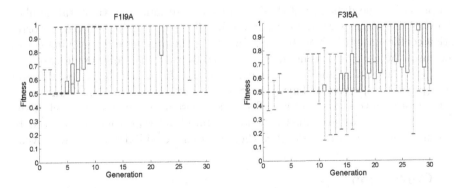

Fig. 3. The mean, minimum and maximum, and first and third quartiles for the 30 generations of the GA for scenarios F1I9A, which is easier to solve and F3I5A, which is more difficult to solve

The Number of Time-Steps. For each of the eleven chosen scenarios this value was compared to the perceived difficulty of the task as determined by the average number of time-steps to complete the scenario when the hBots have a pre-determined behavioural rule set. The correlation coefficient found between these measures of difficulty of scenario was -0.96 (p<.01).

This correlation is expected, scenarios which take longer to complete when the behaviours of the hBots are predetermined would also take longer for the GA to find a suitable solution. These difficult scenarios require greater interaction between the hBots to reach the higher state-levels, which takes time. A drawback of this method is that it requires running the scenario numerous times.

The Number of Identifying States. A more efficient method for determining the difficulty of each scenario is the proportion of identifying states relative to the total number of achievable states at each state-level. An identifying state is a state that is only achievable for the hBots when interacting with the valid object shape and not the invalid object shape. For the eleven chosen scenarios the number of identifying states at each state-level is shown in Table 2.

Table 2. The fraction of identifying states at each state-level for each of the eleven scenarios

Object Shape	Valid	0	0	0	1	1	3	5	5	5	9	9
	Invalid	1	4	9	0	9	5	0	9	4	0	4
Fraction of	State-Level 1	0.00	0.00	0.00	0.02	0.02	0.00	0.02	0.02	0.00	0.00	0.00
Identifying States	State-Level 2	0.02	0.02	0.05	0.10	0.07	0.02	0.12	0.12	0.10	0.07	0.05
per all achievable	State-Level 3	0.05	0.07	0.07	0.22	0.22	0.07	0.27	0.24	0.24	0.10	0.10
states at:	Total	0.07	0.10	0.12	0.34	0.32	0.10	0.41	0.39	0.34	0.17	0.15

Comparing the difficulty of the scenarios as determined by the total number of identifying states with the measured difficulty from running the GAs a correlation coefficient of 0.69 (p<.02) was found. As the number of identifying states decreases the number of generations required to solve the scenario increases.

At state-levels 1, 2 and 3 the correlations are 0.79 (p<.01), 0.75 (p<.01) and 0.63 (p<.05) respectively. This indicates that the number of identifying states at lower state-levels have a larger effect than those at a higher state-level. If the object shape can be identified at a lower state-level then there is no need to identify it at a higher state-level as the object shape will already be destroyed.

5   Conclusion

The difficulty of eleven scenarios was predicted with two methods and compared to the measured difficulty the GA had in solving them. The number of time-steps taken to complete the scenario with a swarm of agents that had pre-programmed behaviours correlated closely with the measured difficulty. The second method calculated the number of identifying states for the pair of object shapes in the scenario. This value gives an indication of how many of the possible states could be used to determine that the object shape being identified is the valid object shape. This method proved to be a marginally less accurate indication of the difficulty of the task scenario but benefits from being considerably quicker.

There were some limitations to the simulation, which should be addressed in future research: using a less common hexagonal lattice, comparisons with existing multi-agent systems are difficult to make; all the agents were modelled with perfect sensors and communication; the system only deals with object shapes that have the values 1, 2 and 3 in their data-chain, ignoring possible shapes that have the values 4, 5 and 6; object shapes that are mirror images of each other appear identical due to the way the hBots sense their surroundings; and currently the number of state-levels would only allow a single hBot to have the equivalent knowledge of five state-level 1 hBots. Although the GA was found to be successful at determining the correct behaviours in these initial scenarios, more complex scenarios need testing. In parallel with this different training methods for the hBots should be considered as this will help identify more efficient self determining solutions for the cooperative object recognition problem.

References

1. Bishop, J., Burden, S., Klavins, E., Kreisberg, R., Malone, W., Napp, N., Nguyen, T.: Programmable parts: A demonstration of the grammatical approach to self-organization. In: 2005 IEEE/RSJ International Conference on Intelligent Robots and Systems (IROS 2005), pp. 3684–3691. IEEE (2005)
2. Gilpin, K., Kotay, K., Rus, D., Vasilescu, I.: Miche: Modular shape formation by self-disassembly. The International Journal of Robotics Research 27(3-4), 345–372 (2008)
3. Gilpin, K., Rus, D.: A distributed algorithm for 2d shape duplication with smart pebble robots. In: 2012 IEEE International Conference on Robotics and Automation (ICRA), pp. 3285–3292. IEEE (2012)
4. Goldstein, S.C., Campbell, J.D., Mowry, T.C.: Programmable matter. Computer 38(6), 99–101 (2005)
5. Hosokawa, K., Shimoyama, I., Miura, H.: Dynamics of self-assembling systems: Analogy with chemical kinetics. Artificial Life 1(4), 413–427 (1994)
6. King, D., Breedon, P.: Robustness and stagnation of a swarm in a cooperative object recognition task. In: Tan, Y., Shi, Y., Chai, Y., Wang, G. (eds.) ICSI 2011, Part I. LNCS, vol. 6728, pp. 19–27. Springer, Heidelberg (2011)
7. Østergaard, E.H., Kassow, K., Beck, R., Lund, H.H.: Design of the atron lattice-based self-reconfigurable robot. Autonomous Robots 21(2), 165–183 (2006)
8. Şahin, E.: Swarm robotics: From sources of inspiration to domains of application. In: Şahin, E., Spears, W.M. (eds.) Swarm Robotics 2004. LNCS, vol. 3342, pp. 10–20. Springer, Heidelberg (2005)
9. White, P., Kopanski, K., Lipson, H.: Stochastic self-reconfigurable cellular robotics. In: Proceedings of the 2004 IEEE International Conference on Robotics and Automation, ICRA 2004, vol. 3, pp. 2888–2893. IEEE (2004)
10. Zykov, V., Mytilinaios, E., Desnoyer, M., Lipson, H.: Evolved and designed self-reproducing modular robotics. IEEE Transactions on Robotics 23(2), 308–319 (2007)

Emergent Diagnoses from a Collective of Radiologists: Algorithmic versus Social Consensus Strategies

Daniel W. Palmer[1], David W. Piraino[2], Nancy A. Obuchowski[2], and Jennifer A. Bullen[2]

[1] John Carroll University, University Heights, OH, USA
dpalmer@jcu.edu
[2] Cleveland Clinic, Cleveland, OH, USA

Abstract. Twelve radiologists independently diagnosed 74 medical images. We use two approaches to combine their diagnoses: a collective algorithmic strategy and a social consensus strategy using swarm techniques. The algorithmic strategy uses weighted averages and a geometric approach to automatically produce an aggregate diagnosis. The social consensus strategy used *visual cues* to quickly impart the essence of the diagnoses to the radiologists as they produced an emergent diagnosis. Both strategies provide access to additional useful information from the original diagnoses. The mean number of correct diagnoses from the radiologists was 50 and the best was 60. The algorithmic strategy produced 63 correct diagnoses and the social consensus produced 67. The algorithm's accuracy in distinguishing normal vs. abnormal patients (0.919) was significantly higher than the radiologists' mean accuracy (0.861; $p = 0.047$). The social consensus' accuracy (0.951; $p = 0.007$) showed further improvement.

1 Introduction

The normal work practice of radiologists includes consulting peers on cases where the medical images provide insufficient, conflicting, or ambiguous information. Ideal consults result in exposure to a potential diagnosis not previously considered [3]. In phase I of our study, we explored the intersection of this idea and Surowiecki's definition of wise crowds [6] to determine the diagnostic effectiveness of a collective of radiologists. Four criteria must be met in order for the crowd to be effective: diversity, independence, decentralization, and a method of aggregating the estimates. In our study, the group of radiologists does not fully meet the diversity criterion, as they all are professional radiologists or fellows training in musculoskeletal radiology. However, they do have differing specialties within the field including sports medicine, tumors, trauma, and emergency medicine. Each radiologist evaluated the images in isolation to meet the independence and decentralization criteria and we developed an automated aggregation algorithm to combine the diagnoses.

M. Dorigo et al. (Eds.): ANTS 2014, LNCS 8667, pp. 222–229, 2014.

In phase II, we sought to improve the collaborative diagnoses through visual stigmergic cues and social decision-making techniques as demonstrated by househunting ants (*Temnothorax albipennis*) [1].

2 Methodology

2.1 Experiment and Data Collection

In phase I, 74 musculoskeletal images with surgical proof or proof by follow-up images were read and evaluated by 12 sub-specialized radiologists (referred to as *readers* in this paper). One third of the images were normal, two thirds were abnormal including moderate and difficult cases. All images were displayed as single images (no image series or volumes) with no window or leveling capabilities (similar to brightness and contrast). Additionally, we provided no patient information or medical history. These conditions severely limit the readers and do not represent actual diagnostic conditions, but do produce more diverse diagnoses. This study was conducted under approval from the Cleveland Clinic's institutional review board; all images were de-identified and the data collected was anonymized prior to analysis. The images were presented in randomized order and each reader gave a differential diagnosis for each abnormality that they found. A diagnosis consists of two parts, a *location* in the image and an *identification* within five broad categories (*Normal* and four abnormality types: *Fracture, Tear, Mass/Mass-like abnormalities* and *Other*). The readers marked a *location* of interest with an enclosing oval and made a differential diagnosis for the *identification* using a multi-region slider bar yielding a vector of certainty percentages for 15 specific abnormality subtypes.

In phase II, nine of the original 12 readers returned to view the collected data, to evaluate the diagnoses, and collectively generate a social consensus diagnosis. We displayed all the *identification* and *location* information for each image and the readers selected which diagnosis they most agreed with.

2.2 Collective Diagnosis Algorithm

We developed a deterministic strategy that separately calculates both the *identification* of the abnormality and its *location* and then combines them into a single diagnosis. To determine the *identification* of the condition in medical image k, we use:

$$A_k = f(index(V_k, \left\{ \max \left\{ \sum_{j=1}^{n_r} c_{ijk} | i \in 1..n_d \right\} \middle| k \in 1..n_m \right\})) \qquad (1)$$

where A_k is the algorithmic diagnosis for image k, n_r is the number of readers, n_d is the number of options for the differential diagnosis and also the size of the diagnosis vector (V_k), n_m is the number of medical images in the study, and c_{ijk} is the certainty of reader i of identification j in image k - found in position

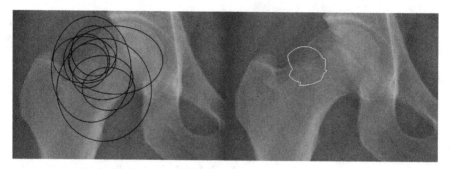

Fig. 1. Individual *locations* (black ovals) on the left, algorithmically combined *location* (white outline) on the right

j of the V_k. The function *index()* returns the position in a vector of a given value and the function $f()$ maps those positions to corresponding abnormalities. To determine the *location* of the abnormality, we overlay all the location ovals and apply a threshold of $> 50\%$. In figure 1, the region outlined on the right was selected by at least seven of the 12 readers. (Color images and additional material can be accessed at http://go.jcu.edu/ants2014)

2.3 Social Consensus Diagnosis

To generate the social consensus, all equivalent diagnoses were collapsed to remove insignificant differences in *location* ovals. Differences in *identifications* or differences in certainty of identical *identifications* remained distinct. Each reader voted for a single diagnosis, and all the votes were tallied producing in a social consensus as defined in Table 1. The first three categories are self-explanatory. A *plurality* exists when no diagnosis exceeds 50% and there is one selected by at least three readers that has a higher percentage of agreement than all others. Any other circumstances constitute a *disagreement*. This includes cases in which no diagnosis has more than two readers agreeing with it, or when ties occur. For the top four categories, the agreed upon diagnosis becomes the social consensus. When a *disagreement* occurs, we take a weighted average of all *identifications* and normalize the certainties.

Table 1. This table shows the definitions and actions for each type of social agreement

Agreement Type	Criteria	Action
Unanimous	100% agreement	Use agreed upon diagnosis
Consensus	≥70% agreement	Use agreed upon diagnosis
Majority	>50% agreement	Use agreed upon diagnosis
Plurality	≤ 50% and ≥ 3 readers agree, no ties	Use agreed upon diagnosis
Disagreement	≤ 2 readers agree or a tie	Average all diagnoses

3   Results

Table 2 displays the outcomes of the 12 individual readers, their (M)ean, the
(A)lgorithmic strategy, and the (S)ocial consensus. The first row shows how
many times the specified diagnosis matched the verified condition. *Mentioned
Correct Diagnosis* indicates that the correct *location* and *identification* are both
included in the diagnosis, but given a lower certainty than other possibilities. An
Incorrect Diagnosis does not include the correct condition. Both the collective
algorithm and the social consensus had more correct diagnoses than any of the
individual readers, but in two instances for social consensus, and in four instances for the algorithmic strategy, readers had the same or fewer incorrect
diagnoses. The main goal of this analysis was to compare the mean of the 12
readers' accuracies with the accuracies of our two strategies. Accuracy was characterized by the area under the receiver operating characteristic (ROC) curve for
each pairwise comparison among the three truth states (normal, fractures/tears,
and mass/mass-like abnormality) (see figure 2). For each reader, nonparametric
accuracy estimates were calculated using the methods in [4]. Additionally, the
accuracies were compared on their ability to distinguish normal patients from
those with an abnormality of any type. Standard errors were calculated using
the methods in [2] for the algorithm and social consensus accuracies and [5] for
the mean of the readers' accuracies. T tests were used to assess the differences
between the mean reader and algorithm/social consensus accuracies for each
comparison and the standard error for these tests was adjusted to account for
the fact that multiple readers read the same images [5]. A significance level of
0.05 was applied. The collective algorithm tied or outperformed 8 of 12 readers
in distinguishing normal patients vs. patients with a trauma, 9 of 12 readers
in distinguishing normal patients vs. patients with a mass, 8 of 12 readers in
distinguishing patients with a mass vs. trauma, and 9 of 12 readers in distinguishing normal patients vs. patients with an abnormality of any type. Table
3 displays the accuracy estimates for the algorithm. The algorithm's accuracy
was significantly higher than the readers' mean accuracy in distinguishing normal patients vs. patients with a mass and in distinguishing normal patients vs.
patients with an abnormality. The readers reached a clear social consensus for
66 images (89%). The social consensus tied or outperformed all 12 readers in
distinguishing normal patients vs. patients with a trauma, 8 of 12 readers in
distinguishing normal patients vs. patients with a mass, 8 of 12 readers in distinguishing patients with a mass vs. trauma, and all 12 readers in distinguishing
normal patients vs. patients with an abnormality of any type. The accuracy

Table 2. Raw Results of the Study

	1	2	3	4	5	6	7	8	9	10	11	12	M	A	S
Correct Diagnoses	47	46	54	38	52	50	56	54	55	60	51	41	50	63	67
Mentioned Correct Diagnosis	16	14	4	3	10	12	13	11	6	10	9	17	11	0	2
Incorrect Diagnoses	11	14	16	33	12	12	5	9	13	4	14	16	13	11	5

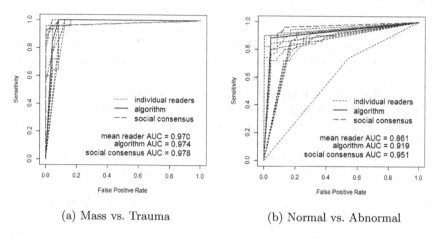

(a) Mass vs. Trauma   (b) Normal vs. Abnormal

Fig. 2. ROC Plots

Table 3. Estimates (standard error) of readers' mean accuracy and algorithm/social consensus accuracy

	Mean reader Accuracy (SE)	Algorithm Accuracy (SE)	Social Consensus Accuracy (SE)
Norm vs. Trauma	0.855 (0.096)	0.899 (0.043)	0.966 (0.024)
Norm vs. Mass(-like)	0.949(0.044)	0.978 (0.021)	0.978 (0.023)
Mass(-like) vs. Trauma	0.970 (0.024)	0.974(0.023)	0.978 (0.022)
Norm vs. Abnorm	0.861(0.093)	0.919 (0.031)	0.951 (0.028)

estimates for the social consensus are displayed in Table 3. Table 4 compares both strategies against the readers' mean performance.

4   Discussion

4.1   Shared Visual Cues for Stigmergic Coordination

Trail blazes, stone cairns and graffiti tags constituting physical examples, and Facebook's likes and Twitter's favored retweets as digital ones, humans employ many kinds of visual stigmergic information. In phase II, the collection of overlaid *location* ovals constitute shared *visual cues*. Readers marked the images independently in phase I, but in phase II, the ovals are aggregated and collectively displayed, propagating information to the readers' as they make their selections. These *visual cues* impart two types of information: the extent of the agreement of the readers and, when there is agreement, a proposed *location* of an abnormality. By engaging the human visual system, the information transfers quickly, setting up expectations in the reader and potentially reducing uncertainty in

Table 4. Estimates (standard error) of readers' mean accuracy and algorithm/social consensus accuracy

	Alg.-Mean reader			Soc. Con.-Mean reader		
	Diff.	95%CI	p	Diff.	95%CI	p
Norm vs. Trauma	0.044	(-0.014, 0.103)	0.120	0.111	(0.031, 0.191)	0.011
Norm vs. Mass(-like)	0.029	(0.002, 0.057)	0.036	0.029	(0.001, 0.056)	0.041
Mass(-like) vs. Trauma	0.004	(-0.008, 0.016)	0.488	0.008	(-0.023, 0.039)	0.587
Norm vs. Abnorm	0.058	(0.001, 0.116)	0.047	0.091	(0.030, 0.1511)	0.007

their diagnoses (see Figure 3). Figure 3A indicates a high level of agreement in a tightly defined *location*. A reader will expect an abnormality there and deem it unlikely that there is another abnormality in the image. Figure 3B indicates some agreement in one *location*, but suggests that there may be other abnormalities. Figure 3C shows very little agreement and dictates careful consideration of the entire image. The single oval in figure 3D indicates strong agreement with a *normal* diagnosis. The readers in our study treated the *visual cues* in three ways: as direction to diagnoses, as confirmation of diagnoses, and, in a few cases, as noise. In cases like 3A and 3D the readers would focus on the marked area and often quickly agree with it. In a case like 3C, some readers would briefly look at the *visual cues*, realize the lack of agreement and then hide the ovals and re-evaluate the image. Once they had a diagnosis, they would reveal the ovals for comparison. Depending on their agreement, they would move on or reconsider. For some images, some readers found the *visual cues* distracting and would re-diagnosed the image without them. In addition to the viewing static *visual cues*, our software provides the reader with the ability to manipulate the visual information. We linked the *locations* with the *identifications* so that the readers can highlight corresponding diagnosis components by changing colors of the ovals, or display individual diagnoses in isolation. These *interactive visual cues* can be used to generate and satisfy queries, and to pair multiple abnormalities identified by the same reader.

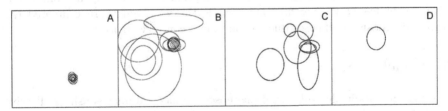

Fig. 3. Visual cues isolated from medical image they relate to

4.2   Examination of Specific Instances of Emergent Diagnoses

Table 5 splits the social diagnoses based on their degree of certainty. We also separate out multiple abnormalities per image and the cases in which the social consensus produced multiple diagnoses for single abnormalities. This increases the number of cases from 74 to 82. Table 5 shows categories that indicate the performance of the social consensus strategy relative to the algorithmic strategy. The headings *same C* and *same I* indicate that the two strategies produced the same diagnosis, either both *Correct* or both *Incorrect*. *I→C* means that where the algorithmic strategy produced an *Incorrect* diagnosis, the social consensus strategy produced a *Correct* one, and *C→I* specifies the reverse. For entry 1^a, five readers specified a 100% certainty in a *normal* diagnosis while six readers didn't mention *normal* at all. All readers produced an incorrect diagnosis and only one those gave a diagnosis that included the correct identification - and that at only a 50% certainty level. The algorithmic strategy produced a diagnosis of *Normal*. During phase II, readers saw the *visual cues* indicating a lack of agreement and more than 70% of them selected the one diagnosis with the correct component and as a result, demonstrated a successful consult.

Table 5. Comparison of Social Consensus to Algorithm by Agreement Type

	same C	same I	I→C	C→I	Other	Totals —
Unanimous	13					13
Consensus	26		1^a			27
Majority	16			1		17
Plurality	7	2	3	2	3	17
Disagreement			4^b	1	3	8
Totals	62	2	8	4	6	82

Entry 4^b specifies four instances in which the phase II readers did not reach agreement. Three of these cases were ties; the fourth one had almost universal disagreement. The algorithmic strategy's approach of inflating the component with the highest certainty fails here; while the weighted average favored the correct diagnosis in all cases due to common secondary choices.

5   Conclusion

We combined multiple, individual diagnoses for medical images using collective algorithmic and social consensus strategies. In both approaches, we accessed and leveraged more useful information from the readers collectively than they could individually. Despite the cost of this approach, it demonstrates that better diagnoses can be obtained and sets the challenge to find efficient ways to reach them. The observations made in this study reinforce the idea that consults should be made without any information provided, to ensure an independent outcomes.

The shared *visual cues* for the *location* worked well, so we plan to also use them for *identification*. The current display shows 12 equal-size rectangles containing a list of the readers' *identifications* and their certainty level. In the next version of the software, boxes will be sized proportionally to the number of readers selecting each one.

Acknowledgements. John Carroll University supported this work with a George Grauel Faculty Fellowship. B. Josipovic, N. Orlando, A. Lanese, and T. Drescher developed the original software. Al Denelsbeck, Mike Kovacina, and Paige Rinker provided valued assistance in preparing this paper. We thank the anonymous reviewers - their feedback improved the paper.

References

1. Brutschy, A., Scheidler, A., Merkle, D., Middendorf, M.: Learning from house-hunting ants: collective decision-making in organic computing systems. In: Dorigo, M., Birattari, M., Blum, C., Clerc, M., Stützle, T., Winfield, A.F.T. (eds.) ANTS 2008. LNCS, vol. 5217, pp. 96–107. Springer, Heidelberg (2008)
2. DeLong, E., DeLong, D., Clarke-Pearson, D.: Comparing the areas under two of more correlated receiver operating characteristic curves: a nonparametric approach. Biometrics 44, 837–845 (1988)
3. Drew, T., Vo, M., Wolfe, J.: The invisible gorilla strikes again sustained inattentional blindness in expert observers. Psychological Science 24(9), 1848–1853 (2013)
4. Obuchowski, N., Goske, M., Applegate, K.: Assessing physicians' accuracy in diagnosing pediatric patients with acute abdominal pain: measuring accuracy for multiple diseases. Statistics in Medicine 20, 3261–3278 (2001)
5. Obuchowski, N., Rockette, H.: Hypothesis testing of diagnostic accuracy for multiple readers and multiple tests: an anova approach with dependent observations. Communication in Statistics – Simulation 24, 285–308 (1995)
6. Surowiecki, J.: The Wisdom of Crowds. Doubleday, New York (2005)

Foraging Agent Swarm Optimization with Applications in Data Clustering

Kevin M. Barresi

Department of Electrical and Computer Engineering
Stevens Institute of Technology, New Jersey, USA
kbarresi@stevens.edu

Abstract. This paper proposes a novel method of swarm optimization called Foraging Agent Swarm Optimization (FASO). FASO is designed to converge on multiple optima in both gradient and point-based search spaces. FASO also operates well in situations where "field optima" are desired, rather than single-point optima. The utility and effectiveness of FASO in a non-gradient search space is demonstrated in the context of data clustering, where we present Foraging Agent Swarm Clustering (FASC). FASC provides several benefits over conventional clustering, such as the ability to automatically determine the number of clusters, and strong performance in both noisy and sparse data sets. FASC is demonstrated to outperform existing methods of clustering in a variety of situations. Positive results by FASC in data clustering suggest that FASO has a promising future in other optimization applications as well.

1 Introduction

Data clustering is a notoriously difficult task in pattern recognition where unlabeled, multi-dimensional data vectors are grouped together in an unsupervised manner. Through cluster analysis, valuable predictions and summarizations of the data set can be made. Data clustering has direct applications in the growing fields of data mining, machine learning, and bioinformatics, and thus, an increasing need for more effective and versatile clustering algorithms exists.

Conventional clustering algorithms sort n-dimensional data points into groups based on positional similarity. Most existing methods are either centroid or density based. Centroid-based clustering algorithms attempt to derive a single "point-centroid" for each cluster. An example of this is the well known k-means method [10]. Centroid based algorithms work well on globular clusters, due to the intrinsic concept of a cluster center. However, k-means performs poorly on arbitrarily shaped clusters and requires the total number of clusters be known *a priori*. This is a major drawback, as this metric is often unknown. Density based clustering algorithms, such as DBSCAN, group points together by a density gradient, and thus excel in creating arbitrarily shaped clusters in noisy environments [5]. However, DBSCAN performance suffers greatly in data sets of varying density. Data clustering operates on the premise of reducing the number

M. Dorigo et al. (Eds.): ANTS 2014, LNCS 8667, pp. 230–237, 2014.

of incorrect point categorizations, while maximizing the number of correct categorizations. In this sense, data clustering can be thought of as an optimization problem.

The ability of swarm algorithms to effectively solve optimization problems has drawn significant attention to the prospect of applying them to data clustering. Earlier attempts utilize basic Particle Swarm Optimization (PSO), with numerous refinements only providing slight performance enhancements [1, 3, 4, 13, 14]. More recent progress has included a Glowworm Swarm Optimization (GSO) based method [2] as well as Bacterial Foraging Optimization (BSO) based clustering [11, 12, 15]. Unfortunately, these methods tend to provide marginal clustering improvements at the expense of significant resource overhead. Furthermore, swarm algorithms generally aim to find one or more "point optima"; the swarm aims to converge on individual points, as opposed to "field optima", where potential solutions exist as a broader area. This quality creates an intrinsic issue that makes clustering via swarm algorithms difficult.

The contributions of this paper are twofold. First, a novel swarm-based optimization algorithm is presented, called Foraging Agent Swarm Optimization (FASO), which excels at finding multiple area optima. The second contribution is a novel clustering scheme named Foraging Agent Swarm Clustering (FASC). As a direct application of FASO in a non-gradient search space, the new method is shown to perform well on a wide range of data sets, outperforming existing clustering techniques and requiring no user configuration.

2   Foraging Agent Swarm Optimization

2.1   Algorithm Description

FASO uses a metaphor of organisms (agents) foraging for food in a user-defined search space. Food, representative of a strong objective function value, is a beneficial commodity that characterizes an optimal solution. Agents are graded by their "happiness", a quantitative approach to marking how optimal an agent's current position is. Happiness is positively impacted by a high objective function value, and negatively impacted by inter-agent overcrowding. In order to effectively move through the search space, agents are first and foremost attracted to other agents with higher happiness values. In this sense, agent happiness is reminiscent of *luciferin* levels of GSO [9]. Agents will move in the direction of the neighboring agent with the highest happiness level. Agents with no superior neighbors are given the opportunity for random movement, allowing for continued searching. The precise nature and magnitude of movement is determined by a vector of ranges. After several cycles of happiness updates, range updates, and movement, groups of agents will become centered around optima. This effectively provides one or more solutions to the provided optimization problem.

2.2   Ranges and Update Mechanisms

Ranges are automatically adapted per agent based on the current position, and guide the swarm towards timely convergence on an appropriate number of optima. Agents rely on one static range per swarm: sensor range (r_s), and two dynamic ranges per agent: foraging range (r_f), and crowding range (r_c). Range magnitudes are bounded by the inequality $r_c < r_f < r_s$.

The sensor range r_s acts as an upper bound for all other ranges, and represents the farthest distance at which an agent can sense objects, including other agents and objective function values. The sensor range is determined by the objective function, giving appropriate scaling based on general gradient observations as well as local search area size. The sensor range can greatly impact that time required for swarm convergence, as well as the number of optima discovered. A large sensor range spanning the entire search space causes the swarm to converge on a single, global optimum. Smaller sensor ranges result in more fine-tuned searching, and increases swarm sensitivity to local optima. As such, a balance between the two is required for effective operation.

The foraging range r_f determines the distance at which other agents and objective function values can be sensed. During the movement stage, this range is used to determine which agents are considered local, and outlines a range of movement. It also determines how far an agent can search for areas of better optima in the current iteration. The foraging range is inversely related to the number of neighboring agents within the previous foraging range. This encourages more localized searching, when a large number of neighbors exists. Formally, the foraging range can be written as follows:

$$r_f^i(t+1) = \alpha + \frac{r_s^i - \alpha}{1 + \beta|A(p_i, r_f^i(t))|}. \tag{1}$$

Where α is a constant representing the minimum acceptable foraging range, and is related linearly to r_s. In practice, setting α equal to 10% of r_s results in strong performance. β is a constant that quantifies the importance of the number of neighboring agents. A higher β value results in a faster fall-off of $r_f^i(t+1)$ in response to a growing number of neighboring agents. Through empirical evaluation, setting β equal to 10 was found to perform well for a wide variety of optimization problems.

The crowding range r_c is the maximum range at which any neighboring agents will negatively impact happiness. In terms of the swarm as a whole, the crowding range affects how closely together agents are located. It is modified to increase or decrease swarm density inside areas of objective function optima. This value can be derived from the foraging range through a simple linear function. Empirical observation has shown that the crowding range should be less than half of the foraging range in order to avoid stifling agent movement.

2.3 Happiness

Happiness is a quantitative measurement bounded between 0 and 1 of how optimal an agent's current position is. Happiness is affected positively by increasing objective function value, and negatively by increased inter-agent crowding, as determined by the crowding range. Formally, the happiness h of agent a_i at position p_i is characterized by the following:

$$h(i) = \frac{O(p_i)}{|A(p_i, r_c^i| + 1}.$$ (2)

Where $O(p_i)$ is the objective function value at position p_i, normalized between 0 (worst) and 1 (best). $A(p_i, r_c^i)$ is a vector of agents whose distance from point p_i is less than r_c^i. In other words, the magnitude of this term is the number of agents located within the crowding range. Clearly, higher objective function values results in higher agent happiness, while a larger number of crowded neighbors results in decreased happiness. In practice, increasing the happiness of an agent is a balancing act between higher objective function values and lower numbers of neighboring agents. Through movement, agents travel to positions representing a "sweet spot" of the two values, resulting in a dense group of agents spread around optima. This behavior is the desired "area optima" effect.

2.4 Movement Rules

Effective movement mechanisms are critical for any swarm based algorithm, as it determines how well agents are able to avoid local optima, while at the same time, maintaining the ability to locate potential areas of interest. FASO agents move with a primary focus on superior local neighbors, with a cascaded series of backup movement cases.

The first step in movement involves searching the local area for neighboring agents. The foraging range r_f is used to generate a list of potential target agents, of which, the agent with the highest happiness score is selected as the candidate target agent. If the candidate's happiness is found to be higher than that of the moving agent, then the agent is moved a small distance towards this neighbor. In this way, agents are indirectly drawn towards areas of optima containing agents, while at the same time searching the paths towards these areas. If no superior neighbors are found, then a "safe" random movement is attempted, where a random position in the search space is selected and evaluated for happiness. If the resulting happiness is greater than or equal to the agent's current happiness, it is moved to this new position. Otherwise, the agent does not move. The random movement is considered safe because it cannot result in the agent being moved to a worse position. If the agent does not have any neighbors, a gradient ascent/descent approach is used. The agent will simply move a small distance in the direction of the gradient unit vector. This has the effect of moving the agent toward optima, in the hopes of either discovering a global optima, or locating new agent neighbors. Neighbor-based movement is characterized by Equation 3, while gradient-based movement by Equation 4:

$$p_i(t+1) = p_i(t) + \delta \left(\frac{p_j(t) - p_i(t)}{||p_j(t) - p_i(t)||} \right).$$ (3)

$$p_i(t+1) = p_i(t) + \delta \left(\frac{\frac{df}{d_1}, \frac{df}{d_2}, \dots, \frac{df}{d_n}}{||\frac{df}{d_1}, \frac{df}{d_2}, \dots, \frac{df}{d_n}||} \right).$$ (4)

Where $p_i(t + 1)$ is the next position of agent a_i, and $p_i(t)$ is the current position. The magnitude of the movement, δ, is determined randomly by the forage range, following the bounding inequality $0 < \delta < \frac{r_i^j}{2}$.

2.5 Sample Results

In order to evaluate the effectiveness of FASO, we tested its ability to find global minima in two multivariate functions: an Ackley function, and a Styblinski-Tang function. FASO was run 50 times for each function, and agent positions were recorded each time after the final iteration was completed.

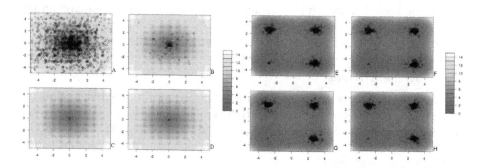

Fig. 1. Agent positions (black) for the Ackley function (A-D) and Styblinski-Tang function (E-H) after 50 instances of FASO, with (A,E) 25 iterations, (B,F) 100 iterations, (C,G) 200 iterations, (D,H) 500 iterations

Overall, results from Figure 1 show that FASO is able to reliably converge on global minima in a small number of iterations, without becoming trapped in local minima. It also confirms that the density of agents per unit area is dependent on the gradient around optima; a smaller slope allows for greater agent spread with less objective function penalty. Agent spread can be controlled by modifying the sensor range; higher sensor ranges result in higher spread, while smaller ranges result in more densely packed agent groups.

3 Foraging Agent Swarm Clustering

Clustering is an application in a non-gradient search space; positions in the space are marked as either containing a data point or not. We define the objective function as the percentage of data points located within an agent's foraging range. In order to account for the non-gradient quality of clustering, we modify the basic happiness function presented in FASO to take into account "data point density" (Equation 5). Data point density is chosen over raw data point count in order to make the happiness score less dependent on foraging range.

$$h(i) = \frac{|O(p_i, r_f^i)|}{\pi r_f^{2i}|A(p_i, r_c^i)|} = \frac{density_{(data)}}{crowding}. \tag{5}$$

FASC operates in three phases: the convergence phase, the consolidation phase, and the assignment phase. The convergence phase uses FASO to superimpose agents over the data set. Agents move towards areas of higher data point density, eventually forming a "net" of coverage around groups of data points. The ability of FASO to converge into "area optima" is apparent in this application, where it is critical for agents to completely cover clusters, rather than converge on the center of a cluster. After a specified number of FASO iterations, the consolidation phase removes "stray" agents that may have become lost in unpopulated areas. While unlikely, this removes the chance of a wondering agent accidentally creating unnecessary clusters. Finally, the assignment phase places data points in clusters formed by agent groups. Using a density based approach similar to that used in the DBSCAN method, clusters of agents are produced by connecting chains of neighboring agents. If the foraging ranges of two agents overlap, they are considered part of the same cluster. With each agent part of a cluster, all data points that fall within an agent's foraging range immediately join that agent's cluster. Points that lie outside of the range of any agent simply join the nearest cluster. These outlying points may alternatively be ignored as noise.

4   Experimental Results

In order to evaluate FASC, four different two-dimensional clustering data sets were used. Together, the data sets are highly representative of most common clustering applications. We used the Aggregation data set [7] as a constant density, non-regular cluster sample, Jain's data set [8] as a multi-density, non-regular cluster sample, and finally S_1 and S_2 [6] as a semi-constant density, regular and non-regular cluster sample with varying degrees of noise. data set sizes ranged from 373 points (Jain) to 5000 points (S_1, S_2).

We compared the effectiveness of FASC to several well known clustering methods, including both centroid (k-means) and density based (DBSCAN) algorithms. All FASC clusters were generated using 100 iterations of FASO, a swarm size of 0.25 to 0.75 times the number of data points, and a sensor range r_s of 0.2 to 0.6 times the average data point to data point distance per data set. With these settings, FASC was able to perform the clustering without any other user configuration.

As shown in Figure 2, FASC performed well on all data sets in terms of cluster detection and point assignment. FASC was able to correctly identify $\frac{38}{39}$ clusters across all data sets. This high level of performance shows that FASC is able work well with clusters of arbitrary size and shape, as well as clusters of varying density: a highly desirable trait.

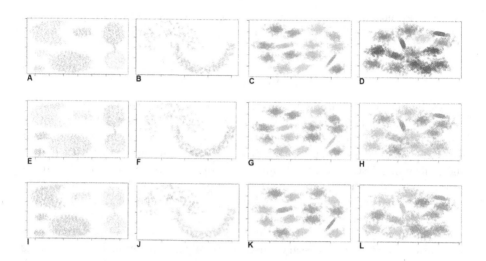

Fig. 2. Clustering results for FASC (A-D), k-means (E-F), and DBSCAN (I-L) on data sets Aggregation (A,E,I), Jain (B,F,J), S_1 (C,G,K), and S_2 (D,H,L)

On the other hand, the conventional clustering methods did not perform as well across all data sets. The k-means method identified $\frac{27}{39}$ clusters, while DB-SCAN identified $\frac{35}{39}$ clusters. As expected, k-means excelled in globular cluster identification, while DBSCAN performed best at identifying density-based clusters. Overall, FASC outperformed both classic methods in the data sets tested.

5   Conclusions

In this paper, a novel swarm optimization algorithm named Foraging Agent Swarm Optimization (FASO) was presented. FASO is designed to excel at converging on one or more optima in both gradient and point-based search spaces. It also operates well in situations where "field optima", rather than single point optima, are desired. Experimental results directly demonstrated that FASO is able to perform well in gradient search spaces containing one or more optima. FASO was also shown to be resilient to local minima that prove problematic to other optimization methods.

The ability of FASO to find "field optima" in point-based search spaces was demonstrated by directly applying FASO to data clustering. The novel clustering method, called Foraging Agent Swarm Clustering (FASC), treats clustering as an optimization problem containing multiple optima. Experimentation showed that FASC is a highly effective clustering technique in a variety of situations. It outperforms both k-means and DBSCAN in terms of cluster quality and autonomous adaptability to numerous clustering situations.

Overall, results for both FASO and FASC are very promising. As a generic swarm algorithm, FASO is able to consistently find optima in a variety of landscapes. Similarly, FASC provides an excellent clustering performance, perhaps paving the path for newer swarm-based clustering methods.

References

1. Alam, S., Dobbie, G., Riddle, P.: An evolutionary particle swarm optimization algorithm for data clustering. In: IEEE Swarm Intelligence Symposium, SIS 2008, pp. 1–6 (September 2008)
2. Aljarah, I., Ludwig, S.: A new clustering approach based on glowworm swarm optimization. In: 2013 IEEE Congress on Evolutionary Computation (CEC), pp. 2642–2649 (June 2013)
3. Cui, X., Potok, T., Palathingal, P.: Document clustering using particle swarm optimization. In: Proceedings 2005 IEEE Swarm Intelligence Symposium, SIS 2005, pp. 185–191 (June 2005)
4. Esmin, A., Coelho, R.: Consensus clustering based on particle swarm optimization algorithm. In: 2013 IEEE International Conference on Systems, Man, and Cybernetics (SMC), pp. 2280–2285 (October 2013)
5. Ester, M., Kriegel, H.P., Sander, J., Xu, X.: A density-based algorithm for discovering clusters in large spatial databases with noise, pp. 226–231. AAAI Press (1996)
6. Fränti, P., Virmajoki, O.: Iterative shrinking method for clustering problems. Pattern Recogn. 39(5), 761–775 (2006)
7. Gionis, A., Mannila, H., Tsaparas, P.: Clustering aggregation. ACM Trans. Knowl. Discov. Data 1(1) (March 2007)
8. Jain, A., Law, M.: Data clustering: A user's dilemma. In: Pal, S.K., Bandyopadhyay, S., Biswas, S. (eds.) PReMI 2005. LNCS, vol. 3776, pp. 1–10. Springer, Heidelberg (2005)
9. Krishnanand, K.N., Ghose, D.: Detection of multiple source locations using a glowworm metaphor with applications to collective robotics. In: Proceedings 2005 IEEE Swarm Intelligence Symposium, SIS 2005, pp. 84–91 (June 2005)
10. MacQueen, J.: Some methods for classification and analysis of multivariate observations (1967)
11. Niu, B., Duan, Q., Liang, J.: Hybrid bacterial foraging algorithm for data clustering. In: Yin, H., Tang, K., Gao, Y., Klawonn, F., Lee, M., Weise, T., Li, B., Yao, X. (eds.) IDEAL 2013. LNCS, vol. 8206, pp. 577–584. Springer, Heidelberg (2013)
12. Olesen, J., Cordero, H.J., Zeng, Y.: Auto-clustering using particle swarm optimization and bacterial foraging. In: Cao, L., Gorodetsky, V., Liu, J., Weiss, G., Yu, P.S. (eds.) ADMI 2009. LNCS, vol. 5680, pp. 69–83. Springer, Heidelberg (2009)
13. Szabo, A., de Castro, L., Delgado, M.: The proposal of a fuzzy clustering algorithm based on particle swarm. In: 2011 Third World Congress on Nature and Biologically Inspired Computing (NaBIC), pp. 459–465 (October 2011)
14. Van Der Merwe, D.W., Engelbrecht, A.: Data clustering using particle swarm optimization. In: The 2003 Congress on Evolutionary Computation, CEC 2003, vol. 1, pp. 215–220 (December 2003)
15. Wan, M., Li, L., Xiao, J., Wang, C., Yang, Y.: Data clustering using bacterial foraging optimization. Journal of Intelligent Information Systems 38(2), 321–341 (2012)

GPU Implementation of Food-Foraging Problem for Evolutionary Swarm Robotics Systems

Kazuhiro Ohkura[1], Toshiyuki Yasuda[1], Yoshiyuki Matsumura[2], and Masaki Kadota[1]

[1] Graduate School of Engineering, Hiroshima University, Hiroshima, Japan
kohkura@hiroshima-u.ac.jp
[2] Faculty of Textile Science and Technology, Shinshu University, Nakano, Japan

Abstract. Evolutionary swarm robotics (ESR) is an artificial approach for developing smart collective behavior in a system of homogenous autonomous robots. Robot behavior is generally controlled by evolving artificial neural networks. ESR has been considered a promising approach for swarm robotics systems (SRSs), because swarm behavior naturally emerges from numerous local interactions among the autonomous robots. In contrast, programming individual robots to display appropriate swarm behavior is extremely difficult. However, even in a simulated SRS, ESR is precluded by a very high computational cost. In this study, we introduce a novel implementation that overcomes the computational cost problem. The method employs parallel problem solving on a graphics processing unit (GPU) and OpenMP on a multicore CPU. To demonstrate the efficiency of the proposed method, we engage an evolving SRS in a food-foraging problem.

1 Introduction

Swarm robotics (SR) [1] investigates the collective behavior of multirobot systems with large redundancy and lack of centralized control. Specifically, since autonomous robots have limited capacity to sense and act in their local environments, they must cooperate in various ways to achieve a given task. Most SR systems are manually programmed [2] [3]. However, the complexity of the robot environment depends on the SR system size, and the complex behavior of large SR systems cannot be adequately generated by a sequence of manually controlled behaviors.

The present study adopts a typical evolutionary robotics approach [4] described in [5]. In this approach, the robot controller is designed by evolving an artificial neural network whose inputs and outputs are sensory inputs and motor outputs, respectively. Typically, the synaptic weights are coordinated with the artificial evolution. This SR approach is frequently called Evolutionary swarm robotics (ESR) [6]. Although ESR achieves an effective and robust collective behavior in a robotic system, the artificial evolution incurs prohibitively large computational cost.

M. Dorigo et al. (Eds.): ANTS 2014, LNCS 8667, pp. 238–245, 2014.

To overcome this problem, we evolve an SR system by the so-called GPU (graphics processing unit) computing technique. GPU computing extends the original role of GPUs (as graphics processing devices) to general-purpose computing. The high-speed architecture of GPUs is specialized for simple and independent calculation. Although GPU performance is degraded in computations involving many conditional branches, GPUs conduct floating-point calculations much more rapidly than CPUs. NVIDIA, a leading GPU-developing company, has launched a parallel computing platform and programming model called CUDA (compute unified device architecture))[1]. GPUs developed by NVIDIA are frequently used for purposes other than graphics [7][8]. For instance, Riegel et al. [9] implemented a Lattice Boltzmann method for numerical fluid mechanics in a GPU, and increased the computational speed (relative to CPU processing) by 7.5 times. Three GPUs improved the processing speed by 18.4 times.

The efficiency of a GPU depends on the number of processes in the algorithm that are suitable for parallel computing. The number of conditional branches is also important. In this study, we expect that GPU computing will accelerate the processing, since conditional branches are restricted to a particular section of the ESR algorithm.

This study considers GPU computing for solving ESR problems. The cooperative foraging problem is a typical problem that can be resolved from an ESR systems approach. Indeed, ESR might be the only way to develop smart collective behavior in this scenario. The study is organized as follows. Section 2 introduces our ESR systems approach to the cooperative foraging problem. Section 3 briefly explains the GPU architecture and the CUDA programming model. The implementation of ESR in a GPU is discussed in Section 4. In Section 5, the proposed method is evaluated in computational experiments. The study concludes with Section 6.

2 Cooperative Foraging Problem

Figure 1(a) shows the initial state of the cooperative foraging problem. The square field covers (5000 × 5000) unit lengths and is surrounded by walls. Initially, members of the robotic swarm are randomly placed in the square nest at the center of the field. The nest is (1000 × 1000) unit lengths. The swarm size (i.e., the number of robots) is set to 8, 16, 24 or 32.

The robotic swarm is required to maximize its collection of circular food sources. Each food source is 600 units in diameter and is randomly placed along the circumference of a circle centered at the field center, as shown in Fig. 1. If a robotic swarm succeeds in transporting a food source to the nest, i.e., when a food source is pushed by the robotic swarm until its center reaches the nest, it disappears, and a new food source is randomly placed on the solid circumference in Figure 1(a). A food source is assumed sufficiently heavy that at least three robots must combine their forces in the same direction to push it toward the nest. The maximum execution time of a single trial is set to 4000.

---

[1] http://www.nvidia.com/cuda/

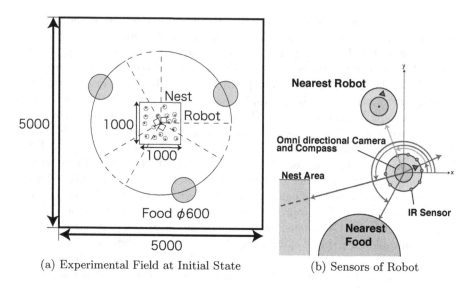

(a) Experimental Field at Initial State   (b) Sensors of Robot

Fig. 1. Cooperative Foraging Problem

Table 1. Point calculation for swarm behavior

A food source reaches the nest	+1000
Distance pushed to the nest	$+1000 \times (1 \ d_{rem} / d_{init})$
A robot is touching a food source	+0.03125 (swarm size = 8) +0.015625 (swarm size = 16) +0.0104167 (swarm size = 24) +0.0078125 (swarm size = 32)

The specifications of the robots are shown in Fig. 1(b). Each robot is 200 units in diameter, and is equipped with six IR sensors at its front and two IR sensors at its back. The sensing range of all detectors is 100 units. At the center of each robot is an omnidirectional camera with a sensing rage of 500 units. Individual robots can detect their nearest neighbor robots and the nearest food source by image processing within the sensing range. A robot receives distance and directional information on nearby objects. It should be noted here that directions are computed by pairs of sine and cosine functions. Each robot is also equipped with a digital compass by which it determines its absolute direction. In addition, all robots are assumed to obtain the absolute direction to the nest. All sensory information is perturbed by 3% Gaussian noise. The right and left wheels of each robot are actuated by separate motors. Therefore, each robot controller receives 18 inputs and generates two outputs. In this study, the controller is assumed to be a fully connected recurrent artificial neural network with four hidden neurons.

The collective behavior of a robotic swarm is evaluated as shown in Table 1. A robotic swarm gains 1000 points when it successfully brings a food source to the nest. If the swarm fails to transport the food to the nest within one time step, its points are calculated as $1000 \times (1 - d_{rem}/d_{init})$, where d_{rem} and d_{init} denote the linear distances between the centers of the nest and the food source in the last and initial states, respectively. If a robot touches a food source during a time step, it receives a small score that depends on the swarm size. Points awarded for touching a food source are 0.03125, 0.015625, 0.0104167, and 0.0078125 for swarm sizes of 8, 16, 24, and 32, respectively. Awarded scores assume that a robotic swarm will gain 1000 points if all of the robots contact food sources in all time steps.

3   CUDA

CUDA is a parallel computing platform and programming model invented by NVIDIA Corp. This integrated developmental GPU environment is widely used because it is well designed and free of charge. The GPU coprocessor executes many threads in parallel. In this environment, the GPU behaves as a device while the CPU and its main memory play host roles. A function executed on a device is called a kernel. A device executes a kernel function under direction by the host.

A device has many streaming multiprocessors (SM), each including numerous CUDA cores. Our objective is efficient use of these CUDA cores. The device can access global memory from all CUDA cores, and each SM holds shared memory, accessible solely from its internal CUDA cores. While shared memory may be speedily accessed from the CUDA cores with low latency, the latency of global memory access is hundreds of cycles. An important consideration is how to simultaneously execute many threads in parallel on CUDA cores while concealing the latency by accessing different threads at different times.

In the CUDA programming environment, threads are organized in two layers. The upper layer comprises grids, which are directly called by kernel functions. The lower layer is composed of thread blocks, typically in a two-dimensional configuration. In the following, thread blocks are simply referred to as blocks. The threads comprising the blocks are also typically arranged in two dimensions. During an execution process, a thread is assigned to a CUDA core, whereas a block is assigned to an SM.

4   GPU Implementation

As mentioned above, threads are configured in a two-layered structure in the CUDA environment. That is, such a structure is suitable for two-layered parallel computing. Therefore, the blocks are used to evaluate individuals in the artificial evolution, since they allow independent update of each individual. The threads in a block are used to calculate the fitness values of individuals in parallel.

Table 2. Settings of GA

Population size	256
Last generation	200
Selection	Tournament (size:2)
	Elitism (size:1)
Crossover rate	0.2
Crossover index	0.35
Mutation rate	0.01
Mutation index	0.25

Table 3. Development environment

CPU	Intel Core i7 2600K (3.40GHz)
GPU	NVIDIA Geforce GTX580
	(512 CUDA cores, 16 SMs)
	(1544MHz)
Main Mem.	8GB
OS	Fedora16 (64bits)
IDE	Nsight Eclipse Edition
	CUDA 5.0

The artificial evolution of the SR system proceeds through the following steps:

1. Initialization: robots and food sources are randomly positioned. All information is initialized.
2. Sensing: a robot retrieves sensory information from its IR sensors, omnidirectional camera and compass.
3. Controller calculation: the obtained sensory information is passed to the robot controller, which provides output information.
4. Collision detection and constraint solving: robots move according to their controller outputs. Collisions with other robots, walls, and food sources are monitored. If robots collide with an object, constraint calculations are executed.
5. Return to step 2 until the last time step is counted.

At Step 1, a block is defined for each individual. The number of threads defined per block equals the swarm size. The robotic swarm is initialized prior to other operations. Step 2 employs two kernel functions. The first kernel controls the sensing by IR sensors. All IR sensors of all robots are assigned to different threads, enabling complete parallel processing of each block. The second kernel controls sensing by omnidirectional cameras and compasses. This kernel function is executed identically to Step 1. At Step 3, the outputs of the robot controller are calculated by the parallel processing adopted in Step 1. At Step 4, the conditional branches for collision detection and constraint solving are unavoidable, this step is thus executed on the CPU. However, we access pinned memory on the host since the pinned memory can be copied to the device memory parallel to other kernel executions. To coordinate the population, the CPU execution is parallelized on OpenMP[2].

Genetic operators are implemented on the GPU as described in [10]. In the following computer simulations, we adopt a real-coded genetic algorithm (GA) [11] with Laplace crossover [12] and power mutation [13]. The pseudorandom number generator used in both GPU and CPU is Xorshift [14].

---

[2] http://openmp.org/wp/

Table 4. Average processing time (s) for swarm size = 8

Operation	Using OpenMP		Using GPU and OpenMP		
	Processing time	Deviation	Processing time	Deviation	Ratio
Op. A	1142.564	98.412	140.913	5.965	8.108
Op. B	102.399	1.109	84.155	0.019	1.217
Op. C	183.802	37.566	193.489	32.672	0.950
Op. D	-	-	111.632	0.060	-
Op. E	0.099	0.003	0.022	0.000	4.476
Op. F	0.592	0.004	41.723	12.504	0.014
Total	1429.456	124.147	571.933	37.888	2.499

Table 5. Average processing time (s) for swarm size = 16

Operation	Using OpenMP		Using GPU and OpenMP		
	Processing time	Deviation	Processing time	Deviation	Ratio
Op. A	3122.674	201.117	296.775	15.443	10.522
Op. B	202.507	0.774	131.170	0.094	1.544
Op. C	642.163	115.369	629.665	100.970	1.020
Op. D	-	-	201.970	0.131	-
Op. E	0.100	0.002	0.022	0.000	4.565
Op. F	1.122	0.004	44.637	11.935	0.025
Total	3968.566	278.123	1304.239	108.911	3.043

5   Computational Experiments

5.1   Settings

The parameters of the real-coded GA are listed in Table 2. The genotype length is 136. In addition to the elitism of size 1, the tournament selection of size 2 is assumed in the natural selection process. The computer environment is summarized in Table 3. Our approach is compared against a typical implementation on a Core i7 CPU multithreaded for population by OpenMP. In the following, the comparison is referred to as the standard method. Similarly, we refer to the GPU implementation of the standard method as the proposed method. The swarm size is set to 8, 16, 24, or 32. Under each condition, the average and standard deviation of the processing time is calculated from 30 runs.

5.2   Results

In all of the computational experiments, the cooperative food foraging problem was solved and robot teams successfully brought one or more food sources to

Table 6. Average processing time (s) for swarm size = 24

Operation	Using OpenMP		Using GPU and OpenMP		
	Processing time	Deviation	Processing time	Deviation	Ratio
Op. A	5885.165	413.318	518.594	38.487	11.348
Op. B	301.589	0.283	179.200	0.033	1.683
Op. C	1806.164	902.308	1667.286	455.317	1.083
Op. D	-	-	298.787	0.133	-
Op. E	0.097	0.001	0.022	0.000	4.471
Op. F	1.662	0.008	54.048	11.619	0.000
Total	7994.676	1109.690	2717.938	482.373	2.941

Table 7. Average processing time (s) for swarm size = 32

Operation	Using OpenMP		Using GPU and OpenMP		
	Processing time	Deviation	Processing time	Deviation	Ratio
Op. A	9612.531	437.086	927.365	65.771	10.365
Op. B	402.251	4.447	237.831	1.158	1.691
Op. C	4211.370	4470.927	3508.710	1243.615	1.200
Op. D	-	-	389.825	0.734	-
Op. E	0.101	0.002	0.022	0.001	4.508
Op. F	2.260	0.033	60.862	11.318	0.037
Total	14228.513	4615.823	5124.614	1264.831	2.777

the nest. The processing times of the various operations are summarized in Tables 4, 5, 6 and 7. Operation A (denoted Op. A) is the sensing activity. The other operations, denoted Ops. B, C, D, E and F, are the ANN calculation, collision detection and constraint solving, data transfer between the GPU and CPU, natural selection and genetic operations, and miscellaneous operations, respectively. Relative to the standard method, the proposed method reduced the total processing time by factors of 2.5, 3.0, 2.9, and 2.8 for swarm sizes of 8, 16, 24, and 32, respectively. The proposed method conferred its greatest benefit in the sensing operation, which was maximized at 11.3 times faster than the standard method. On the other hand, the proposed method only minimally benefitted the ANN calculation (performing only 1.7 times faster than the standard method).

6   Conclusions

In this study, we discussed whether the smart collective behavior in an ESR system can be efficiently implemented by GPU computing. As an example, we implemented the food-foraging problem. In an experimental computer environment, the GPU proved beneficial for parallel processing of ESR, particularly when processing independent sensors equipped on robots. As a next step, we plan to extend the proposed method into a multi-GPU computing environment, which will accommodate a much larger ESR system.

References

1. Şahin, E.: Swarm robotics: From sources of inspiration to domains of application. In: Şahin, E., Spears, W.M. (eds.) Swarm Robotics 2004. LNCS, vol. 3342, pp. 10–20. Springer, Heidelberg (2005)
2. Turgut, A.E., Çelikkanat, H., Gökçe, F., Şahin, E.: Self-organized flocking in mobile robot swarms. Swarm Intelligence 2(2-4), 97–120 (2008)
3. Liu, W., Winfield, A.F., Sa, J.: Modelling swarm robotic systems: A case study in collective foraging. Towards Autonomous Robotic Systems (TAROS 2007), 25–32 (2007)
4. Harvey, I., Husbands, P., Cliff, D., et al.: Issues in evolutionary robotics. School of Cognitive and Computing Sciences, University of Sussex (1992)
5. Şahin, E., Girgin, S., Bayindir, L., Turgut, A.E.: Swarm robotics. In: Swarm Intelligence, pp. 87–100. Springer (2008)
6. Trianni, V.: Evolutionary swarm robotics: evolving self-organising behaviours in groups of autonomous robots, vol. 108. Springer (2008)
7. Stone, S.S., Haldar, J.P., Tsao, S.C., Hwu, W.M., Sutton, B.P., Liang, Z.P., et al.: Accelerating advanced mri reconstructions on gpus. Journal of Parallel and Distributed Computing 68(10), 1307–1318 (2008)
8. Preis, T., Virnau, P., Paul, W., Schneider, J.J.: Accelerated fluctuation analysis by graphic cards and complex pattern formation in financial markets. New Journal of Physics 11(9), 093024 (2009)
9. Riegel, E., Indinger, T., Adams, N.: Implementation of a lattice boltzmann method for numerical fluid mechanics using the nvidia cuda technology. Computer Science-Research and Development 23(3-4), 241–247 (2009)
10. Oiso, M., Matsumura, Y., Yasuda, T., Ohkura, K.: Implementing genetic algorithms to cuda environment using data parallelization. Tehnicki vjesnik/Technical Gazette 18(4) (2011)
11. Goldberg, D.E., et al.: Genetic algorithms in search, optimization, and machine learning, vol. 412. Addison-Wesley, Reading (1989)
12. Deep, K., Thakur, M.: A new crossover operator for real coded genetic algorithms. Applied Mathematics and Computation 188(1), 895–911 (2007)
13. Deep, K., Thakur, M.: A new mutation operator for real coded genetic algorithms. Applied mathematics and Computation 193(1), 211–230 (2007)
14. Marsaglia, G.: Xorshift rngs. Journal of Statistical Software 8(14), 1–6 (2003)

Nature-Inspired Swarm Robotics Algorithms for Prioritized Foraging

Jade Abbott and Andries P. Engelbrecht

Computational Intelligence Research Group, Department of Computer Science,
University of Pretoria, South Africa
{jabbott,engel}@cs.up.ac.za

Abstract. This paper introduces the problem of prioritized foraging. The performance of a naïve foraging algorithm and a desert ant foraging algorithm is evaluated on the prioritized foraging problem. The evaluation is used to motivate the creation of a novel honey bee based foraging algorithm. The performance and adaptability to different environments is evaluated for all algorithms. This paper concludes that a relationship exists between the performance of the existing foraging algorithms and the initial ratio of agents configured to forage prioritized items. The honey bee algorithm was shown to perform well across all configurations.

1 Introduction

An important activity of all natural swarms is foraging for resources [1]. These resources could be food, water, or building materials. In times of stress, the collection of one resource may be prioritized over others - such as water during a drought. Individuals in a natural swarm often adapt behaviour appropriately to enable greater collection of the prioritized item. Item prioritization during foraging exists in real-world robot foraging problems such as search and rescue where some agents move waste material to reach the trapped survivors.

This paper introduces the problem of prioritized foraging. The performance of a naïve foraging algorithm and a desert ant-based foraging algorithm are evaluated on the prioritized foraging problem. The evaluation is used to motivate the creation of an algorithm whose performance is independent of the initial agent configuration. Finally, a novel honey bee based foraging algorithm, with the ability to divide its labourers between items of different priorities, is presented.

Section 2 provides background on foraging, while the problem of prioritized foraging is discussed in section 3. The agents are described in section 4, and section 5 presents the algorithms. Section 6 outlines the experimental setup and the results are reported in 7. Section 8 concludes the paper.

2 Background

In swarm robotics, foraging has become a benchmark problem due to its complex nature involving coordination of numerous sub-tasks. Foraging has a variety of

M. Dorigo et al. (Eds.): ANTS 2014, LNCS 8667, pp. 246–253, 2014.
© Springer International Publishing Switzerland 2014

real-life applications and numerous swarm robotics algorithms have been developed to improve robustness, time and energy efficiency of the foraging process [1]. This section discusses foraging algorithms in nature.

2.1 Ant Foraging

Ant-based algorithms are difficult to replicate in real-life since techniques use pheromone. Agent-based algorithms that use pheromone require agents to be equipped with substance-distributors, beacon-deployers or a complex simulation of pheromone [2].

The desert ant (*Cataglyphis bicolor*) does not make use of pheromone-based communication to forage, as pheromone deposited on desert sand would be blown away by the wind. Instead, desert ants use a technique called path-integration for navigation to relocate food sources that have already been found [3].

Desert ant foraging has been modelled and experiments performed on real agents in [4]. Experiments were performed using a small set of agents in a small arena on different environment types. Two algorithms were evaluated: one based on the desert ant foraging and one with pheromone-like communication. It was shown that communication improved performance; however, the desert-ant algorithm performed reasonably.

2.2 Bee Foraging

Honey bee foraging is made up of three roles [5]: Employed foraging, unemployed foraging and scouting. Scouts explore to locate new food sources. Once a source has been found, scouts return to the hive to communicate information about the source. Scout bees perform a waggle-dance to the hive, containing information about the distance and bearing of the resource, known as recruitment.

Unemployed foragers evaluate the dances of the scout bees. An unemployed bee chooses a location described by the scout bees, and becomes an employed forager. Employed foragers attempt to locate the source, load themselves and return to the hive where unemployed foragers are ready to offload the food. Jansen [6] suggests that unemployed bees become exploring scouts when they do not detect any dancing scout bees. Bee swarms have been used in swarm robotics for problems such as path planning [7] and collective perception [8].

3 Prioritized Foraging

The prioritized foraging problem, is a modified version of the multi-foraging problem [9] with two types of items: prioritized items and non-prioritized items. The goal is to forage all items of the prioritized type. Prioritized items may become trapped among non-prioritized items and thus the non-prioritized items must be removed from the environment, however foraging the non-prioritized item too much may result in a waste of time and energy.

The aim of research in prioritized foraging is to develop an algorithm to efficiently adapt the number of agents foraging prioritized items to those moving non-prioritized items out the way.

4   Agents

Each agent is equipped with a 360°camera to identify an item's type and eight local proximity sensors spaced equally around the circular perimeter of the agent. Both camera and proximity sensors have a depth of view of five times the agent's size. Agents have grippers, to pick up items.

Agents use local broadcast communication, occurring in a radius of five times the agent's size. An agent can forage a single item at a time. The agents do not have a global positioning system capability. All algorithms used in this paper use the same obstacle avoidance and navigation technique.

Item identification and navigation techniques run in a separate thread. An agent will only approach an item when the item has been identified as the agent's allocated item type, otherwise the robot will attempt to move around the item.

5   Algorithm Description

Naïve foraging is discussed in section 5.1, desert ant foraging discussed in section 5.2, and honey bee foraging is discussed in 5.3.

5.1   Naïve Foraging

Naïve foraging includes only the most minimal set of foraging actions. Naïve foraging is a baseline for comparison to evaluate how other techniques compare to a standard model [10,2].

Naïve foraging works as follows: Agents perform a random walk until they find an item. On locating an item, the agent grips the item. If the item has been moved, the agent will continue to explore; otherwise the agent returns the item to the correct sink using a beacon-based homing algorithm.

The following random walk is used: an agent chooses a random direction, σ, and a random distance $m \in (0, M)$ where M is a chosen maximum path length. The agent walks in direction σ for distance m. The agent then chooses new values for σ and m.

5.2   Desert Ant Foraging

Due to the lack of pheromone, desert ant foraging behaviour is a very suitable model for agent foraging. Desert ants use path integration to memorize the location of an existing food source and later to return to the memorized source to find more food. The notion of returning to a previously explored site is known as site fidelity [11]. The desert ant algorithm does not require communication between agents or the dispersal of beacons, and is thus simpler than other many swarm robotics foraging algorithms.

Path integration (PI) is the integration of an ant's odometry such that the ant can maintain a position and heading estimate of where the ant is going, by continuously updating a heading direction vector that points to the starting position [12]. The heading-direction vector is known as the PI vector. The agent control algorithm consists of five states. The agent performs a random walk while performing PI (exploration). On finding an item, the agent loads the item and memorizes the PI vector (loading). The agent uses the path integration vector to move to the sink (homing). When the agent is at the sink, the object is offloaded (offloading). The agent follows the memorized PI vector to the location of the previous item. If another item is found, the item is loaded; otherwise the agent returns to the exploration state (locating).

5.3   Honey Bee Foraging

The presented honey bee algorithm is based on the mathematical model of honey bee foraging in [5]. A portion of the agents are initialized as scouts and the rest as unemployed foragers in a waiting state. State transitions for the honey bee algorithm are described as follows:

1. A scout agent performs a random walk and, upon finding an item and evaluating the site, forages the item by returning the item to the sink using PI.
2. A scout decides to dance, based on the quality of the site. If the estimated quality of the site, μ, is less than the dance threshold, ϕ, the scout agent does not dance and instead continues to forage the site as an employed forager.
3. Otherwise, if the estimated quality of the site, μ, is greater or equal to the dance threshold, ϕ, the scout agent dances for the unemployed foragers.
4. After a dance is complete, the scout agent decides with probability ρ to start exploring again or recruit itself and begins foraging the found site.
5. On detecting a dance, waiting bees become foraging bees with a probability α and switch to foraging the dancer's item type.
6. Employed foragers become unemployed foragers (waiting foragers) if the foraging site has been depleted.
7. Unemployed foragers become scouts if no dances are detected t_{max} time steps.

Site quality, μ_t, for an agent scouting items of type t, is calculated as the estimated density of items of type t in the local vicinity of the found item. The agent has distance sensor values $k_i \in [0,1]$ for $i = 1...n$, where 0 means that nothing is detected in sensor range and 1 indicates that the agent is touching an item and n is the number of distance sensors. The item density of type t, μ_t, is calculated using

$$\mu_t = \frac{1}{n} \sum_{i=1}^{n} k_{i_t} \tag{1}$$

where the sensor value for item type t, k_{i_t}, is calculated using

$$k_{i_t} = \begin{cases} k_i & \text{if item } i \text{ is type } t \\ 0 & \text{otherwise} \end{cases} \tag{2}$$

In times of drought, bees prioritize water over nectar or pollen. Bees are sent out to forage water; however, if they happen to encounter pollen, they will forage it but will not communicate the discovery [5]. The rules for item-type division of labour are based on the behaviour of bees under environmental pressure. An agent foraging the prioritized type, forages a non-prioritized type only if a prioritized item can not be located for max time f_{max}. An agent foraging a non-prioritized item forages the non-prioritized item type until the agent fails to relocate the site or the agent locates a prioritized item, switching back to foraging the prioritized item. An agent foraging a non-prioritized item will not communicate the location of the non-prioritized item site.

6   Experimental Setup

A 2D grid world was used to evaluate the performance of the foraging algorithms. Each agent fits into one grid block and each item takes up one grid block. Items can be picked up, or otherwise form obstacles that agents must navigate around.

The prioritized and non-prioritized sinks were placed beside each other, on a single side of the environment to more accurately represent the type of environment that the problem could be applied to e.g. in mine tunnels, differing from the more common placement at the centre of the environment. The sinks were marked by light beacons that all agents can detect and navigate towards. Agent's beacon homing to the sinks is simulated by each agent evaluating and moving up a light intensity gradient. The colour of the light is used to distinguish between the 2 sinks.

Four different classes of environments were used: uniformly distributed environments, clustered environments with clusters of item types generated by randomly relabelling items in clusters generated by Lumer-Faieta ant cemetery clustering [13] as either prioritized or non-prioritized items, vein environments resembling the natural occurrence of gold and gaussian environments where prioritized items are focused at the environment center.

For each class of environment, the following configurations were tested: The environment grid size, $S = 50, 100, 200$ where S is the width and length of the grid and the percentage p of the grid covered by objects, with $p = 5\%, 20\%, 50\%, 70\%, 90\%$. The ratio of prioritized to non-prioritized items r, is varied, where $r = 0, 0.2, 0.25, 0.333, 0.5, 0.667, 0.75, 0.8, 1$. For each configuration, 30 environments were generated and all algorithms were run on all environments. Honey bee specific parameters were selected as $t_{max} = 200$ timesteps, $f_{max} = 100$ timesteps, $\phi = 0.8$ and $\rho = 0.1$, are based on [5].

Agents were initially configured to forage either the prioritized item or the non-prioritized item with a ratio of $\tau = 0, 0.2, 0.25, 0.333, 0.5, 0.667, 0.75, 0.8, 1$. Different numbers of agents, c, were used with $c = 10, 30, 50, 70, 100$; c is defined as the percentage of cells of the grid size S that are occupied by agents.

All algorithms were run for 10000 time steps, where an agent can move maximum 1 grid cell at a time, to any adjacent cell, with no stopping conditions. For all algorithms, the agents begin randomly placed next to the sink. Due to space constraints, only the percentage of prioritized items foraged, σ, is presented.

7   Results

When comparing two foraging algorithms, pairwise Wilcoxon test was performed, to determine if a statistical difference occurs, at a significance level of 95%. If an algorithm statistically outperforms the other algorithm in the comparison, a win is awarded to that algorithm. The wins are counted per algorithm and are shown in Table 1. The sample set consisted of all data obtained from the experiment over all configurations. The null hypothesis is that the results of the two algorithms come from the same distribution.

Table 1. The overall Pairwise Mann Whitney U wins, averages and standard deviations of for average prioritized item found over time (σ) for each algorithm

Algorithms	Wins	Average	Std Dev
Naive	3	0.528	0.394
Desert Ant	2	0.643	0.387
Honey Bee	1	0.807	0.294

Statistical tests indicate a significant difference between the results of all algorithms. Desert ant foraging performed better than naïve foraging showing the positive effect of site fidelity. The honey bee algorithm out-performed the naïve foraging algorithm and desert ant algorithm indicating the positive effect of communication and adaptivity of the honey bee foraging algorithm. The standard deviation is high for all algorithms due to large variations in the environments provided.

The following hypotheses are addressed: An algorithm foraging a portion of non-prioritized items will have greater performance than an algorithm that does not forage any non-prioritized items. Performance depends on the r and τ. As r increases, the value of τ that yields the greatest value of σ, τ_{best}, will increase approximately linearly for the naïve and desert ant algorithms.

An algorithm with configuration $\tau = 1$, forages only prioritized items. Analysing Table 2, for the naïve and desert ant algorithms, for all values of r where $r \neq 1$, τ_{best} is never equal to 1, proving the hypothesis that the algorithms achieved the best performance when some agents are configured to forage non-prioritized items. Perhaps non-prioritized items are moved out of the way to allow for easier, faster access to prioritized items or allow access to inaccessible prioritized items.

The naïve and desert ant algorithms performed best when τ was slightly greater than r. The existence of the relationship motivates the development of an algorithm that adapts τ to correspond the environment item ratio r.

Analysis of Table 2 indicates that the honey bee foraging algorithm achieves similar performance throughout all configurations for r and τ, highlighting that the performance of the honey bee algorithm is independent of the configuration of τ, resulting in an algorithm that is more flexible and robust. This could mean that the honey bee algorithm could perform well in dynamic environments where

Table 2. The performance, σ, for each foraging algorithm, for each combinations of r and τ. If τ_{best} exists, τ_{best} is provided. The best value of σ is shown in bold.

Algorithm	r	\multicolumn{9}{c}{τ}	τ_{best}
		0	0.2	0.25	0.333	0.5	0.667	0.75	0.8	1	
Naïve	0	1	1	1	1	1	1	1	1	1	
	0.2	0	0.492	0.526	0.567	**0.597**	0.595	0.587	0.577	0.471	0.5
	0.25	0	0.484	0.526	0.557	0.588	**0.595**	0.585	0.575	0.477	0.667
	0.333	0	0.467	0.507	0.544	0.586	**0.596**	0.592	0.584	0.495	0.667
	0.5	0	0.428	0.46	0.508	0.568	0.588	**0.591**	0.589	0.528	0.75
	0.667	0	0.4	0.433	0.487	0.544	0.583	**0.591**	0.593	0.554	0.75
	0.75	0	0.377	0.425	0.47	0.531	0.576	0.585	**0.591**	0.567	0.8
	0.8	0	0.372	0.409	0.455	0.53	0.571	0.584	**0.592**	0.575	0.8
	1	0	0.336	0.375	0.433	0.5	0.552	0.57	0.581	**0.618**	1
Desert Ant	0	1	1	1	1	1	1	1	1	1	
	0.2	0	0.698	0.724	**0.737**	**0.737**	0.712	0.694	0.67	0.519	0.333
	0.25	0	0.678	0.711	0.73	**0.735**	0.715	0.697	0.673	0.530	0.5
	0.333	0	0.65	0.693	0.722	**0.739**	0.725	0.71	0.686	0.562	0.5
	0.5	0	0.596	0.645	0.684	0.729	**0.734**	0.725	0.701	0.621	0.667
	0.667	0	0.554	0.607	0.648	0.706	0.737	**0.738**	0.716	0.675	0.75
	0.75	0	0.533	0.587	0.63	0.691	0.731	**0.739**	0.72	0.703	0.75
	0.8	0	0.523	0.577	0.62	0.682	0.725	0.736	**0.74**	0.718	0.8
	1	0	0.488	0.543	0.588	0.654	0.702	0.718	0.726	**0.758**	1
Honey Bee	0	1	1	1	1	1	1	1	1	1	
	0.2	**0.687**	**0.687**	0.686	0.686	0.686	0.685	0.686	0.685	**0.687**	
	0.25	0.678	**0.679**	0.678	0.678	**0.679**	**0.679**	0.678	0.677	**0.679**	
	0.333	**0.674**	**0.674**	**0.674**	**0.674**	**0.674**	**0.674**	0.673	**0.674**	**0.674**	
	0.5	0.668	**0.669**	0.668	0.668	0.668	0.668	0.668	0.668	**0.669**	
	0.667	0.671	0.671	0.671	0.671	0.671	**0.672**	0.671	0.671	0.671	
	0.75	0.672	**0.673**	0.671	0.671	0.672	**0.673**	0.672	**0.673**	**0.673**	
	0.8	0.674	0.674	0.674	0.674	0.674	**0.675**	**0.675**	**0.675**	**0.675**	
	1	**0.691**	0.69	**0.691**	0.69	**0.691**	**0.691**	0.69	0.69	0.69	

agents and items can be destroyed. This robustness of the honey bee algorithm may be attributed to the division of labour strategy used by the algorithm.

However, according to Table 2, the desert ant algorithm performs better than the honey bee algorithm, for particular configurations of r and τ. This indicates that, if the value of r is known for a particular environment, then it is beneficial to use desert ant foraging and choose τ appropriately. A possible reason why the desert ant algorithm performs better when optimally configured for a particular environment than the honey bee algorithm is that the honey bee algorithm takes time to adapt to the environment, while the desert ant algorithm with optimal configuration has no division of labour overhead and may outperform the honey bee algorithm under those circumstances.

8   Conclusions and Future Research

This paper concludes that an algorithm that forages a portion of non-prioritized items will have greater performance than an algorithm that does not forage any non-prioritized items and that the performance of the naïve and desert ant foraging algorithms is dependant on the ratio of agents initially configured to forage the prioritized item and the ratio of prioritized items in the environment. The honey bee algorithm was shown to perform well across all of the initial configurations; however, the desert ant algorithm still outperformed the honey bee algorithm for certain environment object ratios and agent forage type ratios.

Future work will also evaluate the performance of the honey bee algorithm in dynamic environments and will include an evaluation of other performance measures and an in depth discussion of the scalability of the algorithms.

References

1. Winfield, A.F.: Foraging robots. In: Meyers, R. (ed.) Encyclopedia of Complexity and Systems Science, pp. 3682–3700. Springer (2009)
2. Hoff, N.R., Sagoff, A., Wood, R.J., Nagpal, R.: Two foraging algorithms for robot swarms using only local communication. In: 2010 IEEE International Conference on Proceedings of Robotics and Biomimetics (ROBIO), pp. 123–130. IEEE (2010)
3. Collett, M., Collett, T.S., Bisch, S., Wehner, R.: Local and global vectors in desert ant navigation. Nature 394(6690), 269–272 (1998)
4. Hecker, J.P., Letendre, K., Stolleis, K., Washington, D., Moses, M.E.: *formica ex machina*: Ant swarm foraging from physical to virtual and back again. In: Dorigo, M., Birattari, M., Blum, C., Christensen, A.L., Engelbrecht, A.P., Groß, R., Stützle, T. (eds.) ANTS 2012. LNCS, vol. 7461, pp. 252–259. Springer, Heidelberg (2012)
5. Seeley, T.D.: The wisdom of the hive: the social physiology of honey bee colonies. Harvard University Press (2009)
6. Janson, S., Middendorf, M., Beekman, M.: Searching for a new home—scouting behavior of honeybee swarms. Behavioral Ecology 18(2), 384–392 (2007)
7. Lin, J.H., Huang, L.R., et al.: Chaotic bee swarm optimization algorithm for path planning of mobile robots. In: Proceedings of the 10th WSEAS International Conference on Evolutionary Computing, World Scientific and Engineering Academy and Society (WSEAS), pp. 84–89 (2009)
8. Schmickl, T., Möslinger, C., Crailsheim, K.: Collective perception in a robot swarm. In: Şahin, E., Spears, W.M., Winfield, A.F.T. (eds.) SAB 2006 Ws 2007. LNCS, vol. 4433, pp. 144–157. Springer, Heidelberg (2007)
9. Balch, T.: The impact of diversity on performance in multi-robot foraging. In: Proceedings of the Third Annual Conference on Autonomous Agents, pp. 92–99. ACM (1999)
10. Østergaard, E.H., Sukhatme, G.S., Matari, M.J.: Emergent bucket brigading: a simple mechanisms for improving performance in multi-robot constrained-space foraging tasks. In: Proceedings of the Fifth International Conference on Autonomous Agents, pp. 29–30. ACM (2001)
11. Switzer, P.V.: Site fidelity in predictable and unpredictable habitats. Evolutionary Ecology 7(6), 533–555 (1993)
12. Ronacher, B.: Path integration as the basic navigation mechanism of the desert ant cataglyphis fortis (forel, 1902)(hymenoptera: Formicidae). Myrmecological News 11, 53–62 (2008)
13. Lumer, E.D., Faieta, B.: Diversity and adaptation in populations of clustering ants. In: Proceedings of the Third International Conference on Simulation of Adaptive Behavior: from Animals to Animats, pp. 501–508. MIT Press (1994)

Particle Swarm Optimisation with Enhanced Memory Particles

Ian Broderick and Enda Howley

Discipline of Information Technology, National University of Ireland, Galway, Ireland
{i.broderick1,enda.howley}@nuigalway.ie

Abstract. Particle swarm optimisation (PSO) is a general purpose optimisation algorithm in which a population of particles are attracted to their past success and the success of other particles. This paper introduces a new variant of the PSO algorithm, PSO with Enhanced Memory Particles, where the cognitive influence is enhanced by having particles remember multiple previous successes. The additional positions introduce diversity which aids exploration. Balancing the need for exploitation with this additional diversity is achieved through the use of a small memory and by using Roulette selection to select a single position from memory to use when calculating particles' velocities. The research shows that PSO EMP performs better than the Standard PSO in most cases and does not perform significantly worse in any case.

1 Introduction

The particle swarm optimisation algorithm was developed by James Kennedy and Russell Eberhart [1] as a result of their work on modelling the behaviour of flocking birds. In PSO particles are driven by two influences, a social influence and a nostalgic influence. The social influence encourages particles to emulate the behaviour of the most successful particles in the swarm, while the nostalgic influence tends to pull them back to their own past successes. The social influence depends on the interactions and connections between particles. Much research has been conducted on the social component, by varying the topology in which the particles are connected [6] or how the social influence is calculated [3,4]. The focus of this paper will be on the nostalgic or cognitive influence on the particles with the goal of answering two main research questions: (1) Can the performance of the Standard PSO be improved by having particles remember additional previous good positions? (2) What is the most efficient way to utilise this additional information?

The rest of this paper will be organised as follows. Section 2 will summarize the previous research that is relevant to this paper. Section 3 will introduce the new PSO EMP algorithm. The experimental data will be presented in Section 4. Finally, Section 5 will give the conclusions reached as a result of this research.

M. Dorigo et al. (Eds.): ANTS 2014, LNCS 8667, pp. 254–261, 2014.

2   Background Research

2.1   Particle Swarm Optimisation

In PSO optimisation there is a population of agents, known as particles. These particles move around a solution space, defined by a fitness function, and evaluate their fitness at each location they visit. Particles are defined by a location within the solution space (x_t) and a velocity (v_t). At each iteration every particle calculates its velocity, moves to a new location and evaluates its fitness.

$$x_{t+1} = x_t + v_t \qquad (1)$$

$$v_{t+1} = \chi \left(v_t + r_1 c_1 \left(p_{id} - x_t \right) + r_2 c_2 \left(p_{gd} - x_t \right) \right) \qquad (2)$$

Where r_1 and r_2 are random numbers between 0 and 1. c_1 and c_2 are acceleration coefficients. χ is the constriction coefficient.

$$\chi = \frac{2}{\left| 2 - \varphi - \sqrt{\varphi^2 - 4\varphi} \right|}, \varphi = c_1 + c_2 \qquad (3)$$

Selecting $c_1 = c_2 = 2.05$ gives $\varphi = 4.1$ and $\chi \approx 0.72984$. Setting these values guarantees convergence [11]. The use of these values with Eq.1 and Eq.2 has become the Standard PSO [2]. p_{id} is the position where the particle achieved its best fitness. p_{gd} is the best position achieved by any particle.

2.2   Social Influence

Mendes et al. introduced the Fully Informed Particle Swarm (FIPS) [3]. In the FIPS a particle uses information from all of its neighbours when calculating the velocity, not just the best performer. This is done by splitting the c_2 acceleration constant between the particles in the neighbourhood. Two different methods for doing this were examined. In the Standard FIPS all neighbours contribute the same weight to the velocity equation. In the weighted FIPS the contribution of each neighbour is weighted by its previous fitness. Jordan et al. introduced a variant of FIPS, the ranked FIPS [4]. The ranked FIPS uses a ranking system to weight the influence of each particle in the neighbourhood.

Both the FIPS and ranked FIPS attempt to improve the social component of the velocity equation by dividing c_2 term. This paper will examine the cognitive component, dividing the c_1 term using methods similar to those used by the FIPS and ranked FIPS.

2.3   Enhancing Memory in PSO

This section summarises some of the research conducted into the use of internal and external memories in PSO to make better use of the available information. An internal memory stores information within the population, while an external memory stores information separate to the population.

Yin et al. introduced the Cyber Swarm Algorithm which adopted the idea of using a Reference Set in a PSO [7]. A Reference Set is an external memory storing elite mutually diverse solutions taken from the entire swarm that is dynamically updated during the optimisation process. The members of the Reference Set are used when calculating the particles' velocities. This approach was found to improve on the Standard PSO.

Hu et al. proposed the use of an external memory to store all Pareto optimal solutions to improve the performance of PSO in MOP [8]. The external memory is then used when selecting the neighbourhood best. Chao et al. used both an internal and external memory to enhance the performance of PSO for MOP [9]. Each particle's extended memory stores a set of mutually non-dominating non-inferior pBest solutions. A random pBest is chosen from memory when a particle is calculating its velocity. An external memory is also used to store all Pareto optimal solutions.

The research in this paper will use an internal memory to store additional best positions for each particle.

3   PSO with Enhanced Memory Particles

In a typical PSO implementation each particle remembers only the best position it has been to, forgetting all other positions. This discards a lot of information that could potentially be useful. In PSO EMP particles remember a number of their best positions. These additional remembered positions are used when calculating the particles' velocities. The additional positions should encourage swarm diversity which may lead to improved performance.

The new PSO EMP algorithm will replace the individual best term in the velocity equation with a new EMP term (Eq. 4). A number of different approaches to calculating this term have been explored. The approaches typically involve a division of the c_1 acceleration coefficient. This division does not affect the convergence of the algorithm as long as the sum of the terms gives the whole c_1. The proposed methods of dividing the c_1 term are outlined below.

$$v_{t+1} = \chi * (v_t + r_1 EMP_{\text{term}} + r_2 c_2 (p_{gd} - x_t)) \tag{4}$$

3.1   Equal Influence EMP

The simplest method for calculating the EMP term is to give equal influence to all remembered positions when calculating the velocity of a particle. This is done by splitting the c_1 acceleration coefficient into N equal parts, where N is the number of remembered positions. The velocity equation then becomes:

$$v_{t+1} = \chi * \left(v_t + r_1 \sum_{j=1}^{N} \frac{c_1}{N} (p_{id,j} - x_t) + r_2 c_2 (p_{gd} - x_t) \right) \tag{5}$$

Where $p_{id,j}$ is the jth best position remembered by the particle. This approach is similar to the approach used in the FIPS for dividing the c_2 term [3].

3.2 Power Law Distribution EMP

The second method for calculating the EMP term divides c_1 among the remembered positions using the ranking of these positions according to their fitness. The best remembered position will have twice the influence of the second best, the second twice that of the third, and so on down the ranking. In this way the influences are determined by a power law distribution.

$$v_{t+1} = \chi * \left(v_t + r_1 c_1 \sum_{j=1}^{N} rank_j \left(p_{id,j} - x_t \right) + r_2 c_2 \left(p_{gd} - x_t \right) \right) \qquad (6)$$

$$rank_j = \frac{2^{j-1}}{2^N - 1} \qquad (7)$$

Where a rank of $j = N$ is the best and $j = 1$ the worst. This method is similar to the approach used by Jordan et al. in the ranked FIPS [4].

3.3 Roulette EMP

The final approach for determining the EMP term is inspired by Roulette Wheel selection [10]. Roulette selection will be used in the proposed PSO EMP to select a single position from memory to be used in the velocity update equation. Since fitnesses in PSO tend to be quite similar toward the end of the optimisation process, the probability of a position from memory being chosen will be calculated by rank as given in Eq. 7 above. This method will mean that the best position a particle remembers will have twice the probability of being chosen as the second best position, and so on.

4 Experimental Results

This section will outline the experiments carried out on the PSO EMP algorithm to select the optimum memory size and method of calculating the EMP term. The algorithm's performance will then be compared to the Standard PSO.

A suite of 32 test functions is used in the experiments. The first seven test functions are functions that are widely used by researchers when testing new PSOs. These are the Sphere, Rosenbrock, Ackley, Griewank , Rastrigin, Schaffer and Griewank10 functions. The remaining 25 functions were proposed by Suganthan et al. [5]. All functions in the test suite are tested in 30 dimensions, except for Schaffers f6, which is a two dimensional problem, and Griewank10, the 10 dimensional version of the Griewank function.

The following parameters are used in all experiments. The maximum number of iterations is set to 10,000 iterations. A swarm of 50 particles randomly initialised within the function bounds is used. χ is set to 0.72984 and $c_1 = c_2 = 2.05$. Particles do not evaluate their fitness outside of the function bounds, except in functions f_7 and f_{25}, where the global optimum is outside of the initialisation region. Results for each experiment are averaged from 25 independent runs.

4.1 Experiment 1: Memory Size

The first factor that needs to be determined is the number of positions that a particle should remember. This experiment tested the three EMP approaches on seven test functions varying the number of remembered positions. Each EMP approach was tested using the gBest, lBest and von Neumann topologies. The performance measures used were the mean best fitness achieved and the percentage of runs which achieved the goal for each function. Table 1 below shows the results for the Roulette EMP PSO with a gBest topology (GEMP). It can be seen that a memory of size 2 yields the best results across the test functions, in terms of both the mean best value achieved and the success rate. Similar results were found for other combinations of swarm topology and EMP approach.

Table 1. GEMP ($M = MemorySize$)

	$M = 2$	$M = 3$	$M = 4$	$M = 5$
Sphere	**5.35E-069 (2.44E-068)** 1.00	2.41E-029 (6.559E-029) 1.00	8.02E-015 (1.74E-014) 1.00	1.35E-009 (2.04E-009) 1.00
Rosenbrock	**1.76E+001 (2.32E+000)** 1.00	2.64E+001 (1.47E+001) 1.00	3.20E+001 (1.75E+001) 1.00	2.57E+001 (2.18E+000) 1.00
Ackley	**7.55E-015 (0.00E+000)** 1.00	4.52E-014 (4.25E-014) 1.00	3.18E-004 (1.55E-003) 1.00	3.04E-003 (1.03E-002) 0.96
Griewank	**2.90E-002 (2.42E-002)** **0.84**	3.94E-002 (3.23E-002) 0.76	4.44E-002 (4.59E-002) 0.72	5.13E-002 (4.97E-002) 0.64
Rastrigin	**2.35E+001 (6.07E+000)** **1.00**	6.03E+001 (2.24E+001) 0.88	1.15E+002 (2.17E+001) 0.32	1.37E+002 (4.59E+001) 0.24
Schaffer	0.00E+000 (0.00E+000) 1.00	0.00E+000 (0.00E+000) 1.00	0.00E+000 (0.00E+000) 1.00	0.00E+000 (0.00E+000) 1.00
Griewank10	1.21E-002 (1.63E-002) 0.96	**4.01E-003 (7.54E-003)** 1.00	1.15E-002 (9.67E-003) 1.00	1.61E-002 (1.48E-002) 1.00

It is expected that remembering additional good positions should aid performance by keeping the particles more diverse. However, too many remembered positions will hinder convergence. Remembering two positions seems to give the desired balance between diversity and convergence. This result matches the finding of Yin et al. [7] for the Cyber Swarm where it was found that more than three guiding solutions blurred the guidance information and so degraded performance. Mendes [3] also obtained the best results for the FIPS using the Ring and von Neumann topologies without self, where particles are influenced by 2-4 solutions. A memory size of two will be adopted for all further experiments.

4.2 Experiment 2: EMP Approach

The second factor to be determined experimentally is which of the approaches to determining the EMP term is the most effective. In this experiment the three approaches discussed above are compared using the first seven test functions. The criteria used to evaluate performance will be the best mean value achieved and the proportion of runs meeting the goal. Table 2 below shows the results for each EMP approach for a global topology. It can be seen that using the Roulette approach to determine the EMP term yields the best results in terms of the

Table 2. EMP Approach (Global Topology)

	Roulette	Equal	Power Law
Sphere	**4.47E-068 (1.38E-067)** **1.00**	3.31E-005 (1.49E-004) 1.00	1.77E-005 (5.87E-005) 1.00
Rosenbrock	**1.88E+001 (8.62E+000)** **1.00**	2.41E+001 (1.15E+001) 1.00	2.55E+001 (7.50E+000) 1.00
Ackley	**7.55E-015 (0.00E+000)** **1.00**	3.18E+000 (2.57E+000) 0.04	3.89E+000 (2.48E+000) 0.08
Griewank	**4.15E-002 (4.51E-002)** **0.80**	2.93E-001 (4.21E-001) 0.44	6.45E-001 (1.26E+000) 0.36
Rastrigin	**2.28E+001 (5.83E+000)** **1.00**	4.60E+001 (1.76E+001) 1.00	4.88E+001 (1.91E+001) 0.96
Schaffer	0.00E+000 (0.00E+000) 1.00	0.00E+000 (0.00E+000) 1.00	0.00E+000 (0.00E+000) 1.00
Griewank10	**1.97E-002 (2.86E-002)** **0.84**	1.28E-001 (1.02E-001) 0.24	1.61E-001 (2.27E-001) 0.40

mean best value achieved and the success rate. Similar results were obtained for the ring and von Neumann topologies.

As in the first experiment the most important factor seems to be adding diversity without hindering convergence. The Roulette EMP approach introduces the least variance. It will behave like the standard PSO in most iterations, as the best position has the highest chance of being selected. However, the occasional selection of the second best position may help keep the particles diverse and aid exploration. As the Roulette EMP gives the best final solution with consistent performance, this is the method that will be adopted for further testing.

4.3   Experiment 3: Comparison to Standard PSO

This experiment will compare the developed EMP algorithm with the Standard PSO, using the global (gBest), ring (lBest) and von Neumann topologies, to see if the proposed algorithm gives the desired performance increases. The Roulette EMP with a memory size of two and a global topology, as this was the found to perform best, will be compared to the Standard PSO across the full range of 32 test functions.

Table 3 below shows the mean best values achieved by each PSO and the standard deviation across the suite of test functions. The Roulette EMP outperforms the Standard PSO on a large number of the test functions. It is also the best performer in 11 of the 32 functions, while being the worst performer in only 5 functions. The most notable performance of the Roulette EMP PSO is on the Sphere and Ackley functions, where the mean best values achieved are substantially better than the value achieved by the Standard PSO. A t-test was also conducted to test the statistical significance of these results. If the two tailed p value if less than 5% the difference is deemed significant. Table 4 shows the results of the t-test. These results show that PSO EMP can yield significant performance improvements in some cases and only decreases the performance significantly in a relatively small number of cases.

Table 3. Comparison to Standard PSO

	Roulette Mean (std)	gBest Mean (std)	lBest Mean (std)	vonNeumann Mean (std)
Sphere	**4.47E-068 (1.38E-067)**	2.30E-008 (1.02E-007)	4.82E-004 (5.34E-004)	7.68E-005 (1.23E-004)
Rosenbrock	**1.88E+001 (8.62E+000)**	2.22E+001 (1.42E+000)	2.39E+001 (2.48E+000)	2.57E+001 (7.92E-001)
Ackley	**7.55E-015 (0.00E+000)**	1.91E-001 (9.31E-001)	1.65E-002 (9.98E-003)	3.63E-003 (3.71E-003)
Griewank	4.15E-002 (4.51E-002)	4.54E-002 (5.48E-002)	**1.37E-002 (1.10E-002)**	1.82E-002 (2.02E-002)
Rastrigin	**2.28E+001 (5.83E+000)**	4.34E+001 (1.43E+001)	1.02E+002 (2.52E+001)	8.50E+001 (2.94E+001)
Schaffer	0.00E+000 (0.00E+000)	0.00E+000 (0.00E+000)	3.00E-004 (5.56E-004)	2.05E-004 (4.13E-004)
Griewank10	1.97E-002 (2.86E-002)	7.19E-002 (8.30E-002)	**1.82E-002 (1.86E-002)**	3.11E-002 (3.01E-002)
f1	**-4.50E+002 (5.57E-014)**	-4.50E+002 (1.45E-008)	-4.50E+002 (3.39E-004)	-4.50E+002 (5.99E-005)
f2	-4.46E+002 (4.55E+000)	**-4.50E+002 (3.40E-001)**	1.92E+002 (1.60E+002)	9.06E+001 (2.61E+002)
f3	2.24E+006 (6.34E+005)	**1.28E+006 (6.32E+005)**	5.18E+006 (1.60E+006)	4.45E+006 (1.72E+006)
f4	**5.91E+002 (5.33E+002)**	7.40E+002 (1.82E+003)	8.21E+003 (1.89E+003)	8.80E+003 (2.71E+003)
f5	**3.67E+003 (6.81E+002)**	4.09E+003 (7.90E+002)	4.26E+003 (7.52E+002)	4.11E+003 (8.38E+002)
f6	**4.18E+002 (3.76E+001)**	5.17E+002 (1.81E+002)	5.42E+002 (7.42E+001)	5.25E+002 (5.71E+001)
f7	-1.80E+002 (1.41E-002)	-1.80E+002 (1.80E-002)	-1.79E+002 (9.21E-002)	-1.79E+002 (2.50E-001)
f8	-1.19E+002 (6.39E-002)	-1.19E+002 (6.86E-002)	-1.19E+002 (6.51E-002)	-1.19E+002 (5.75E-002)
f9	**-3.04E+002 (7.53E-002)**	-2.75E+002 (2.06E-001)	-2.15E+002 (3.06E+001)	-2.47E+002 (2.40E+001)
f10	-1.53E+002 (3.99E+001)	**-2.48E+002 (2.46E+001)**	-2.41E+002 (2.46E+001)	-2.52E+002 (1.92E+001)
f11	1.14E+002 (3.61E+000)	**1.12E+002 (4.15E+000)**	1.16E+002 (2.55E+000)	1.16E+002 (2.71E+000)
f12	1.45E+004 (1.30E+004)	**8.91E+003 (9.84E+003)**	1.77E+004 (9.59E+003)	1.30E+004 (6.57E+003)
f13	**-1.26E+002 (1.29E+000)**	-1.24E+002 (2.34E+000)	-1.19E+002 (2.21E+000)	-1.20E+002 (3.13E+000)
f14	-2.87E+002 (3.08E-001)	**-2.88E+002 (5.97E-001)**	-2.87E+002 (1.37E-001)	-2.86E+002 (2.02E-001)
f15	5.60E+002 (2.50E+002)	4.48E+002 (8.32E+001)	**3.69E+002 (3.85E+001)**	4.03E+002 (1.15E+002)
f16	3.18E+002 (1.60E+002)	3.47E+002 (1.42E+002)	**2.65E+002 (2.38E+001)**	2.83E+002 (9.07E+001)
f17	4.32E+002 (1.21E+002)	3.82E+002 (1.68E+002)	**3.27E+002 (1.59E+001)**	3.46E+002 (1.40E+001)
f18	9.30E+002 (0.00E+000)	**9.19E+002 (4.12E+001)**	9.32E+002 (0.00E+000)	9.29E+002 (0.00E+000)
f19	9.34E+002 (0.00E+000)	9.31E+002 (2.77E+001)	9.32E+002 (0.00E+000)	**9.29E+002 (0.00E+000)**
f20	1.06E+003 (0.00E+000)	7.71E+002 (1.50E+002)	7.17E+002 (0.00E+000)	**6.94E+002 (0.00E+000)**
f21	8.60E+002 (0.00E+000)	1.00E+003 (2.64E+002)	8.60E+002 (0.00E+000)	8.60E+002 (0.00E+000)
f22	**1.30E+003 (0.00E+000)**	1.34E+003 (2.98E+001)	1.32E+003 (0.00E+000)	1.33E+003 (0.00E+000)
f23	8.94E+002 (0.00E+000)	9.85E+002 (2.04E+002)	8.94E+002 (0.00E+000)	8.94E+002 (0.00E+000)
f24	4.60E+002 (0.00E+000)	5.04E+002 (2.15E+002)	4.60E+002 (0.00E+000)	4.60E+002 (0.00E+000)
f25	4.73E+002 (0.00E+000)	4.78E+002 (5.41E+000)	4.75E+002 (0.00E+000)	**4.72E+002 (0.00E+000)**
Best	**11**	7	5	3
Worst	5	10	12	5
Better		20	20	18
Worse		11	8	11

Table 4. Statistical Comparison to Standard PSO

	gBest	lBest	vonNeumann
Statistically Better	10	14	15
Statistically Same	16	14	13
Statistically Worse	6	4	4

5   Conclusions

Two research questions were posed at the beginning of this paper which the research has attempted to address. In terms of the first question the research has found that the new PSO EMP can yield significant performance improvements over the Standard PSO. Experiment 2 addressed the second research question and found that the most important factor for PSO EMP seems to be balancing diversity and convergence, and therefore exploration and exploitation. The most successful variants were those which only gave a little extra diversity, i.e. Roulette EMP with M=2. However, this little extra diversity seems to give improved exploration. The other variants introduced too much variance leading to particles being pulled in too many directions and preventing them from exploiting known good areas. Further research is required to obtain an optimal exploration exploitation balance for PSO EMP.

References

1. Kennedy, J., Eberhart, R.: Particle swarm optimization. In: Proceedings of IEEE International Conference on Neural Networks, vol. 4, pp. 1942–1948. IEEE Press (November/December 1995)
2. Bratton, D., Kennedy, J.: Defining a standard for particle swarm optimization. In: IEEE Swarm Intelligence Symposium, pp. 120–127 (April 2007)
3. Mendes, R., Kennedy, J., Neves, J.: The fully informed particle swarm: simpler, maybe better. IEEE Transactions on Evolutionary Computation 8(3), 204–210 (2004)
4. Jordan, J., Helwig, S., Wanka, R.: Social interaction in particle swarm optimization, the ranked fips, and adaptive multi-swarms. In: Proceedings of the 10th Annual Conference on Genetic and Evolutionary Computation (GECCO 2008), pp. 49–56. ACM, New York (2008)
5. Suganthan, P.N., et al.: Problem definitions and evaluation criteria for the CEC 2005 special session on real-parameter optimization. Technical report, Nanyang Technological University, Singapore and KanGAL Report Number 2005005 (2005)
6. Kennedy, J., Mendes, R.: Population structure and particle swarm performance. In: Proceedings of the 2002 Congress on Evolutionary Computation (CEC 2002), pp. 1671–1676. IEEE Computer Society, Washington, DC (2002)
7. Yin, P.-Y., Glover, F., Laguna, M., Zhu, J.-X.: Cyber Swarm Algorithms – Improving particle swarm optimization using adaptive memory strategies. European Journal of Operational Research 201, 377–389 (2010)
8. Hu, X.H., Eberhart, R.C., Shi, Y.H.: Particle swarm with extended memory for multi-objective optimization. In: Proceedings of the IEEE Swarm Intelligence Symposium (SIS 2003), pp. 193–197 (2003)
9. Zhou, C., Zhang, G.-A., Zhou, H.: Extended Individual Memory Based Multi-objective Particle Swarm Optimization. In: International Conference on Future Computer and Communication (ICFCC), Wuhan, pp. 390–394 (2010)
10. Sivaraj, R., Ravichandran, T.: A review of selection methods in genetic algorithm. Int. J. Eng. Sci. Tech. 3, 3792 (2011)
11. Clerc, M., Kennedy, J.: The particle swarm: explosion stability and convergence in a multi-dimensional complex space. IEEE Trans. Evolution. Comput. 6(1), 58–73 (2002)

Sorting in Swarm Robots Using Communication-Based Cluster Size Estimation

Hongli Ding and Heiko Hamann

Department of Computer Science,
University of Paderborn, Paderborn, Germany
`hongli.ding@uni-paderborn.de, heiko.hamann@uni-paderborn.de`

Abstract. Inspired by sorting behaviors of social insects, we are interested in sorting by robot swarms using only local information and hence achieving high degrees of robustness and scalability. In this work, we propose a gossip-based sorting method which allows two swarms of simple homogeneous autonomous robots to sort themselves in two not pre-assigned areas. Key feature of this method is the estimation of cluster sizes based on communication that allows to determine the local majority. In a series of simulation experiments, we show the effectiveness of the approach and investigate the influence of different swarm sizes.

1   Introduction

Recent research shows that social insects sort their brood in sophisticated patterns. These well-organized brood sorting patterns emerge spontaneously from dynamic interactions during the process of depositing and removing brood. During this sorting process, no specified spatial plans or any global representation is required, nor any hierarchical decisions are made [2]. By interacting among other individuals and with the environment, individuals act following their own goals and knowledge about the environment. The collective behavior on the group level emerges from the sum over all individual decisions, actions, and the interactions among individuals and the environment. The sorting system of social insects has many attractive features such as scalability, flexibility, and robustness. Abstract models based on sorting behaviors of social insects have been applied in many areas such as search, collective sorting, data mining, numeric data analysis, and graph partitioning [4]. Inspired by how ants and honey bees sort their broods, we are interested in how to implement these natural sorting behaviors and strategies in a swarm of robots. We simplify the sorting task to sorting robots instead of objects. This preliminary work aims at sorting robots of different classes which can be considered as robots carrying objects of different types. Hence, our algorithm still aims to sort objects.

Object sorting by swarm robots is a complex task which involves mechanism such as self-organization, collective decision making, and pattern recognition. Abstract models of sorting objects by a group of minimalist homogeneous robots were proposed by Deneubourg et al. [2]. These models are based on simple rules

M. Dorigo et al. (Eds.): ANTS 2014, LNCS 8667, pp. 262–269, 2014.

which are used to determine the probabilities of picking up and dropping down objects. The major drawback of these models is the complexity of the procedure to obtain the local object density. Being limited to the minimal sensors of real robots imposes a challenge in achieving sorting similar to that observed in ants. This method was extended to sort more than three types of objects [5]. Inspired by [2], an approach using an overhead camera to identify robots, object positions and orientations, and global data about the entire arena was proposed [8]. It is unclear how this approach could be transferred to use local sensors only. This approach was extended by adding a forward-facing camera on each robot, which not only allows robots to share image data with their neighbors, but also allows robots to estimate cluster sizes [7]. Both approaches [7,8] achieve good sorting results while it is hard to apply these approaches to real robots, in particular swarm robots, due to the applied complex and sophisticated sensors.

In this paper, we describe a gossip-based sorting method which aims to sort different robots by using only local information and simple onboard sensors. Similar to previous studies, our algorithm is based on simple behavioral rules. In addition, we examine how the system performance changes with different swarm densities (different swarm sizes on a constant area).

This paper is organized as follows. Sec. 2 describes the scenario and objective of this work. A global approach is proposed in this section. Sec. 3 focuses on the proposed sorting algorithm. In Sec. 4, we report the experimental setups and the simulation results. We conclude the paper and outline the future work in Sec. 5.

2   Scenario Description

In a given space as shown in Fig. 1a, the ground is divided into three areas: two black areas and one white area. Green and red robots are initially randomly distributed in the white part of the arena. These two groups of robots have to sort themselves in the two black areas using only local information (shown in Fig. 1b).

In this scenario, the allocation of the two black areas to the two robot groups is not predetermined. The robot swarms decide collectively on this allocation at runtime during the sorting process by interacting with other robots and the environment. Each agent's individual decision is based on local information only. The mechanism is implemented as a self-organizing system which requires no global information like spatial plans, hierarchical decisions, or message broadcasting. We investigate how swarm robots can make correct decisions and self-organize themselves to achieve this complex sorting task under these strict constraints.

The two groups of robots are randomly distributed in the white area initially as shown in Fig. 1a. At the beginning, all robots walk randomly in the white area by avoiding obstacles, especially other robots. When a robot detects a black ground, it stops while continuing to communicate with other robots in its neighborhood. At the same time, each stopped robot identifies other stopped robots in its neighborhood based on a unique robot identity (UID). They communicate and transfer the UIDs of stopped neighbors to other stopped robots and count

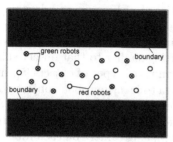 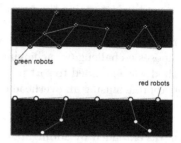

(a) initial setup (red and green robots positioned within the white area)   (b) expected result (green robots share one black area and red robots share the other black area)

Fig. 1. Initial setup and expected result of the simulation experiment

the number of stopped robots of both groups at the boundary. When one group of stopped robots represents the majority at this boundary, then the minority robots leave this boundary and search for an unoccupied boundary. With time going on, the number of robots representing the majority is expected to increase at this boundary. Robots position themselves in a line formation at the boundaries to form, what we call a *robot barrier*. It allows the same robots to pass and prevents the passage of other robots. For example, if the barrier is composed of a certain number of red robots which is above a threshold, then the red robots are allowed to cross this barrier. They hence allocate this black area and stay inside of it. In contrast, green robots are not allowed to pass this barrier. The control algorithm is organized in several modules as follows:

- *Count robots*: Stopped robots count the number of different robots staying at the boundary by communicating with each other.
- *Minority robots leave*: After counting the number of different robots at a boundary, the robots know whether they represent the minority. If they are the minority, they leave this boundary for a random walk in the white area.
- *Barrier formation*: When a moving robot reaches the communication range of a stopped robot of the same color, it moves on a circular trajectory around the stopped robot in order to position itself in a neighboring position at the boundary. It stops once a black ground is detected, and consequently becomes a part of the robot barrier. The circular trajectory enforces a certain distance between robots in the robot barrier.
- *Pass robot barrier*: When the number of majority robots on the boundary exceeds a threshold, robots of the same color are allowed to pass the barrier and position themselves at an appropriate spot within the black area.

3   Gossip-Based Sorting Algorithm

In this work, the gossip-sorting method is based on gossip communication to count the number of robots in a swarm. The idea of gossip communication is that

robots exchange their local knowledge in pairs when robots reach the communication range of each other. Each robot that is stopped at a boundary and that is
in the communication range of other robots, communicates with each neighbor,
and they mutually exchange their local knowledge. In this work, robots exchange
information about the number of robots of each group at the boundary that the
robot either can perceive itself or about which it has received information via
past gossiping. After several gossip communication iterations between robots,
every robot knows the number of different robots at the same boundary.

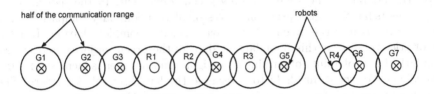

Fig. 2. Robot barrier under construction

Table 1. Example of the gossip-based communication

	Direct line communication	Gossip communication	After gossip communication
G1	G1	- -	- -
G2	G2, G3	G3 says: G2,G3,R1 are in the barrier	G2 knows G2,G3,R1
G3	G2, G3, R1	R1 says: G3, R1,R2 are in the barrier	G3 knows G2,G3,R1,R2
R1	G3, R1, R2	G3 says: G2,G3,R1,R2 are in the barrier R2 says: R1,R2,G4 are in the barrier	R1 knows G2,G3,R1,R2,G4
⋮	⋮	⋮	⋮

Fig. 2 is an example of the gossiping method which shows a robot barrier
under construction. This robot barrier consists of two groups of robots: green
labeled with UIDs G1, G2, etc., and red labeled with UIDs R1, R2, etc. We
assume that the communication is robust (subject to future work, see Sec. 5).
Robots are assumed to communicate only with neighbors in the line of sight,
that is, messages are neither addressed to a particular robot (i.e., no multi-hop
communication) nor sent through obstacles including other robots. In Fig. 2,
robot G1 cannot communicate with robot G2, because they are not in each
others communication range. Robot G2 cannot communicate with robot R1,
because multi-hop messaging is not available (robot G3 would need to serve as
relay).

Table 1 shows how robots exchange messages based on gossip communication.
As shown in Fig. 2, robot G1 is isolated. Similarly, Robot G5 and R4 cannot
communicate with each other. In consequence, the robot barrier shown in Fig. 2

consists of three sub-groups: $S_1 = \{G1\}$, $S_2 = \{G2, G3, R1, R2, G4, R3, G5\}$, and $S_3 = \{R4, G6, G7\}$. Obviously, sub-group S_1 will consider itself as majority robot until the approach of other red robots. Each robot in sub-group S_2 knows about all other robots from the same sub-group after the gossip-based communication. Each robot has a counter for red robots n_r and a counter for green robots n_g. In this case, green robots are the majority ($n_g = 4$, $n_r = 3$). Consequently, red robots leave the barrier and start a random walk into the white area until they find another place to stay. Similarly for sub-group S_3, green robots are the majority and hence red robot leaves the barrier. In this case, all red robots leave the boundary and there are only green robots left. In the sequel, other green robots have a higher probability to stop at this boundary compared to red robots, even before the green robots have formed a complete barrier. If a red robot tries to stop at this boundary, it detects the green majority and leaves. Similarly, the probability of forming a barrier at another area is increased for the red robots. We set a threshold to determine the length of robot barriers (The threshold is set to 5 for all experiments in this work). Once the majority group size at a barrier reaches this threshold, other robots of that kind are allowed to pass the barrier through two robots of the same color. Hence, this robot group claims the respective black area.

4   Simulation Environment and Results

In this section, we give a brief description of our simulation environment. Results of our sorting method for different swarm densities are shown. Videos of our simulations are available online[1].

4.1   Simulation Environment

Experiments were conducted using the foot-bot robot [1] in the ARGoS simulator [6]. ARGoS is designed to simulate complex experiments involving large swarms of robots of different types. It allows to transfer robot controllers from the simulation directly to real robots without any modification [6].

In the following simulation experiments, we use an arena of 3 meters by 3 meters, divided to three areas (two blacks, one white). Two groups of robots (green and red) are randomly distributed in the white area. The communication range of each robot is set to 60 cm, which is the minimal communication range required to allow a robot to pass between two robots in communication distance (requirement for passing barriers). With this communication range the robot covers 12.6% of the total arena. We use only basic onboard sensors: IR sensors, range and bearing sensors.

4.2   Simulation Results

Second Swarm as Disturbance. In this experiment, our sorting method is tested for different swarm sizes $N \in \{10, 16, 20, 26, 30, 40\}$ (i.e., different swarm

---

[1] See https://www.youtube.com/user/SortingRobots

densities because the area is constant) forming groups of red and green robots in the following way: $(n_r, n_g) \in \{(2,8), (4,12), (5,15), (6,20), (8,22), (10,30)\}$. The swarms are composed of approximately 25% of red robots and 75% of green robots. Initially, both red and green robots are uniformly positioned in the white area. For each parameter setting, 10 runs were done. The simulation ends when either all robots have stopped in black areas or one group of robots has stopped in black areas while the other group of robots moves in the white area. We define the sorting rate as the percentage of robots sorted in black areas (i.e., robots positioned correctly in the neighborhood of their own kind). When both kinds of robots are found within the same black area, we consider the group of robots which represents the majority as sorted. Fig. 3a shows the sorting rates and the required time for the different swarm sizes.

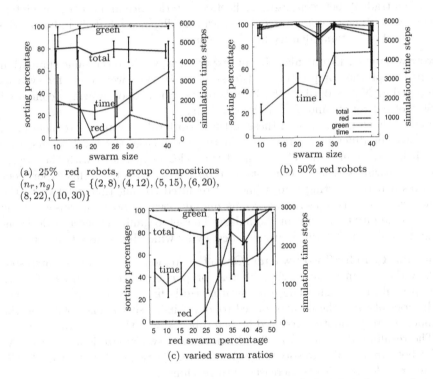

(a) 25% red robots, group compositions $(n_r, n_g) \in \{(2,8), (4,12), (5,15), (6,20), (8,22), (10,30)\}$

(b) 50% red robots

(c) varied swarm ratios

Fig. 3. Results of the simulations, sorting rate and simulation time (mean values of 10 runs, error bars give standard deviation)

From Fig. 3a, we can see that the disturbance from approximately 25% of red robots has no significant influence on the sorting results of green robots, while the low sorting rate of red robots decreases the total sorting rate. The relatively small numbers of red robots induce a small probability to occupy any boundaries. However, the total sorting rate is still bigger than 75%.

As seen in Fig. 3a, the swarm size has little influence in these experiments. Only the required time indicates a trend to increased times for bigger swarms. Furthermore, there is a trend indicating that a swarm size of $N = 20$ might be optimal in terms of the system's convergence time.

The total sorting rate is influenced by several conditions. The total swarm density is too low to guarantee a good cooperation among robots. The sizes of the two robot groups are of importance. The green robots are enough to occupy both boundaries, hence, the red robots have a limited chance to occupy any boundary. Many of the experiments end with the red robots being trapped in the white area. Therefore, the red robots have a small chance to sort themselves.

The curve of required time decreases at first, probably because the low swarm density provides insufficient opportunities for cooperation among individuals. The robots spend much time searching or waiting for other robots. We therefore observe a tradeoff between positive effects of interference and obstructive interference which has commonly seen in swarm intelligence [3]. This indicates the existence of an optimal density for this scenario.

Two Swarms of Equal Size. In this experiment, we keep the same simulation setups except the composition of two swarms. 10 runs were done for swarms composed of $N \in \{10, 16, 20, 26, 30, 40\}$ robots. Each robot swarm is composed of 50% of red robots and 50% of green robots.

Fig. 3b shows the results for these settings. Both robot swarms sort themselves efficiently for different swarm sizes. In the worst case, the total sorting rate is still higher than 85%. The required time clearly increases with the swarm size. Comparing to the experiments with only 25% red robots, this experiment achieves a better sorting rate, while the required sorting time relative to the same swarm size is longer. This seems mainly because both swarms have the same probability to occupy any boundary. The self-organized collective decision-making process about which group is occupying which black area takes time.

Optimal Ratio for Two Swarms. Given the previous results, we are interested in how the swarm behaves with different proportion for two swarms. What is the optimal proportion? The total swarm size is fixed to $N = 20$ robots. For each series of runs, the number of red robots is increased by one robot and the number of green robots is reduced by one robot (5% of the total swam size).

The results are given in Fig. 3c. When the red swarm size is $n_r \leq 20\% \cdot N$, red robots are not able to sort themselves at all. For ratios of $n_r > 20\% \cdot N$, the sorting rate for red robots increases until reaching 100%.

In contrast to the red robots, the number of green robots reduces from $95\% \cdot N$ to $50\% \cdot N$. For all the cases $n_g > 50\% \cdot N$, they represent the majority, hence, they have a bigger probability to occupy one of the boundaries to sort themselves well. The case for both swarms having the same quantity ($n_r = n_g = 10$) has the same result as shown in Fig. 3b for $N = 20$: a 100% sorting rate.

5 Conclusion

Sorting by swarm robots is a complex task. Recent research in sorting by swarm robots relies often on sophisticated sensors and complex image processing methods. We have proposed a gossip-based sorting method based exclusively on local information without using any sophisticated sensors. Our method allows different robot swarms to sort themselves efficiently by using the mechanism of self-organization. Comparing to previous research, we use a smaller communication range and simpler sensors to achieve sorting. The relationship between system performance, swarm size, and different compositions of competing swarms are studied in this work.

We plan to extend this work to the actual task of sorting objects by considering the two kinds of robots as robots carrying different types of objects. Instead of having the robots stop at the black areas they drop the carried object, pick a new one, and try to find the appropriate black area for it. Our findings about the influence of ratios of robot types will be applied, for example, by trying to guarantee that both kinds of objects are carried by two robot groups of approximately equal size at all time. The influence of our assumption of robust communication will be investigated by simulating unreliable communication. This might introduce difficulties for the gossiping method which might require additional methods. We also plan the obvious follow-up work to implement this method on real robots.

References

1. Bonani, M., Longchamp, V., Magnenat, S., Rétornaz, P., Burnier, D., Roulet, G., Vaussard, F., Bleuler, H., Mondada, F.: The marXbot, a miniature mobile robot opening new perspectives for the collective-robotic research. In: International Conference on Intelligent Robots and Systems, pp. 4187–4193 (2010)
2. Deneubourg, J.L., Goss, S., Franks, N., Franks, A.S., Detrain, C., Chrétien, L.: The dynamics of collective sorting: robot-like ants and ant-like robots. In: Proceedings of the First International Conference on Simulation of Adaptive Behavior: From Animals to Animats, pp. 356–363. MIT Press, Cambridge (1990)
3. Hamann, H.: Towards swarm calculus: Urn models of collective decisions and universal properties of swarm performance. Swarm Intelligence 7(2-3) (2013)
4. Handl, J., Knowles, J., Dorigo, M.: On the Performance of Ant-based Clustering. In: Design and Application of Hybrid Intelligent Systems, pp. 204–213. IOS Press, Amsterdam (2003)
5. Melhuish, C., Wilson, M., Sendova-Franks, A.: Patch sorting: Multi-object clustering using minimalist robots. In: Kelemen, J., Sosík, P. (eds.) ECAL 2001. LNCS (LNAI), vol. 2159, pp. 543–552. Springer, Heidelberg (2001)
6. Pinciroli, C., Trianni, V., O'Grady, R., Pini, G., Brutschy, A., Brambilla, M., Mathews, N., Ferrante, E., Caro, G.D., Ducatelle, F., Birattari, M., Gambardella, L.M., Dorigo, M.: ARGoS: a modular, parallel, multi-engine simulator for multi-robot systems. Swarm Intelligence 6, 271–295 (2012)
7. Vardy, A.: Accelerated patch sorting by a robotic swarm. In: 9th Conference on Computer and Robot Vision (CRV) (2012)
8. Verret, S., Zhang, H., Meng, M.Q.H.: Collective sorting with local communication. In: IROS, pp. 2687–2692. IEEE (2004)

Using Fluid Neural Networks to Create Dynamic Neighborhood Topologies in Particle Swarm Optimization

Stephen M. Majercik

Computer Science Department, Bowdoin College, Brunswick, Maine, USA
smajerci@bowdoin.edu

Abstract. Fluid Neural Networks (FNNs) are a model of interacting mobile automata. The automata move on a lattice, affecting each other's motion in a way that can result in clusters of automata that change over time, making FNNs a potential basis for dynamic neighborhood topologies in Particle Swarm Optimization. We describe Fluid Neural Network Particle Swarm Optimization (FNN-PSO), a PSO algorithm that uses a dynamic neighborhood mechanism based on FNNs, and we report promising results from experiments indicating that FNN-PSO can outperform both the standard PSO algorithm and PCGT-PSO, a PSO algorithm based on partially connected grid topologies [3], over a range of neighborhood topologies and influence models.

1 Introduction

The Particle Swarm Optimization (PSO) algorithm, introduced by Kennedy and Eberhart [5], is one of the most successful swarm based optimization techniques. In this algorithm, a swarm of particles flies through the solution space of the objective function, seaching for the optimum function value, each particle guided by its *personal best*, the best solution it has found do far and the *global best*, the best solution the entire swarm has found so far. The algorithm, however, tends to converge to a local optimum. For this reason, many PSO variants use a neighborhood topology that specifies smaller, overlapping neighborhoods, thus replacing the global best with multiple *local (neighborhood) bests* that help diversify the search.

Dynamic topologies, in which the neighborhoods of particles change over time, have also been used to address this issue. We propose Fluid Neural Network Particle Swarm Optimization (FNN-PSO), a PSO algorithm in which dynamic neighborhoods are constructed using *fluid neural networks*, a type of neural network in which mobile automata move and interact with each other [8]. Our work generalizes the PSO algorithm based on *partially connected grid topologies* of [3]. Experiments show that FNN-PSO is competitive with both that algorithm and a standard PSO algorithm.

In Section 2, we describe fluid neural networks and FNN-PSO. We describe the results of our experiments in Section 3. In Section 4, we discuss related work, and we conclude with ideas for future work in Section 5.

M. Dorigo et al. (Eds.): ANTS 2014, LNCS 8667, pp. 270–277, 2014.

2   PSO and FNN-PSO

2.1   Standard PSO

In the standard PSO algorithm (S-PSO), a swarm of particles iteratively searches a d-dimensional solution space. Each particle i remembers the best solution it has found so far (its *personal best*, or *pbest*), p_i, and the best solution found so far by the particles in i's neighborhood (the *neighborhood best*, or *nbest*), n_i. On each iteration, the velocity v_i of particle i is updated (but confined to a range $[V_{min}, V_{max}]$ for each component of v_i, where V_{min} and V_{max} are the minimum and maximum values of the search space), such that its motion is biased toward both p_i and n_i, and the new velocity is used to update its position x_i. The update equations for the constriction coefficient variant of S-PSO are:

$$v_i \leftarrow \chi(v_i + U(0, \phi_1) \otimes (p_i - x_i) + U(0, \phi_2) \otimes (n_i - x_i)) \qquad (1)$$

$$x_i \leftarrow x_i + v_i \qquad (2)$$

where:

- ϕ_1 and ϕ_2, the *acceleration coefficients* that scale the attraction of particle i to p_i and n_i, respectively, are equal and have the value 2.05,
- $U(0, \phi_i)$ is a vector of real random numbers uniformly distributed in $[0, \phi_i]$,
- $\otimes$ is component-wise multiplication, and
- χ is a constriction coefficient (approximatley 0.7298).

2.2   FNN-PSO

Fluid neural networks were used by Miramontes to model the chaotic dynamics in *Leptothorax* ant colonies [6] and were explored further in [8]. A fluid neural network (FNN) is a set of n automata, or *neurons*, that occupy positions on a $d \times d$ lattice. The number of neurons and lattice size are such that $n < d \times d$; thus, neurons can move. Time is discrete. Neurons have an activation level and are *active* if and only if their activation level exceeds a specified threshold. $S_i(t)$, the activation level of neuron i at time t, depends on the activation levels of i's neighbors and is determined by the following formula:

$$S_i(t) = \tanh \left(g \times \sum_{j \in N(i)} J_{ij} S_j(t-1) - \theta_i \right) \qquad (3)$$

where g is the *gain*, $N(i)$ is the set of neurons in the eight positions adjacent to i and i itself, J_{ij} is a *coupling matrix*, θ_i is the threshold of particle i, and θ is the global threshold. If $S_i(t) > \theta$, neuron i becomes active. Neurons may also *spontaneously activate* with probability p_a (the *spontaneous activation probability*). In this case, a neuron's activation level is set to S_a (the *spontaneous activation level*). In [8], and in our work, $\forall i, j \; J_{ij} = 1$ (i.e. J has no impact on the dynamics), $\theta_i = 0$, and $\theta = 10^{-16}$.

The operation of FNN-PSO is similar to that of S-PSO with a von Neumann (or Moore) topology. In these topologies, the particles are thought of as being on a toroidal grid and a particle's neighbors are those particles in the spaces to the north, south, east, and west of it (von Neumann) or in the eight spaces adjacent to it (Moore). In FNN-PSO, the toroidal grid is replaced by the lattice of an FNN and the PSO particles occupy this lattice and move in the same way as the neurons of an FNN. The neighbors of particle i are those particles in the locations specified by the von Neumann (or Moore) topology (relative to i), but, because the density of the FNN is less than 1.0, some of those spaces may be unoccupied, reducing the number of particles in i's neighborhood. In addition, because the particles move, a particle's neighbors will change over time. See Algorithm 1 for pseudocode.

Algorithm 1. FNN-PSO

---

1 **Inputs:**
2   n, the size of the swarm
3   f, the function to be optimized (minimized)
4     $iters_{max}$, the maximum number of iterations
5     $d, g, J, \theta_i, \theta, p_a$, and S_a (parameters specifying an FNN)
6 **Outputs:**
7   x^*, the position of the minimum function value found
8     $f(x^*)$, the value or the function at that position
9 **for** $i \leftarrow 1 \ldots n$ **do**
10   Initialize particle i with:
11   - a random position x_i and random velocity v_i in the solution space,
12   - a random position on the FNN lattice, and
13   - a random FNN activation level in $[0.0, 1.0]$;
14 **while** $iteration\ i < iters_{max}$ **do**
15   **for** $i \leftarrow 1 \ldots n$ **do**
16      $p_i \leftarrow$ position of the best solution particle i has found so far;
17      $N(i) \leftarrow$ particles currently in particle i's neighborhood in the FNN;
18      $n_i \leftarrow$ position of the particle in $N(i)$ with the lowest function value;
19      $v_i \leftarrow$ velocity of particle i updated using p_i and n_i and Equation 1;
20      $x_i \leftarrow$ position of particle i updated using Equation 2;
21   Calculate $f(x_i)$ and update $pbest$, $nbest$, and x^*;
22 **return** x^* and $f(x^*)$

FNN-PSO generalizes the PSO algorithm of [3], in which a particle's neighbors are determined by a partially connected grid topology. (We will refer to this algorithm as PCGT-PSO.) Particles move on a 2-dimensional torus and, like FNN-PSO, a particle's neighbors at time t are those particles that are in the locations specified by the standard topology being used, either von Neumann or Moore. In both FNN-PSO and PCGT-PSO, the particles move, but in FNN-PSO, the grid is a lattice, rather than a torus. In both, on each iteration, each particle moves randomly to one of the eight adjacent cells, if any of those cells are empty. The major difference between FNN-PSO and PCGT-PSO is that particles in FNN-PSO have an activation level that changes depending on the activation levels of their neighbors and determines whether they are active and can move, whereas the particles in PCGT-PSO always move.

3 Experimental Results

We compared the performance of FNN-PSO to that of the S-PSO algorithm described in Section 2.1 and PCGT-PSO on six standard benchmark functions: Sphere, Rosenbrock, Ackley, Griewank, Rastrigin, and Penalized Function P8. (See [2] for the function definitions.) Sphere and Rosenbrock are uni-modal functions, while Ackley, Griewank, Rastrigin, and Penalized Function P8 are multi-modal functions with many local optima. The optimum (minimum) value for all of these functions is 0.0. We used asymmetric initialization and randomly shifted the location of the optimum away from the center of the search space in order to mitigate the tendency of PSO algorithms to converge to the center. We tested each of these functions in 30 dimensions, allowing 10,000 iterations. The FNN parameters are described below. We ran each algorithm 50 times and computed the mean and standard deviation of the lowest function value found.

We tested each algorithm on combinations of two topologies—von Neumann and Moore—and two neighborhood influence models—neighborhood best (*nbest*) and *fully informed particle swarm* (FIPS). In *nbest*, a particle is influenced only by the *pbest* of the fittest particle in the neighborhood. In FIPS, a particle is influenced by the *pbest* of every particle in its neighborhood.

We fixed the following parameters to the same values as those used in [8]: the threshold for an individual neuron i ($\forall\, i\ \theta_i = 0$), the global activation threshold ($\theta = 10^{-16}$), and the coupling matrix ($\forall\, i,j\ J_{ij} = 1$). We varied the following parameters in our experiments: the lattice size (d), the gain (g), the spontaneous activation level (S_a), and the spontaneous activation probability (p_a).

Our working hypothesis was that the dynamics of FNNs with high *information transfer* would provide a good basis for dynamic neighborhood creation. Information transfer is used in [8] as a measure of the propagation of activation events (when a particle becomes active) through the lattice. At one extreme—very low density and/or gain—activation events die out almost immediately. At the other extreme—very high density and/or gain—activation events spread indiscriminately, becoming noise. In between these two regimes (referred to as "the edge of chaos" in [8]), information spreads but individual neurons remain in the same state for a long time. Thus, although not every neuron is in the neighborhood of every other neuron at each time step, there are waves of activation that travel across the lattice, so that a neuron in one place is affected by a neuron becoming active at a distant location. In a sense, the waves of activation would produce "waves of neighborhoods" forming and dissolving over time.

We ran exploratory tests over a broad range of FNN parameter settings: lattice sizes $\{7 \times 7,\ 8 \times 8,\ 9 \times 9,\ 10 \times 10\}$, densities $\{0.10,\ 0.12,\ 0.14,\ \ldots,\ 0.90\}$, gains {0.1, 0.2, 0.3, 0.4, 0.5}, spontaneous activation levels {0.1, 0.2, 0.3, 0.4}, and spontaneous activation probabilities {1e-6, 1e-5, 1e-4, 1e-3, 1e-2, 1e-1}, for a total of 19,680 parameter settings. We tested the performance of FNN-PSO using parameter settings with high information transfer and found that FNN-PSO outperformed S-PSO and PCGT-PSO in many cases (with respect to the minimum function value found). Subsequent tests with parameters that had low information transfer did almost as well, however, suggesting that our information transfer hypothesis was incorrect, and we do not report those results.

The activity patterns (the change in number of active neurons over time) of the most successful runs had a characteristic appearance: cyclic, varying sharply between no neurons active and all neurons active, and spending approximately the same amount of time in both states. And, since tests indicated that neurons are actually moving about 99% of the time they are active, this is the movement pattern as well. We conjectured that it is beneficial to have periods during which neighborhoods are static alternating with periods in which neighborhoods are changing, and we explored the performance of FNN-PSO given different activity patterns as a function of gain (maintaining density at 0.5).

These experiments indicated that lower gains yielded better performance. We compared the performance of one of these parameter settings (40 particles on a 9 × 9 lattice, a gain of 1e-4, a spontaneous activation level of 0.2, and a spontaneous activation probability of 1e-3) to that of S-PSO with 40 particles and PCGT-PSO with 40 particles on a 9 × 9 torus (see Table 1). For each of the 24 cases (six functions for each of the four topology/influence models), the best result is shown in bold-face. FNN-PSO had the best performance in ten cases, S-PSO in five cases, and PCGT-PSO in seven cases. (The total is less than 24 due to ties, which were not counted.) In each case, we compared the best algorithm to the next best algorithm using a 2-tailed Mann-Whitney U-test. The difference was significant at the 0.05 level or less in nine of the ten FNN-PSO cases in which FNN-PSO was top-ranked (90%), four of the five cases in which S-PSO was top-ranked (80%), and four of the seven cases in which PCGT-PSO was top-ranked (57%). Of the ten cases in which FNN-PSO was ranked first, the difference was significant in nine of them (90%). Of the five cases in which S-PSO was ranked first, the difference was significant in four of them (80%). Of the seven cases in which PCGT-PSO was ranked first, the difference was significant in four of them (57%). Instances where the difference was significant are underlined in Table 1. We ran the same tests using neighborhoods in which a particle was excluded from its own neighborhood: of the 24 cases, FNN-PSO had the best performance in 15 cases (nine of them significant), S-PSO in seven cases (five of them significant), and PCGT-PSO in one case (which was significant).

4   Related Work

A number of researchers have investigated the benefits to be obtained from using dynamic topologies. Mohais et al. looked at two methods for randomly restructuring neighborhoods, one in which a single edge is randomly reconfigured on each iteration, and one in which all neighborhoods are randomly recreated periodically [7]. This approach, when used with a FIPS information sharing approach, outperformed a number of PSO variants. Since the gain and density parameters of the FNN in FNN-PSO control the rate at which neighborhoods are reconfigured, the FNN mechanism may provide a useful way to interpolate between these two appoaches.

Table 1. Mean and Standard Deviation of Minimum Function Value Obtained (Best performance in bold-face and underlined if statistically significant)

| Function | Von Neumann Topology | | | | | |
| | Neighborhood Best | | | FIPS | | |
	S-PSO	PCGT-PSO	FNN-PSO	S-PSO	PCGT-PSO	FNN-PSO
Sphere	7.57e-30	**3.55e-30**	4.04e-30	0.0	1.69e+00	0.0
	3.06e-29	**1.22e-29**	2.86e-29	0.0	7.21e-01	0.0
Rosenbrock	**8.20e+00**	1.86e+01	1.37e+01	**8.32e-01**	2.95e+01	4.86e+00
	2.89e+00	3.25e+00	2.25e+00	**2.39e+00**	8.10e-01	1.07e+01
Ackley	**3.99e+00**	4.15e+00	4.01e+00	5.09e+00	4.67e+00	**3.33e+00**
	8.06e+00	8.15e+00	8.09e+00	7.41e+00	6.64e+00	**7.70e+00**
Griewank	**4.38e-03**	9.89e-03	4.83e-03	4.55e-02	9.80e-01	**2.86e-03**
	6.93e-03	1.23e-02	7.15e-03	7.43e-02	4.01e-02	**5.30e-03**
Rastrigin	4.81e+01	5.50e+01	**4.00e+01**	5.79e+01	2.03e+02	**3.28e+01**
	1.40e+01	1.75e+01	**1.29e+01**	3.37e+01	1.16e+01	**2.32e+01**
Penal P8	6.65e-02	4.35e-02	**2.70e-02**	3.49e+00	6.38e+00	**1.87e-02**
	2.51e-01	1.15e-01	**7.78e-02**	5.94e+00	1.67e+00	**9.07e-02**

| Function | Moore Topology | | | | | |
| | Neighborhood Best | | | FIPS | | |
	S-PSO	PCGT-PSO	FNN-PSO	S-PSO	PCGT-PSO	FNN-PSO
Sphere	1.45e-29	**1.12e-29**	1.62e-29	1.96e-29	0.0	0.0
	4.89e-29	**4.12e-29**	5.53e-29	1.38e-28	0.0	0.0
Rosenbrock	**7.09e+00**	1.83e+01	9.58e+00	2.10e+01	3.81e-03	**3.13e-03**
	2.61e+00	1.66e+00	2.18e+00	1.02e+01	1.48e-03	**1.85e-02**
Ackley	2.79e+00	**1.23e+00**	3.23e+00	7.51e+00	**3.76e+00**	4.46e+00
	6.97e+00	**4.92e+00**	7.41e+00	8.10e+00	**8.11e+00**	7.71e+00
Griewank	1.38e-02	9.00e-03	**6.89e-03**	2.66e-02	**2.96e-04**	1.05e-02
	1.61e-02	1.19e-02	**1.05e-02**	4.24e-02	**1.46e-03**	1.70e-02
Rastrigin	5.12e+01	5.74e+01	**4.43e+01**	8.20e+01	1.63e+02	**4.91e+01**
	1.44e+01	2.34e+01	**1.16e+01**	6.24e+01	7.93e+00	**2.53e+01**
Penal P8	1.46e-01	**5.19e-02**	5.40e-02	4.97e+00	**3.08e-32**	8.89e-01
	3.63e-01	**1.47e-01**	1.56e-01	6.68e+00	**1.07e-31**	1.54e+00

Akat and Gazi compared three approaches to creating dynamic neighborhoods [1] and raised the issue of the effect of *information flow topology* on the performance of the PSO algorithm. In the general case, the parameter determining neighborhood composition is different for each particle, resulting in nonreciprocal neighborhoods, which can be represented as directed graphs. If these digraphs are strongly connected *over time*, i.e. if there is a fixed interval I such that the union of the digraphs over every interval of iterations of that length is strongly connected, then information flow in the swarm will be preserved and every particle eventually has access to the information gathered by every other particle. We are currently investigating whether FNN-PSO provides this type of strong connectedness over time.

Wang and Xiang proposed a dynamic ring topology in which particles are connected unidirectionally based on their personal bests [9]. Particles are ordered by personal bests and have neighbors that are either 1) all better, in which case one is chosen based on a variant of a fitness-distance ratio ("learn from far and better ones"), or 2) all worse, in which case a weighted average of the neighborhoods is used ("centroid of mass"). Their algorithm was competitive with a number of standard PSO variants.

García-Nieto and Alba tested a variant of the S-PSO algorithm in which the neighborhood for each particle on each iteration is constructed by choosing k other particles, or informants, randomly [4]. They tested the algorithm over a range of values for k and found evidence for a quasi-optimal neighborhood size in the range of 6-8 neighbors. The neighborhood creation mechanism of FNN-PSO is somewhat similar, in that it is constructing neighborhood sets based on the random motion of the particles in the FNN space. A maximum neighborhood set size (12) could be obtained by using a topology in which the neighborhood is composed of those locations whose Manhattan distance from the particle is less than or equal to two, and the expected neighborhood set size could be varied from 0 to 12 by changing the lattice size. We are currently running tests to determine whether FNN-PSO performance peaks when the expected neighborhood set size falls in the range that García-Nieto and Alba found to be quasi-optimal.

5   Discussion and Further Work

FNN-PSO is a promising PSO algorithm, but our results are not sufficient to recommend a particular topology, influence model, or set of FNN parameters. Of the nine cases in which FNN-PSO's top-ranked performance was statistically significant, five of them used the von Neumann topology and four of them used the Moore topology, so the algorithm does not appear to be sensitive to the underlying topology. The algorithm did seem to be somewhat sensitive to the influence model used; FIPS was the influence model in six of the nine cases where FNN-PSO was top-ranked. The better performance of the FIPS model may be due to the motion of the particles; particles may be in neighborhoods only briefly, and the FIPS model allows them to influence each other no matter how brief that period is. We are currently testing whether the advantage conferred by FIPS disappears if particles move more slowly (by reducing the gain and/or raising the activation threshold).

Our results indicate that good performance is achieved by FNN parameters that produce relatively long periods during which no particles are moving in the FNN and neighborhoods are static, punctuated by short periods, initiated by the spontaneous activation of a particle, during which most of the particles are moving. Further testing is needed to determine whether a constant, but very slow, rate of mixing might be better than periods of complete inactivity punctuated by brief periods of mixing.

We have described some current work in Section 4 investigating the temporal connectedness of the swarm in FNN-PSO and the effect of expected neighborhood size. There are other interesting avenues to explore. The coupling matrix, J, can modulate the impact two particles have on each other's activation level. Currently, all elements of J, are set to 1.0, so it has no impact. There is potential to use this matrix to allow particles to affect each other differently depending on their fitness-distance ratio, e.g. two particles with a high ratio should affect each other more strongly since this would increase both their activation levels, making it more likely that they would both be active and, thus, in a neighborhood together.

Finally, it might be beneficial for each particle to have its own gain parameter, adjusted dynamically based on the fitness of the particle, e.g. when the particle's fitness increases, its gain might be increased slightly (and decreased when the particle's fitness remained the same). This would allow particles with higher fitness to stay active longer, influencing other particles in the swarm more strongly.

References

1. Akat, S., Gazi, V.: Particle swarm optimization with dynamic neighborhood topology: Three neighborhood strategies and preliminary results. In: Swarm Intelligence Symposium, SIS 2008, pp. 1–8. IEEE (2008)
2. Bratton, D., Kennedy, J.: Defining a standard for particle swarm optimization. In: Swarm Intelligence Symposium, SIS 2007, pp. 120–127. IEEE (2007)
3. Fernandes, C., Rosa, A., Laredo, J., Cotta, C., Merelo, J.J.: Performance and scalability of particle swarms with dynamic and partially connected grid topologies. In: Proceedings of the 5th International Joint Conference on Computational Intelligence, pp. 47–55 (2013)
4. García-Nieto, J., Alba, E.: Empirical computation of the quasi-optimal number of informants in particle swarm optimization. In: Proceedings of the Genetic and Evolutionary Computation Conference, GECCO 2011, pp. 147–154 (2011)
5. Kennedy, J., Eberhart, R.: Particle swarm optimization. In: Proceedings of IEEE, pp. 1942–1948 (1995)
6. Miramontes, O.: Order-disorder transitions in the behavior of ant societies. Complexity 1(3), 56–60 (1995)
7. Mohais, A.S., Mendes, R., Ward, C., Posthoff, C.: Neighborhood re-structuring in particle swarm optimization. In: Zhang, S., Jarvis, R.A. (eds.) AI 2005. LNCS (LNAI), vol. 3809, pp. 776–785. Springer, Heidelberg (2005)
8. Solé, R.V., Miramontes, O.: Information at the edge of chaos in fluid neural networks. Physica D 80, 171–180 (1995)
9. Wang, Y.X., Xiang, Q.L.: Particle swarms with dynamic ring topology. In: IEEE Congress on Evolutionary Computation, pp. 419–423 (2008)

A Low-Cost Real-Time Tracking Infrastructure for Ground-Based Robot Swarms

Alan G. Millard[1], James A. Hilder[1,2], Jon Timmis[2], and Alan F.T. Winfield[3]

[1] Department of Computer Science,
York Robotics Laboratory, University of York, UK
millard@cs.york.ac.uk
[2] Department of Electronics, York Robotics Laboratory, University of York, UK
{james.hilder,jon.timmis}@york.ac.uk
[3] Bristol Robotics Laboratory, University of the West of England, Bristol, UK
alan.winfield@uwe.ac.uk

Optical tracking systems are used in many research laboratories for monitoring and recording the movements of mobile robots. The data gathered by such systems is invaluable for offline post-experiment analysis, for example, measuring the area coverage of a robot swarm. These systems can also be used to provide robots with online feedback about their current position and orientation, for the purposes of indoor localisation. Unfortunately, obtaining precise and reliable tracking data often comes at the cost of expensive equipment.

For many research laboratories, commerical motion capture systems such as Vicon[1], comprising even a small number of cameras, are prohibitively expensive. The cost of developing a custom-built system such as IRIDIA's Arena Tracking System [6] may also be too great. Open-source solutions such as SwisTrack [1] and AprilTag [5] provide a cheap alternative, however, the precision of the tracking data obtained using visible-light cameras is often inferior to that aquired from commercial motion capture systems that use infrared cameras and retroreflective markers. OptiTrack[2] is a recent competitor in the motion capture market that offers a cost-effective commercial solution, whilst still delivering precise and reliable tracking data. We have recently built an OptiTrack system at the York Robotics Laboratory, which is capable of simultaneously tracking up to 32 ground-based robots in real-time within a 2.5m square arena. The cost of the OptiTrack hardware and software totalled just under $6,000 (USD), so this may potentially provide an affordable solution for other research laboratories.

We currently use the system to track swarms of e-puck robots that are each augmented with a Linux Extension Board, which improves their processing and memory resources, and enables Wi-Fi communication. Figure 1 provides an overview of our experimental infrastructure. Three Flex 13 cameras are connected via USB to an OptiHub 2 — a custom USB hub that allows the cameras to synchronise with each other. This, in turn, connects via USB to the OptiTrack server (a Windows machine with tracking software installed), allowing tracking data to be obtained from the cameras. The server processes the tracking data, logs it for post-experiment analysis, and makes it available to e-pucks and users

---

[1] www.vicon.com

[2] www.naturalpoint.com/optitrack

M. Dorigo et al. (Eds.): ANTS 2014, LNCS 8667, pp. 278–279, 2014.

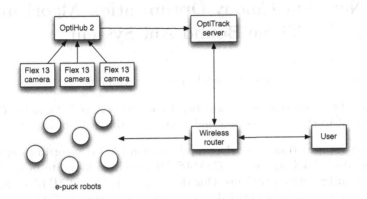

Fig. 1. Overview of the tracking infrastructure, showing data flow between system components

in real-time via the wireless LAN. Each robot is assigned a static IP address, and connects to the LAN via a wireless router. This allows networked computers to connect to any robot using the SSH protocol. For further details please see [2], which contains information we believe will be of use to the wider swarm robotics research community, and will assist others in developing low-cost solutions for tracking ground-based robot swarms in real-time.

In future, we will use this tracking infrastructure to implement exogenous fault detection as proposed in [3]. This fault detection approach has already been shown to work for single-robot systems [4], and will soon be extended for use with robot swarms.

References

1. Lochmatter, T., Roduit, P., Cianci, C., Correll, N., Jacot, J., Martinoli, A.: Swis-Track - A Flexible Open Source Tracking Software for Multi-Agent Systems. In: Proceedings of the 2008 IEEE/RSJ International Conference on Intelligent Robots and Systems (IROS), pp. 4004–4010 (2008)
2. Millard, A.G., Hilder, J.A., Timmis, J., Winfield, A.F.T.: A Low-cost Real-time Tracking Infrastructure for Ground-based Robot Swarms. Tech. Rep. YCS-2014-489, University of York (2014)
3. Millard, A.G., Timmis, J., Winfield, A.F.T.: Towards Exogenous Fault Detection in Swarm Robotic Systems. In: Natraj, A., Cameron, S., Melhuish, C., Witkowski, M. (eds.) TAROS 2013. LNCS (LNAI), vol. 8069, pp. 429–430. Springer, Heidelberg (2014)
4. Millard, A.G., Timmis, J., Winfield, A.F.T.: Run-time Detection of Faults in Autonomous Mobile Robots Based on the Comparison of Simulated and Real Robot Behaviour. In: Proceedings of the 2014 IEEE/RSJ International Conference on Intelligent Robots and Systems (IROS) (to appear, 2014)
5. Olson, E.: AprilTag: A robust and flexible visual fiducial system. In: 2011 IEEE International Conference on Robotics and Automation (ICRA), pp. 3400–3407 (2011)
6. Stranieri, A., Turgut, A.E., Francesca, G., Reina, A., Dorigo, M., Birattari, M.: IRIDIA's Arena Tracking System. Tech. Rep. TR/IRIDIA/2013-013, Université Libre de Bruxelles (2013)

A New Ant Colony Optimization Algorithm: Three Bound Ant System

Nikola Ivkovic[1] and Marin Golub[2]

[1] Faculty of Organization and Informatics, University of Zagreb, Varazdin, Croatia
[2] Faculty of Electrical Engineering and Computing, University of Zagreb, Croatia

Since their introduction, ant colony optimization (ACO) algorithms and especially the MAX-MIN ant system (MMAS) [4] have been found to be well suited to various optimization problems. Our theoretical analysis of MMAS allowed us to create a new algorithm, entitled 'three bound ant system (TBAS)', which has lower computational complexity while retaining and even improving the quality of constructed solutions. In this paper, TBAS is briefly explained and experimentally explored in terms of algorithmic speedup and solution quality. Some distinctive characteristics of TBAS with regard to MMAS are:

- three pheromone bounds (the lower bound τ_{LB}, the upper bound τ_{UB}, and the contraction bound $\tau_{CB} = \omega \cdot \tau_{UB}$),
- occasional pheromone contractions instead of regular evaporations,
- a unique pheromone reinforcement procedure, and
- the lower pheromone bound equal to the initial pheromone value.

In the solution construction procedure, TBAS uses a well-known random-proportional rule in a manner identical to that in MMAS (and many other ACO algorithms). TBAS uses all the parameters that MMAS does, although some parameters are employed somewhat differently (e.g. parameter ρ), with TBAS also using the additional parameter ω, where $\tau_{LB}/\tau_{UB} \le \omega \le 1$.

The pheromone update procedure of TBAS starts with pheromone reinforcement, which is followed by the pheromone contraction procedure, provided that a pheromone trail outgrows the upper pheromone bound τ_{UB}. During the pheromone reinforcement it is necessary to select one or more solutions whose components will be assigned an additional pheromone value. For that purpose, different strategies like iteration best, κ-best or max-κ-best can be used [2]. All the components of the selected solution s^{bs} are reinforced in the TBAS according to expression (1), where in the first iteration $Q_0 = 1$, and in the subsequent i-th iteration Q_i is modified by $Q_{i+1} = Q_i/(1 - \rho)$. The pheromone contraction multiplies all the pheromone trails and Q_i with $\omega' = \tau_{UB}/\tau_{max} \cdot \omega$ while ensuring that every pheromone trail τ_c remains inside the interval $[\tau_{LB}, \tau_{CB}]$.

$$\tau_c = \tau_c + \frac{Q_i}{f(s^{bs})}, \forall c \in s^{bs} \tag{1}$$

Owing to different pheromone update procedures, TBAS has lower computational complexity than MMAS. In the case when precomputed values of $\tau_c^\alpha \cdot \eta_c^\beta$ are stored in a lookup table (LUT) to speed up the algorithm, the TBAS has an additional advantage in that it is often sufficient to recompute pheromone

M. Dorigo et al. (Eds.): ANTS 2014, LNCS 8667, pp. 280–281, 2014.

trails only for components affected by the pheromone reinforcement. The actual speedup of TBAS over MMAS, for an equal number of iterations and equal common parameters, depends on the optimization problem, algorithm implementation and parameters, among others, and can vary from negligible to highly significant. According to [3], the time spent in the pheromone update procedure can amount to 90% of the total running time of the algorithm. In order to demonstrate a possible speedup of TBAS over MMAS we chose an optimization problem and parameter settings that we expected would yield a significant speedup. Experiments were conducted on instances of ATSP, with sizes ranging from 53 to 5000, available at http://www2.research.att.com/~dsj/chtsp/ and http://comopt.ifi.uni-heidelberg.de/software/TSPLIB95/. Both algorithms shared the same source code, except for the pheromone update procedure, as well as the same parameters ($\alpha = 1.3$, $\beta = 4$, $\rho = 0.02$, with the favorite nodes list size set at 40, etc.), with the exception of $\omega = 0.03125$ which only exists in TBAS. The experiments conducted on an HP 6830s laptop showed that the speedup ratio in the case when the number of ants $m = 10$ is between 2 and 18, and for $m = 1000$ the speedup ratio was between 1.01 and 1.2 (i.e. TBAS was 1% to 20% faster than MMAS).

To explore TBAS in terms of solution quality, experimental comparison of TBAS and MMAS was conducted on 55 instances of the quadratic assignment problem (QAP) [4,1] from QAPLIB, with sizes ranging from 15 to 256 locations. The algorithms were compared without local optimization and with 2-opt local optimization after 10000 iterations, in each category with two parameter settings. For MMAS the parameters were set at commonly recommended values, while the reinforcement strategy was tuned by comprehensive experimentation. For TBAS the parameters were copied from MMAS, after which the parameters ω and $\vartheta = \tau_{LB}/\tau_{UB}$ were tuned based on p^{QAP} probability [1]. The experiments were repeated 100 times, from which median values were used. In both categories TBAS clearly outperformed MMAS, which was confirmed by a very high level of statistical significance obtained by Friedman test and various post hoc procedures. The newly proposed TBAS has a special theoretical relationship with MMAS. In addition, we experimentally proved that it can compete with MMAS both in terms of algorithmic speedup and solution quality.

References

1. Ivkovic, N., Golub, M., Malekovic, M.: A pheromone trails model for MAX-MIN ant system. In: Hao, J.K. (ed.) 10th Conf. on Artificial Evolution, pp. 35–46 (2011)
2. Ivkovic, N., Malekovic, M., Golub, M.: Extended trail reinforcement strategies for ant colony optimization. In: Panigrahi, B.K., Suganthan, P.N., Das, S., Satapathy, S.C. (eds.) SEMCCO 2011, Part I. LNCS, vol. 7076, pp. 662–669. Springer, Heidelberg (2011)
3. Oliveira, S.M., Hussin, M.S., Stützle, T., Andrea, R., Dorigo, M.: A detailed analysis of the population-based ant colony optimization algorithm for the TSP and the QAP. In: Krasnogor, N., Lanzi, P.L. (eds.) 13th Conf. on Genetic and Evolutionary Computation, pp. 13–14. ACM (2011)
4. Stützle, T., Hoos, H.H.: MAX-MIN ant system. Future Generation Comp. Syst. 16, 889–914 (2000)

An Adaptive Bumble Bees Mating Optimization Algorithm for the Hierarchical Permutation Flowshop Scheduling Problem

Yannis Marinakis and Magdalene Marinaki

School of Production Engineering and Management, Technical University of Crete, Chania, Greece
marinakis@ergasya.tuc.gr, magda@dssl.tuc.gr

In the last few years, a new algorithm, the Bumble Bees Mating Optimization (BBMO) algorithm, has been proposed. BBMO is based on the mating behaviour of the bumble bees. In this extended abstract, we present a procedure to optimize the parameters of the Bumble Bees Mating Optimization algorithm during the iterations of the algorithm and, thus, the user does not need to give any parameters for the algorithm. The outcome of the algorithm, besides the solution of the problem, is the best calculated parameters for each instance. We use as basis the last version of the BBMO algorithm, the CNTBBMO, where no transformation in continuous values is needed [1]. The algorithm is used for the solution of a hierarchical version of the Permutation Flowshop Scheduling Problem where two different criteria, the makespan and the total flow time are optimized in lexicographical order. In this extended abstract, we select an a priori approach where two different optimization criteria are selected and are optimized in lexicographical order. The proposed algorithm is based on the Combinatorial Neighborhood Topology Bumble Bees Mating Optimization (CNTBBMO) [1]. Generally, a BBMO algorithm has three kinds of bumble bees in the colony, the queen bee, the worker bees and the drones (males). Each bee is a solution of the problem and it is represented via the permutation of the jobs. The main phases of the algorithm are: *Initialization, Drones' Selection, Offspring' Creation, Feeding the new queens, Mutation phase, Mating phase, Next Iteration* [1]. The most important and novel part of the algorithm is the optimization of the parameters during the iterations of the algorithm. Initially, random values of the parameters are given taking into consideration not to exceed some specific bounds. For example, w_1 should always be less than w_2 (w_1, w_2 are two parameters necessary for CNTBBMO [1]). The parameters that are optimized are the number of bees, the number of iterations, the number of VNS (local search) iterations, the w_1 and w_2, the u_{bound} and the l_{bound} [1]. The upper bounds (u_{bound}) and lower bounds (l_{bound}) should always have positive values and the u_{bound} should be greater than the l_{bound}. Another value that is selected as a threshold value is a value corresponding to the number of consecutive iterations with no improvement in the results of the best solution. The algorithm was tested on the 90 benchmark instances of Taillard. The results of the algorithm are presented in Table 1.

1. Marinakis, Y., Marinaki, M., (2014), Combinatorial neighborhood topology bumble bees mating optimization for the vehicle routing problem with stochastic demands, *Soft Computing*, DOI: 10.1007/s00500-014-1257-1.

M. Dorigo et al. (Eds.): ANTS 2014, LNCS 8667, pp. 282–283, 2014.

Table 1. Analytical presentation of the results of the proposed algorithm

	Ta20X5	Ta20X10	Ta20X20	Ta50X5	Ta50X10	Ta50X20	Ta100X5	Ta100X10	Ta100X20
				Primary Objective: Makespan					
				Values Optimization					
$AB1$	1224.6	1517	2241.1	2738.9	3035.5	3837	5254.7	5690.3	6536.8
$AB2$	14647.2	21168.3	34771.6	70328.8	92615.4	131748.3	253083.6	312416.9	406070.5
$AP1$	1231.2	1536.7	2272.1	2753.8	3090.9	3917.4	5272.3	5754.5	6656.9
$AP2$	14915.1	21384.2	35089.2	72190.4	94969.1	133221.4	261002.3	321608.2	415993.1
$Q1$	0.18	0.22	0.27	0.09	1.75	3.43	0.20	1.11	3.79
$Q2$	5.18	5.79	5.66	5.84	7.53	8.29	4.89	7.46	7.62
$QB1$	0.73	1.54	1.66	0.64	3.62	5.60	0.53	2.25	5.70
$QB2$	7.11	6.90	6.60	8.63	10.27	9.51	8.17	10.64	10.25
$BKSN$	8	3	0	4	0	0	2	0	0
LVI	0.00	0.00	0.09	0.00	0.49	2.70	0.00	0.38	3.44
UVI	1.14	0.50	0.60	0.32	2.98	4.17	0.58	1.52	4.18
				Parameters optimization					
$iter$	182.3	142.3	139.7	209.6	209.8	188.1	251.2	304.6	242.5
NB	122	75.2	100.3	175.6	106.9	85.1	213.9	191.6	138.6
LB	2.60	2.63	2.68	2.71	2.82	2.82	2.77	2.83	2.81
UB	5.20	5.12	5.27	4.96	5.37	5.54	5.68	5.35	5.54
w_{1opt}	2.40	2.25	2.59	2.40	2.31	2.38	2.28	2.35	2.30
w_{2opt}	2.95	2.87	3.23	3.15	2.86	2.94	2.68	2.83	2.87
LS	129.4	130	143.4	147.3	149.2	148.3	146.7	149.3	149.3
				Ten Runs 1st Objective					
$A10$	1226.1	1518.8	2242.8	2740.6	3037.2	3838.8	5256.4	5692	6538.6
$AQ10$	0.30	0.35	0.35	0.15	1.81	3.48	0.23	1.14	3.82
$stdev$	1.38	1.47	1.34	1.46	1.41	1.49	1.34	1.48	1.56
var	1.95	2.22	1.83	2.17	2.03	2.27	1.83	2.24	2.45
				Ten Runs 2nd Objective					
$A10$	14648.9	21170.2	34773.3	70330.6	92617.3	131750.3	253085.4	312418.4	406072.3
$AQ10$	5.19	5.80	5.66	5.84	7.53	8.30	4.89	7.46	7.62
$stdev$	1.57	1.49	1.39	1.42	1.54	1.31	1.54	1.36	1.49
var	2.47	2.24	1.99	2.13	2.39	1.76	2.41	1.90	2.25
				Primary Objective: Total Flow Time					
	Ta20X5	Ta20X10	Ta20X20	Ta50X5	Ta50X10	Ta50X20	Ta100X5	Ta100X10	Ta100X20
				Values Optimization					
$AB1$	13945.4	20027.4	32929.7	67197.1	87716.7	123589.3	244732.7	297899.3	387194
AB	1323.4	1667.9	2450	2873.3	3309	4124.6	5402.3	5904	6904.2
$AP1$	14044.2	20200.5	33162.9	68192.1	89226.8	125458.9	249053.4	303514.6	393525.2
$AP2$	1325.3	1682.7	2463.4	2884.4	3316.4	4160.6	5425.2	5978.7	6991.7
$Q1$	0.10	0.12	0.06	1.11	1.83	1.59	1.42	2.48	2.62
$Q2$	8.55	10.26	9.63	5.04	10.94	11.18	3.02	4.92	9.63
$QB1$	0.80	0.98	0.76	2.61	3.58	3.12	3.21	4.42	4.29
$QB2$	8.66	11.21	10.22	5.42	11.19	12.16	3.45	6.26	11.02
$BKSN$	6	5	6	0	0	0	0	0	0
LVI	0.00	0.00	0.00	0.71	0.95	1.19	0.81	1.63	2.06
UVI	0.39	0.39	0.19	1.38	2.25	2.09	2.09	3.42	3.24
				Parameters optimization					
$iter$	125.9	152.3	151.1	198.5	182.8	197.6	316.6	261.8	250.2
NB	108.6	124.7	146.3	103.9	87.7	111.9	129.9	104.6	126.2
LB	2.65	2.74	2.78	2.83	2.72	2.86	2.84	2.78	2.85
UB	5.25	5.36	5.40	5.63	5.47	5.62	5.36	5.39	5.30
w_{1opt}	2.41	2.44	2.34	2.20	2.42	2.64	2.34	2.37	2.06
w_{2opt}	2.93	3.00	2.99	3.01	2.93	3.12	2.97	2.98	2.88
LS	135.4	148.1	148.2	149.2	149.4	149.3	149.5	149.1	149.1
				Ten Runs 1st Objective					
$A10$	13947.3	20029.3	32931.3	67198.8	87718.7	123591.2	244734.5	297901	387195.9
$AQ10$	0.11	0.13	0.06	1.11	1.83	1.59	1.42	2.48	2.62
$stdev$	1.38	1.42	1.46	1.38	1.49	1.49	1.41	1.49	1.44
var	1.94	2.03	2.16	1.95	2.25	2.26	2.01	2.25	2.10
				Ten Runs 2nd Objective					
$A10$	1325.3	1669.87	2451.92	2874.98	3310.91	4126.3	5404.18	5905.62	6906.32
$AQ10$	8.70	10.39	9.72	5.10	11.00	11.23	3.06	4.95	9.66
$stdev$	1.53	1.52	1.42	1.51	1.56	1.44	1.60	1.46	1.40
var	2.37	2.32	2.06	2.31	2.46	2.12	2.62	2.18	1.99

Gene Expression in DNA Microarrays: A Classification Problem Using Artificial Bee Colony (ABC) Algorithm

Beatriz A. Garro[1], Roberto A. Vazquez[2], and Katya Rodríguez[1]

[1] Instituto en Investigaciones en Matemáticas Aplicadas y en Sistemas
Universidad Nacional Autónoma de México, México, D.F.
{beatriz.garro,katya.rodriguez}@iimas.unam.mx
[2] Intelligent Systems Group, Facultad de Ingeniería,
Universidad La Salle, México, D.F.
ravem@lasallistas.org.mx

1 Introduction

Analyzing the information captured in a DNA microarray to diagnose a diseases using computation intelligence techniques is complex due to the samples number is much lower than the genes number. Moreover, many genes could be irrelevant to diagnose a specific diseases; for that reason it is necessary to study techniques that allow to select the most relevant genes. There are many works that present different feature selection techniques whose results are applied to classify DNA microarrays data [1], [2], [3]. In this paper, we describe how ABC algorithm could be applied to select the best set of genes from a DNA microarray in order to classify the gene expression with the aim to diagnose efficiently a disease.

2 Proposed Methodology

The problem to be solved can be defined as follows: Giving a set of input patterns $\mathbf{X} = \{\mathbf{x}_1, \ldots, \mathbf{x}_p\}, \mathbf{x_i} \in \mathbb{R}^n, i = 1 \ldots, p$ and a set of desired classes $\mathbf{d} = \{d_1, \ldots, d_p\}, d \in \mathbb{N}$, find a subset of genes $G \in \{0, 1\}^n$ such that a function defined by $(F(\mathbf{X}|_G, \mathbf{d}))$ is minimized.

The food source's position represents the solution to the problem which is defined with an array $I \in \mathbb{R}^n$. Each individual $I_q, q = 1, \ldots, NB$ is binarized using a threshold level th in order to select the best set of genes that compose the gene expression defined as $G^k = T_{th}(I^k), k = 1, \ldots, n$; component whose values is set to 1, indicates that this gene will be selected. The proposed fitness function is described in Equation 1, where tng is the total number of gene expressions to be classified, D is any distance measure, K is the number of classes and $\mathbf{c}$ is the center of each class.

$$F(\mathbf{X}|_G, \mathbf{d}) = \frac{\sum_{i=1}^{p}\left(\left|\underset{k=1}{\overset{K}{\arg\min}}\left(D\left(\mathbf{x}_i|_G, \mathbf{c}^k\right)\right) - d_i\right|\right)}{tng}. \tag{1}$$

M. Dorigo et al. (Eds.): ANTS 2014, LNCS 8667, pp. 284–285, 2014.

3   Experimental Results

The proposed methodology was tested using a bench-mark high-dimensional biomedical DNA microarray data set: leukemia ALL-AML database [2]. In order to validate statistically the experimental results, the proposed methodology was executed 30 runs for each distance measure.

Table 1 shows the results obtained with the proposed methodology. For the case of average accuracy, Manhattan distance was a little better than Euclidean distance obtaining a Tr. cl. of 100% and a Te. cl. of 77.7.9%. For the case of average number of genes, Manhattan distance found more genes to diagnose the disease in less iterations. Finally, the best accuracy was achieved by Euclidean distance providing a Te. cl. of 82.4% using only four genes in 446 iterations.

Table 1. Average behavior of the proposed methodology using distance classifiers

Dist.	Accuracy		# of Genes	# of iter.	Best acc.		# of genes	# of iter
	Tr. cl.	Te. cl.			Tr. cl.	Te. cl.		
Euc.	0.98 ± 0.01	0.76 ± 0.03	2618.3	933.5	1.00	0.82	4	446
Man.	1.00 ± 0.00	0.77 ± 0.01	3563.8	58.8	1.00	0.79	3557	59

Tr. cl. = Training classification rate, Te. cl. = Testing classification rate.

4   Conclusions

The feature selection task presents good results using the ABC algorithm. In this process, in order to binarize each population individual a threshold of 0.5 was setting. On the other hand, a gene number reduction was evident between original problem and the genes founded by ABC algorithm obtained a good testing classification. Finally, similar results were obtained when the proposed methodology was compared against [2].

Acknowledgments. The authors thank DGAPA-UNAM and ULSA for the economic support under grant number IN107214 and I-61/12, respectively. Beatriz Garro thanks CONACYT for the posdoctoral scholarship provided.

References

1. Deegalla, S., Boström, H.: Classification of microarrays with knn: Comparison of dimensionality reduction methods. In: Yin, H., Tino, P., Corchado, E., Byrne, W., Yao, X. (eds.) IDEAL 2007. LNCS, vol. 4881, pp. 800–809. Springer, Heidelberg (2007)
2. Golub, T.R., et al.: Molecular classification of cancer: class discovery and class prediction by gene expression monitoring. Science 286(5439), 531–537 (1999)
3. Tang, E.K., et al.: Feature selection for microarray data using least squares svm and particle swarm optimization. In: CIBCB 2005, pp. 1–8 (2005)

Morphology Learning via MDL and Ants

Päivi Suomalainen

Attido Oy, Espoo, Finland
paivi.suomalainen@attido.com

1 Introduction

In computational linguistics, data preprocessing is an important step. For example, word segmentation into morphemes is widely used in various learning tasks, such as text classification, information extraction, part-of-speech (PoS) tagging and machine translation. In fact, our morphology learner is one part of our future information extraction software, which will be tailored for extracting relevant marketing information from the web, for example, from twitter. Working with text written in a language having a rich morphology, it is very important to have the words in their base form when training and testing a classifier in order to increase the performance of the classifier.

In order to achieve a good segmentation, we utilize the important feature of the minimum description length (MDL) principle [4], the ability to avoid under- and over-fitting. The heuristic we use with our simple MDL-based cost function is *ant colony optimization* (ACO) [3]. ACO is a nature inspired meta-heuristic used in many different kinds of optimization problems. Our algorithm iteratively splits the words in our corpus into stems and suffixes based on a pheromones and heuristic information, and after that, combines some of the words based on the MDL cost function. As far as we know, no ACO-based search heuristic has been proposed to tackle the problem of morphology learning until now.

2 Algorithm

Our aim is to minimize the combined code-length $l(\mathcal{M}) + l(\mathcal{D}|\mathcal{M})$, that is, the length of the model $\mathcal{M}$ and the length of the data $\mathcal{D}$ given the model. First, we encode the model in straightforward manner and use the probabilities of the morphemes in the data in order to encode it.

Our algorithm begins by creating a colony with one ant for each word-frequency pair in our data. Each ant has its own pheromone trails. That is, each ant has pheromone information for all possible segmentations of its word into half. All splits are considered as equal at the initial phase of the algorithm. Our algorithm iterates over a splitting phase, combining phase, and over a phase that we call *stealing*. Each iteration begins with going through all the ants, and each ant splits its word using a probabilistic rule based on pheromone information and on the frequency and probability information of the minimal-cost model so far. After going through the whole colony, each ant has a possibility to combine the segmented word or to steal one character from the suffix to the stem,

M. Dorigo et al. (Eds.): ANTS 2014, LNCS 8667, pp. 286–287, 2014.
© Springer International Publishing Switzerland 2014

if no combination was made. These actions are made only if the action results decrease on the current code length. After that, each ant updates its pheromone trails according to a function of the code-length of the current solution, and after updating the trails an evaporation phase of the pheromones takes place. And finally, the current code-length is compared to the all-time minimum, and the all-time minimum is updated if the current code-length is smaller than all-time minimum. This is repeated until the all-time minimum code-length has not been updated in fifty iterations.

3   Experiments

The data set for our experiments consists of the English and Finnish versions of the *Acquis communautaire*, i.e., the rights and obligations that European Union countries share. The data are extracted from the DGT-TM-2012 corpus [1]. We test our approach against the Morfessor [1] and the Morfessor Categories-MAP [2] algorithms. In our experiments, the language we use is Finnish, which is a language having a rich morphology. We evaluate five hundred randomly selected words in order to compare the different algorithms.

We compared our ACO-based heuristic against a greedy iterative one that exploits the frequencies of the suffixes of the words when splitting them. Our @tico algorithm achieved much more compression than the greedy heuristic. After that, we concentrated on comparing our @tico algorithm against the Morfessor and the Morfessor Categories-MAP algorithms. We collected five hundred words randomly, and analyzed them with respect to the results of @tico and the two Morfessor algorithms. Our @tico algorithms outperformed the MDL-based Morfessor algorithm, but not the Morfessor Categories-MAP algorithm. The latter uses a more sophisticated probabilistic modeling than our straightforward MDL-model, so we did not expect our simple approach to achieve as good results as the Morfessor Categories-MAP algorithm.

Acknowledgments. We would like to thank Timo and Mikko Jääskelainen for providing us the chance to work with this problem.

References

1. Creutz, M., Lagus, K.: Unsupervised discovery of morphemes. In: Proceedings of the Workshop on Morphological and Phonological Learning of ACL 2002, Philadelphia, Pennsylvania, pp. 21–30 (2002)
2. Creutz, M., Lagus, K.: Inducing the morphological lexicon of a natural language from unannotated text. In: Proceedings of the International and Interdisciplinary Conference on Adaptive Knowledge Representation and Reasoning (AKRR 2005), Espoo, Finland (2005)
3. Dorigo, M., Stützle, T.: Ant Colony Optimization. MIT Press (2004)
4. Grünwald, P.: The Minimum Description Length Principle. MIT Press (2007)

---

[1] http://ipsc.jrc.ec.europa.eu/index.php?id=197

Parallelizing Solution Construction in ACO for GPUs

Noriyuki Fujimoto and Shigeyoshi Tsutsui

Graduate School of Science, Osaka Prefecture University, Osaka, Japan
{fujimoto,tsutsui}@mi.s.osakafu-u.ac.jp

1   Introduction

We present a GPU implementation of an ant colony optimization (ACO) algorithm called the Cunning Ant System (cAS) [3]. Although local search can be effectively combined to ACO [4], such local search algorithms are heuristics that inherently depend on the target problem and there are many problems for which such a heuristic is not known. Hence, we intend to evaluate ACO as a meta-heuristic which can be applied to problems without effective local search, and does not use local search in this paper.

2   Parallelization of the Solution Construction

ACO has parallelism among solutions, but in order to extract high performance of a GPU, parallelizing processing for each solution is desirable. Fig. 1 shows a pseudo code of our solution construction. The lines four to six can be replaced with the prefix-sums of tau[index[i]*n+vw[i..n-1]] into rw[i..n-1]. We implemented prefix-sums with the warp-shuffle instructions [2]. The line eight was implemented with the atomic instruction atomicMin(). We also parallelized a simple for loop with embarrassing parallelism in our sampling code.

3   Experiments

Table 1 shows the performance of the proposed GPU algorithm and the corresponding CPU algorithm for several problem instances from the QAPLIB benchmark library [1]. QAPLIB provides the optimal values of the objective function for these problem instances. For each problem instance, the measurement was conducted 25 times consecutively and the average value of the 25 trials was adopted.

4   Conclusion

The effect of parallelizing the solution construction in ACO has been experimentally shown on the NVIDIA CUDA GPU architecture. If no local search heuristic is known for the target problem, our approach seems to be effective.

Future work includes performance improvement of the proposed implementation and extensive applications to other problems.

M. Dorigo et al. (Eds.): ANTS 2014, LNCS 8667, pp. 288–289, 2014.

```
1    Generate a random permutation of {0,1,2,...,n-1} into idx[];
2    Initialize v[], vw[], copyLength;
3    for (int i = copyLength; i < n; i++) {
4      rw[i] = tau[idx[i] * n + vw[i]];
5      for (int j = i+1; j < n; j++)
6        rw[j] = rw[j-1] + tau[idx[i] * n + vw[j]];
7      float ptr = a random real in [0, rw[n-1]];
8      int pos = min value j in rw[i..n-1] s.t. ptr < rw[j];
9      int node = vw[pos]; vw[pos] = vw[i]; v[idx[i]] = vw[i] = node;
10   }
```

Fig. 1. The solution construction in a pseudo code

Table 1. Results of ACO on a GPU in case that population size is equal to $4n$, the maximum number of evaluations is $200000n$ where n is the number of the locations. The algorithm terminates when an optimal solution is found by any thread or the number of evaluations arrives at the maximum value. (GPU programs run on: NVIDIA GeForce TITAN GPU, 3.1GHz Intel Core i7-3770S CPU, Windows 7 Professional SP1, Visual Studio 2008 Professional, CUDA 5.5, CPU programs run on: 2.67 GHz Intel Xeon X5550 CPU, Linux 2.6.27.29 (Fedora10 x86_64), gcc 4.3.2)

problem instance	CPU		GPU (with prefix-sums)			GPU (without prefix-sums)			speedup by prefix-sums
	time (sec)	error ()	time (sec)	error ()	speedup (to CPU)	time (sec)	error ()	speedup (to CPU)	
tai35b	22.9	1.9	22.5	1.1	1.0	21.1	1.1	1.1	0.94
tai40b	28.3	2.1	23.9	2.4	1.2	26.8	2.4	1.1	1.12
tai50b	59.0	3.4	39.4	1.6	1.5	40.5	1.7	1.5	1.03
tai60b	97.8	1.4	54.4	0.9	1.8	60.7	0.9	1.6	1.12
tai80b	232.2	2.4	100.0	2.6	2.3	117.1	2.6	2.0	1.17
tai100b	432.0	1.2	174.6	1.4	2.5	202.4	1.3	2.1	1.16
tai150b	1410.3	3.2	480.6	2.9	2.9	621.5	3.0	2.3	1.29

References

1. Burkard, R.E., Çela, E., Karisch, S.E., Rendl, F.: QAPLIB - a quadratic assignment problem library (2002), www.seas.upenn.edu/qaplib
2. Demouth, J.: Shuffle: Tips and tricks (2013), http://on-demand.gputechconf.com/gtc/2013/presentations/S3174-Kepler-Shuffle-Tips-Tricks.pdf
3. Tsutsui, S.: cAS: Ant colony optimization with cunning ants. Parallel Problem Solving from Nature, 162–171 (2006)
4. Tsutsui, S., Fujimoto, N.: Aco with tabu search on a gpu for solving qaps using move-cost adjusted thread assignment. In: Proceedings of the Genetic and Evolutionary Computation Conference (GECCO), pp. 1547–1554 (2011)

Solving Resource-Constraint Project Scheduling Problems Based on ACO Algorithms

Antonio Gonzalez-Pardo and David Camacho

Computer Science Department, Escuela Politécnica Superior,
Universidad Autónoma de Madrid, Spain
{antonio.gonzalez,david.camacho}@uam.es

1 Introduction

Constraint Satisfaction Problems (CSP) belongs to this kind of traditional *NP-hard* problems with a high impact in both, research and industrial domains. CSP problems are represented using triples (X, D, C) where X represents a set of variables that needs to be assigned with a particular value (V), which must satisfy a set of constraints (C) [1,2].

The possible utilization of ACO algorithms to solve CSP problems requires the design of the decision graph where the ACO is executed. The classical approach [3,4,5] builds a graph where the nodes represent the variable/value pairs ($< variable, value >$) and the edges connect those nodes whose variables X are different. One of the problem with this representation is that problems composed by many variables or by variables that could be assigned with many different values, become really difficult to model due to the size of the resulting graph. Another limitation with a full-connected approach is that continuous problems cannot be represented (this is a hot topic research problem in CSP), and only those problems with a finite set of values for the variables are allowed.

This work uses the representation proposed in [6] to solve *Resource-Constraint Project Scheduling Problems* (RCPSP). This new model creates a node for each given variable. In this way, given a problem composed by N variables where each variable can be assigned with a value from a set of M different values $(d_i \in D)$, will be modelled into a graph composed by N nodes (instead of $N * M$ nodes created in the classical approaches). The restrictions of the problem are represented in the edges of the graph. Two nodes will be connected if there is, at least, one restriction that involve the variables contained in each node.

The simplification of the graph entails a change in the behaviour of the ants, that using the proposed model they have to select the value that is assigned to the variable encoded in the node.

2 Experimental Results

The dataset used has been extracted from PSPLib library[1], several single-mode datasets with different number of jobs have been selected. 2040 problems have

---

[1] http://www.om-db.wi.tum.de/psplib/main.html

M. Dorigo et al. (Eds.): ANTS 2014, LNCS 8667, pp. 290–291, 2014.

Table 1. Makespan and accuracy obtained using the Single-Mode problems from PSPLib dataset for the 2040 analysed problems

Instance	PSPLib makespan	Makespan Without Obl.	Accuracy Without Obl.	Makespan With Obl.	Accuracy With Obl.
j30.sm	58.99 ± 14.08	59.98 ± 14.77	58.33 %	60.0 ± 14.81	58.33 %
j60.sm	79.8 ± 17.44	82.79 ± 20.70	52.71 %	82.83 ± 20.75	52.29 %
j90.sm	94.94 ± 20.59	99.1 ± 25.64	52.08 %	**99.07 ± 25.63**	**52.29 %**
j120.sm	122.19 ± 39.78	135.73 ± 49.22	8.0 %	135.76 ± 49.28	8.0%

been analysed from 4 datasets that contains instances of Single-Mode projects composed by 30 jobs (*j30.sm*), 60 jobs (*j60.sm*), 90 jobs (*j90.sm*) and 120 jobs (*j120.sm*). The ACO algorithm has been executed 10 times for each problem and the evaporation rate has been fixed to 5%. The ACO algorithm executed for the dataset *j30.sm* is composed by 100 ants during 400 steps whereas for the instances *j60.sm*, *j90.sm* and *j120.sm*, the ACO is composed by 200 ants executing during 500 steps.

Table 1 shows, for each RCPSP dataset, the average of the minimum makespan published in PSPLib, the average minimum makespan obtained using our approach, and the accuracy of the algorithm for each instance. This accuracy is computed as the number of problems that our algorithm is able to obtain the minimum makespan published divided by the total number of problems that compose the dataset.

The results obtained are really close to the best makespan obtained in the literature, our algorithm is able to find these best solutions in more than the 50% of the problems analysed. As it could be expected, the quality of the makespan, and the accuracy, decreases with the complexity of the problems.

Acknowledgments. This work has been partially supported by the Spanish Ministry of Science and Innovation under grant TIN2010-19872 (ABANT) and Savier project (Airbus Defence & Space project, FUAM-076914).

References

1. Tsang, E.P.K.: Foundations of constraint satisfaction. Computation in cognitive science. Academic Press (1993)
2. Barto, L., Kozik, M.: Constraint satisfaction problems solvable by local consistency methods. J. ACM 61(1), 1–19 (2014)
3. Hoseini Semnani, S., Zamanifar, K.: The power of ants in solving distributed constraint satisfaction problems. Appl. Soft Comput. 12(2), 640–651 (2012)
4. Khan, S., Bilal, M., Sharif, M., Sajid, M., Baig, R.: Solution of n-queen problem using aco. In: IEEE 13th International Multitopic Conference, INMIC 2009 (2009)
5. Solnon, C.: Ants can solve constraint satisfaction problems. IEEE Transactions on Evolutionary Computation 6, 347–357 (2002)
6. Gonzalez-Pardo, A., Camacho, D.: A new csp graph-based representation for ant colony optimization. In: 2013 IEEE Conference on Evolutionary Computation (CEC 2013), vol. 1, pp. 689–696 (2013)

Author Index